MATERIALS SCIENCE AND TECHNOLOGIES

SILICON CARBIDE: NEW MATERIALS, PRODUCTION METHODS AND APPLICATIONS

MATERIALS SCIENCE AND TECHNOLOGIES

Additional books in this series can be found on Nova's website
under the Series tab.

Additional E-books in this series can be found on Nova's website
under the E-books tab.

MATERIALS SCIENCE AND TECHNOLOGIES

SILICON CARBIDE: NEW MATERIALS, PRODUCTION METHODS AND APPLICATIONS

SOFIA H. VANGER
EDITOR

Nova Science Publishers, Inc.
New York

For permission to use material from this book please contact us:
Telephone 631-231-7269; Fax 631-231-8175
Web Site: http://www.novapublishers.com

Additional color graphics may be available in the e-book version of this book.

LIBRARY OF CONGRESS CATALOGING-IN-PUBLICATION DATA
Silicon carbide : new materials, production methods, and applications /
editor, Sofia H. Vanger.
p. cm.
Includes index.
ISBN 978-1-61122-312-5 (hardcover)
1. Silicon carbide. 2. Nanostructured materials. I. Vanger, Sofia H.
TP261.C3S46 2010
620.1'93--dc22
2010038651

Published by Nova Science Publishers, Inc. ✦ New York

CONTENTS

PREFACE

Silicon Carbide (SiC) is well known for its excellent material properties, high durability, high wear resistance, light weight and extreme hardness. This combination of properties makes them ideal candidates for tribological, semiconductor and MEMs, and optoelectronic applications. However, SiC is also known for its low fracture toughness, extreme brittleness and poor machinability. This book presents topical research data in the study of silicon carbide, including the etching and thin film formation of silicon carbide using highly reactive gases; production and characterization of SiC particles; microstructure of silicon carbide nanowires; ductile regime material removal of silicon carbide; limitation of SiC in the liquid-state processing of Al-MMC; and the effects of ion implantation in silicon carbide.

Chapter 1 – Etching and film formation of silicon carbide performed using highly reactive gases of chlorine trifluoride (ClF_3) and monomethylsilane (MMS, SiH_3CH_3), respectively, are reviewed. Silicon carbide etching achieving high etch rate is possible using chlorine trifluoride gas at the temperatures higher than 770 K. The etch rate is 10-20 μm/min and 5 μm/min, for polycrystalline 3C-silicon carbide and single-crystalline 4H-silicon carbide, respectively. Si-face of 4H-silicon carbide shows pitted surface after etching at low temperatures of 570 – 1270 K; the surface maintains smooth appearance after etching at 1570 K and at its concentration of 1 %. The C-face of 4H-silicon carbide shows the similar trend, and is entirely very smoother than that of Si-face. Most of the etch pits formed near 700 K at the Si-face and C-face 4H-silicon carbide correspond to dislocations revealed by the X-ray topograph.

Chapter 2 – In this paper, production and characterization of SiC particles reinforced Al alloy matrix composites produced by squeeze casting has been investigated. Al-4.5Cu-3Mg alloy was used for matrix alloy and 5, 10, 15 %vol. SiC particles reinforced composites have been produced at 100 MPa pressure and 500 ^{o}C die temperature applied at casting operations. The effects of volume fraction of SiC on microstructure, mechanical properties have been investigated. Also gravity cast matrix alloy has been produced to determine effect of pressure on microstructure and properties.

The results showed that, the composites have homogeneously distributed porosity free SiC particles. Squeeze casting also resulted in refining microstructure, porosity and other casting defects such as shrinkage cavities were minimized due to pressure application during solidification. The porosity of unreinforced alloys decreased from 6.63% to 1.08%. In composites, increasing volume fraction of particles caused to increase porosity of composite. Hardness of matrix alloy was increased from 94 BHN to 100 BHN by squeeze casting method. In composites, hardness increased with increasing volume fraction of SiC. The highest value was recorded at 15 %vol. SiCp, 139 BHN.

Squeeze casting method, also, improved tensile properties of matrix alloy, the UTS values increased from 195 MPa to 247 MPa and elongation from 1.78% to 3.30%. The elongation of composites decreased from 1.53% to 0.45% with increasing volume fraction of SiC from 5 to 15 percent.

Chapter 3 - Silicon carbide nanowires were produced from carbon nanotubes and nanosize silicon powder in a tube furnace at temperatures between 1100°C and 1350°C. Diameters of nanowires were controlled by selecting precursors, adjusting temperature, vapor pressure of silicon, and sintering time. The mechanism of SiC nanowires formation remains controversial, but when metals were found in carbon nanotube precursors, author were able to identify the growth mechanism as the vapor-liquid-solid, or VLS, process. In the TEM images 'flowers' were observed at the end of nanowires, and EDS analysis confirmed the presence of heavy metals at these ends. But formation of SiC nanowires obtained from highly oriented pyrolytic graphite and carbon nanotubes without metal catalysts was possible assuming the vapor-solid process, the VS mechanism. At the present time the possibility that the VS process contributes to the growth of SiC nanowires cannot be excluded from nanotube-based samples.

All SiC nanowires possess large quantities of planar defects, among which twins appear to be most abundant. Twins were identified in TEM images. By analyzing the geometry of crystallographic planes in the adjacent crystallites and from electron diffraction patterns recorded in the bright and dark images, it was possible to identify the angle between crystallographic planes as 141 degrees. Other planar defects frequently reported to be present in SiC nanowires are stacking faults. Stacking faults can affect x-ray diffraction patterns. A small shoulder that accompanies the [111] reflection is often considered the evidence of stacking faults. Distinguishing stacking faults from twins in TEM images is not a simple task. Besides, such an analysis does not allow for quantitative evaluation of the defects. In this paper author report a method of simultaneous analysis of profiles of several x-ray diffraction lines. This analysis allowed us to determine the crystallite sizes and the population of twins. It appears by this analysis that the population of stacking faults in silicon carbide nanowires is low. This result is confirmed by the analysis of TEM images.

All nanowires have crystalline cores and are coated with amorphous layers. The thickness of that layer varies from 1 to 10 nm, and its composition depends on properties of the precursors and the history of post production treatment. EDS was employed to analyze the chemical composition of that layer and FTIR to characterize its molecular structure. SiO and C=C groups were found in the amorphous layer of some specimens. Treatment of the specimens in acids and its affect on the composition of that layer is discussed.

Chapter 4 – Advanced ceramics such as silicon carbide (SiC) are increasingly being used for industrial applications. These ceramics are hard, strong, inert, and light weight. This combination of properties makes them ideal candidates for tribological, semiconductor and MEMS, and optoelectronic applications respectively. Manufacturing SiC without causing surface or subsurface damage is extremely challenging due to its high hardness, brittle characteristics and poor machinability. Often times, severe fracture can result when trying to achieve high material removal rates during machining of SiC due to its low fracture toughness. One of the most critical steps before machining SIC is to establish the Ductile to Brittle Transition (DBT) depth, also known as the critical depth of cut (DOC). The DBT depth is determined via scratch testing either with a diamond stylus or a single crystal diamond cutting tool. In this chapter, a study on the DBT of a single crystal SiC wafer is discussed. This study also demonstrates that ductile regime Single Point Diamond Turning (SPDT) is possible on SiC to improve its surface quality without imparting any form of surface or sub surface damage. Machining parameters such as depth of cut and feed used to carry out ductile regime machining are discussed. Sub-surface damage analysis was carried out on the machined sample using non-destructive methods such as Optical Microscopy, Raman Spectroscopy and Scanning Acoustic Microscopy to provide evidence that this material removal method (i.e., SPDT) leaves a damage-free surface and subsurface. Optical microscopy was used to image the improvements in surface finish whereas Raman spectroscopy and scanning acoustic microscopy was used to observe the formation of amorphous layer and sub-surface imaging in the machined region respectively. All three techniques complement the initial hypothesis of being able to remove a nominally brittle material in the ductile regime.

Chapter 5 – One-dimensional (1D) SiC materials have drawn increasing attentions for their use in fundamental research, and for potential nano-technological applications in electronics and photonics operating in harsh environment. In the present chapter, author firstly report the fabrication methods and some novel properties of 1D SiC nanostructures. Then the results of computer simulation on the mechanical properties of SiC nanowires are presented. For perfect SiC nanowires, under axial tensile strain, the bonds of the nanowires are just stretched before the failure of nanowires by bond breakage. Under axial compressive strain, the collapse of the SiC nanowires by yielding or column buckling mode depends on the length and diameters of the nanowires. The nanowires collapse through a phase transformation from crystal to amorphous structure in several atomic layers under torsion strain. The effects of twinning and amorphous coating on the mechanical behavior are also presented. Amorphous layer coating can induce brittle to ductile transition in SiC nanowires. And the critical strain of the nanowires can be enhanced by the twin-stacking faults.

Chapter 6 – By virtue of its physicochemical, thermal and mechanical properties, silicon carbide has been widely recognized as a leading candidate for the manufacture of aluminum matrix composites. However, when designing Al composites reinforced with SiC, consideration of the processing route is of prime importance. Particularly, when Al/SiC composites are processed by the liquid state route, there always exists the potential of affecting the SiC reinforcements, going from a mild sensitizing to a severe degradation and eventually to a reactive dissolution, accompanied by the formation of the deleterious aluminum carbide (Al_4C_3) phase. And even though in many cases the composite may exhibit a good appearance, the presence of Al_4C_3 might sabotage its performance in terms of time, stress level and atmospheric conditions. The latter is on account of the susceptibility of Al_4C_3 of reacting with humidity in the atmosphere. In the worst case scenario the composite collapses. Certainly, a number of processing factors determine whether the reinforcements will be threatened or whether they will ultimately keep their properties and exhibit their full potential. Based on the current literature and on the processing→microstructure→properties paradigm, in this contribution, an analysis of the potentialities and limitations of SiC in liquid aluminum, and their consequences, is made. Likewise, suggestions for preventing SiC degradation are outlined.

Chapter 7 – The current status of the effects of ion implantation into SiC is reviewed with a focus on the implantation of helium ions and its consequences on the microstructure of SiC.

The helium implantation results in a strain profile build-up as a consequence of the generation and accumulation of both point defects and helium-related complexes. At high fluence, the damage accumulation leads to amorphization that can be avoided by performing implantation at elevated temperature. Upon post-implantation annealing, the point defect recovery leads to the relaxation of the elastic strain and the helium-related complexes evolve into extended defects. The evolution of the strain and defects as a function of implantation temperature and annealing temperature is discussed with regard to the mobility of involved species. Moreover, all the defects are associated with a change in volume resulting in surface expansion or swelling of the implanted material that must be taken into account for the applications of SiC. Finally, the effects of implantation on the mechanical properties of SiC are discussed with regard to the microstructure changes.

Chapter 8 – Over the years the synthesis of silicon carbide nanomaterials (SiCNM) has been thrust area of research and development for its unique and potential properties. Numerous efforts have been made to generate SiCNM at the large scale and with enhanced purity. SiCNM have been developed in the form of nanosphere, nanorods, nanowires, nanofibers, nanowhiskers and nanotubes using different techniques and methodologies. SiC is particularly attractive due to its excellent mechanical properties, high chemical resistivity, and its potential application as a functional ceramic or a high temperature semiconductor. This chapter deals with preparation technologies of silicon carbide based nanomaterials. An overview of some important applications have also been discussed.

Chapter 9 – Conversion of carbides to carbon by means of selective removal of metal atoms from carbide lattice allows obtaining porous nanocarbon materials whose phase composition and structure are determined by the properties of initial carbide and conversion conditions. There are no published data on the synthesis of pore-free carbon materials and fullerites by this method. The fundamentally different method for the conversion of SiC to carbon phases by chlorinated derivatives of methane is used. The main point of the method is substitution of carbon atom for silicon atom within the carbide lattice. In the present paper, the unique possibilities of proposed conversion method are demonstrated with silicon carbide (nanowhiskers 3C -SiC, n-type and semi-insulating platelet-shaped single crystals of 6 H -SiC polytype facetted by polar (0001) planes) as an example. By changing the process conditions one can control crystallization of various carbon allotropes which differ in density, and also obtain pore-free transparent mechanically strong coatings exhibiting high adhesion strength, low-porosity materials, and films of fullerite C 42 of different thickness. The above method can be applied for large-scale production of thin pore-free coatings on SiC, for synthesis C 42 -{0001} SiC hetero-structures, and for creation of new strategy in producing fullerites through conversion of carbides with different structure by chlorinated and fluorinated derivatives of methane and other organic compounds.

In: Silicon Carbide: New Materials, Production ... ISBN: 978-1-61122-312-5
Editor: Sofia H. Vanger

Chapter 1

ETCHING AND THIN FILM FORMATION OF SILICON CARBIDE USING HIGHLY REACTIVE GASES

Hitoshi Habuka[1]
Department of Chemical and Energy Engineering, Yokohama National University, 79-5 Tokiwadai, Hodogaya, Yokohama, Kanagawa 240-8501, Japan

Etching and film formation of silicon carbide performed using highly reactive gases of chlorine trifluoride (ClF_3) and monomethylsilane (MMS, SiH_3CH_3), respectively, are reviewed. Silicon carbide etching achieving high etch rate is possible using chlorine trifluoride gas at the temperatures higher than 770 K. The etch rate is 10-20 μm/min and 5 μm/min, for polycrystalline 3C-silicon carbide and single-crystalline 4H-silicon carbide, respectively. Si-face of 4H-silicon carbide shows pitted surface after etching at low temperatures of 570 – 1270 K; the surface maintains smooth appearance after etching at 1570 K and at its concentration of 1 %. The C-face of 4H-silicon carbide shows the similar trend, and is entirely very smoother than that of Si-face. Most of the etch pits formed near 700 K at the Si-face and C-face 4H-silicon carbide correspond to dislocations revealed by the X-ray topograph.

Polycrystalline 3C-silicon carbide thin film is formed on silicon surface using monomethylsilane gas at the temperatures lower than 1070 K. During the film formation, silicon is produced by thermal decomposition in gas phase and at substrate surface, and is incorporated into the silicon carbide film. Such the excess silicon can be significantly reduced by the addition of hydrogen chloride (HCl) gas. Although the silicon carbide film formation saturates within 1 minute due to the surface termination by C-H bonds, it can start again after annealing at 1270 K in ambient hydrogen. Monomethylsilane molecule can chemisorb on the silicon surface to produce silicon carbide film, even at room temperature, after the silicon surface is cleaned at 1370 K and is cooled down in hydrogen ambient. Such the low temperature film formation is expected to be possible, because the hydrogen terminated silicon surface has silicon dimers. The silicon carbide film formed at room temperature can maintain its surface after the etching using hydrogen chloride gas at 1070 K.

---

[1] e-mail: habuka1@ynu.ac.jp.

1. INTRODUTION

1.1. Silicon Carbide

Silicon is one of the major elements in earth's crust. Carbon is one of the most familiar elements for animals, plants, human life and industry. Silicon and carbon have a strong covalent bond between them to form the excellent material of silicon carbide (SiC).

Single-crystalline 4H-silicon carbide is a fascinating wide band-gap semiconductor material [1-3], suitable for high power and high temperature electronic devices [4] because of its remarkable properties, such as high electron mobility, high thermal conductivity, high chemical stability, high mechanical hardness, high break down electric field and small dielectric constant [4, 5]. Additionally, many researchers have reported the stability of silicon carbide micro-electromechanical systems (MEMS) under corrosive conditions using acid and alkaline chemical reagents [6-9]. Polycrystalline 3C-silicon carbide is widely used for various purposes, such as dummy wafers and reactor parts, in silicon semiconductor device production processes, because gas emission from its surface is significantly small.

Following the history of semiconductor materials production technology [10], the electronics devices manufacturing process needs an easy and cost effective technique, such as wet and/or dry cleaning, for achieving and maintaining clean surface of the substrate materials. Here, it should be noted that the suitable properties of silicon carbide often provided difficult problems. Its chemically and mechanically stable nature often makes it very difficult to prepare the entire surface in the wafer production process, such as surface polishing and removal of any damaged layer. Applicable chemical reagents and processes should be found and developed for silicon carbide material production.

In various industries other than semiconductor devices, silicon carbide thin film has been used for coating various materials [11], such as carbon, in order to protect them from corrosive and high temperature environment. For producing silicon carbide film by chemical vapor deposition (CVD), very high temperature, higher than 1500 K [12, 13], is required. By the current technology, various materials having the melting point lower than the CVD temperature cannot be coated with silicon carbide. Thus, the low temperature silicon carbide CVD film coating [14-22] will enable and create significant kinds of application in various industries.

1.2. Dry Etching

Wet and dry etching methods of silicon carbide have been studied by many researchers [5, 23-33], using various gases and various wet etchants. However, the largest etch rate reported was nearly 1 μm/min, except for that assisted by ultraviolet light [30]. Here, chlorine trifluoride gas is very reactive even at low temperatures and has a very strong capability to etch various materials, such as silicon [35] without plasma assistance. Favorably, it has a very low global warming potential (GWP) [11].

In Section 2, details of polycrystalline 3C-silicon carbide etching using chlorine trifluoride gas [36, 37] are shown, particularly focusing on the etch rate, gaseous products, surface chemical bonds and the surface morphology of the silicon carbide. Next, in Section 3, the dry etching of single-crystalline 4H-silicon carbide using chlorine trifluoride gas [38-41] over the wide temperature range of 570-1570 K is reviewed, particularly about the etch rate, surface chemical reaction rate constant, surface morphology and etch pits.

In Section 3, the etch rate and rate constant of fused silica (silicon dioxide, SiO_2) using chlorine trifluoride gas [42] is also explained. Its information is necessary for designing the safe-process and reactor for chlorine trifluoride gas by suppressing fused silica etching, because the current semiconductor device manufacturing processes widely use fused silica, as a material of process tube and various parts.

1.3. Film Deposition

In Section 4, a technology using monomethylsilane gas for achieving low-temperature silicon carbide film formation on silicon surface [43-45] is reviewed. In order to enable an industrial use of the silicon carbide film deposition technology using monomethylsilane gas, the following issues are explained.

(i) The thermal decomposition behavior of monomethylsilane gas in gas phase and at surface,
(ii) The method of adjusting the film composition (*e.g.*, Si content),
(iii) The surface chemical reaction rate (stop and recover).
(iv) The necessary condition to achieve very low temperature silicon carbide film deposition.
(v) The capability of silicon carbide film for protecting substrate materials from the corrosive environment.

2. Polycrystalline 3C-Silicon Carbide Etching Using Chlorine Trifluoride Gas

The polycrystalline 3C-silicon carbide surface is etched using chlorine trifluoride gas at the concentration between 10 and 100 % in ambient nitrogen. From its results, the detail of the etch rate, gaseous products, surface chemical bonds and the surface morphology are clarified.

2.1. Reactor and Processes Using Chlorine Trifluoride Gas

In order to etch silicon carbide by chlorine trifluoride gas, the horizontal cold-wall reactor shown in Figure 2.1 is used. This reactor consists of a gas supply system, a quartz chamber and infrared lamps. A 30 mm wide x 40 mm long x 0.2-1 mm thick 3C-silicon carbide substrate manufactured using chemical vapor deposition (CVD) (Admap Inc., Tokyo) is held horizontally on the bottom wall of the quartz chamber. This polycrystalline substrate is widely used in thermal oxidation furnaces in semiconductor silicon device manufacturing because of its very high purity. [46]. Because the grain boundary and the various crystal planes existing in a polycrystalline material generally tend to cause non-uniform etching to give a very coarse surface, the polycrystalline substrate is suitable to evaluate the etching nature and performance, such as smoothing or roughening.

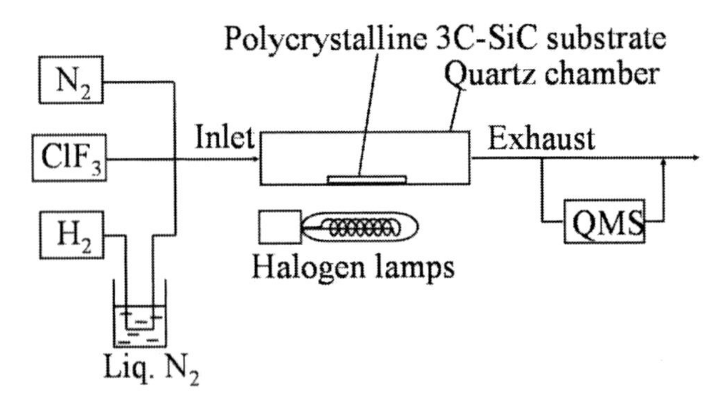

Figure 2.1. Horizontal cold-wall reactor used for etching polycrystalline 3C-silicon carbide substrate.

The silicon carbide substrate is heated by infrared rays from the halogen lamps through the quartz chamber walls. The electric power provided to the infrared lamps is adjusted based on the temperatures measured in ambient nitrogen (without chlorine trifluoride gas) prior to etching.

The gas supply system introduces chlorine trifluoride gas, nitrogen gas and hydrogen gas. Hydrogen gas is used to remove the silicon oxide film on the silicon carbide substrate surface, the same as those on the silicon surface [35]. This reactor has a small cross section above the substrate in order to achieve a very high consumption efficiency of the chlorine trifluoride gas. The height and width of the quartz chamber are compactly designed to be 10 mm and 40 mm, respectively, similar to the chamber in our various studies [35, 47].

The etching using chlorine trifluoride gas is carried out following the process shown in Figure 2.2. This mainly consists of three steps:

(a).cleaning the silicon carbide substrate surface by removing the silicon oxide film on it by baking in ambient hydrogen at 1370 K for 10 min,
(b).changing the gas from hydrogen to nitrogen, and
(c).etching the silicon carbide substrate surface using chlorine trifluoride gas.

During step (a), hydrogen gas is introduced at atmospheric pressure into the reactor at a flow rate of 2 slm. Water vapor and the other impurities in the hydrogen gas are removed by passing it through a liquid nitrogen trap (77 K) at the entrance of the reactor. The etch rate and the change in surface roughness after Step (a) are negligible. Next, in step (b), the quartz chamber and the silicon carbide substrate are cooled to room temperature. The hydrogen gas present in the quartz chamber must be sufficiently purged with nitrogen gas to avoid an explosive reaction between hydrogen and chlorine trifluoride. During step (c), the silicon carbide substrate is heated and adjusted to temperatures between 670 K and 970 K. The silicon carbide substrate is etched by chlorine trifluoride (>99.9 %, Kanto Denka Kogyo Co., Ltd., Tokyo) at a flow rate of 0.1-0.25 slm without further purification and without dilution. In order to evaluate the gaseous products, part of the exhaust gas from the reactor is fed into a quadrupole mass spectra (QMS) analyzer, as shown in Figure 2.1.

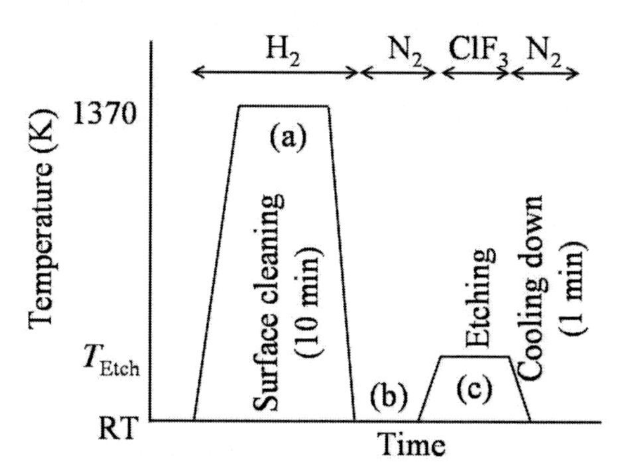

Figure 2.2. Process for cleaning and etching polycrystalline 3C-silicon carbide surface.

The silicon carbide etch rate using chlorine trifluoride gas is evaluated from the decrease in the weight of the silicon carbide substrate. Therefore, the measured etch rate is the average value over the entire top surface of the silicon carbide substrate.

The surface morphology of the polycrystalline 3C-silicon carbide substrate before and after the etching is observed using an optical microscope. The root-mean-square (RMS) surface roughness and the average roughness, R_a, are measured. In order to evaluate the condition of the chemical bonds of the silicon carbide surface before and after the etching, X-ray photoelectron spectra (XPS) are obtained.

2.2. Etch Rate

The etch rate of the polycrystalline 3C-silicon carbide substrate surface is shown in Figure 2.3, which was obtained using 100% chlorine trifluoride gas at various gas flow rates in the temperature range of 670 to 970K at atmospheric pressure. The squares, circles and triangles show the etch rate at the chlorine trifluoride gas flow rate of 0.2, 0.1 and 0.05 slm, respectively. As shown in Figure 2.3, the etch rate at the substrate temperature less than 670K is quite low; its value is less than 1 μm/min. However, with the increasing substrate temperature, the etch rate significantly increases particularly near 720 K. At the substrate temperature of 770 K, the etch rate at the flow rate of 0.2 slm becomes 20 μm/min; it remains constant at substrate temperatures greater than 770 K.

As shown in Figure 2.3, the etch rate changes with the flow rates. The etch rates are 25, 10 and 5 μm/min at the flow rates of 0.2, 0.1 and 0.05 slm, respectively. For each chlorine trifluoride gas flow rate, the trend in the flat etch rate at temperature greater than 770K is maintained.

Figure 2.4 shows that the etch rate increases with the chlorine trifluoride gas flow rate between 0.1 and 0.25 slm at 770 K. The etch rate is 40 μm/min at a chlorine trifluoride gas flow rate of 0.25 slm. This etch rate is comparable to the highest value which has been achieved by a photoassisted wet etching process [30] and is greater than the etch rates for the existing silicon carbide dry etching methods.

Since a flow rate of 0.25 slm was the maximum for the reactor system used, etch rates at higher than 0.25 slm were not evaluated. However, the increasing trend is expected to be maintained at the higher flow rates of chlorine trifluoride.

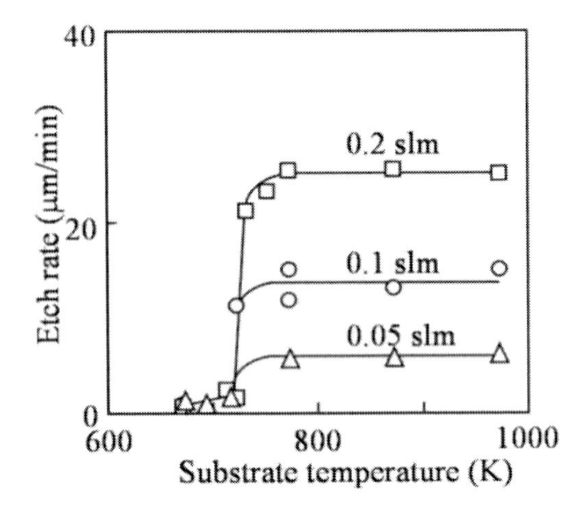

Figure 2.3. Etch rate of the polycrystalline 3C-silicon carbide substrate surface by chlorine trifluoride gas (100%) at atmospheric pressure in the temperature range between 670 and 970 K. Square: 0.2 slm, circle: 0.1 slm, and triangle: 0.05 slm.

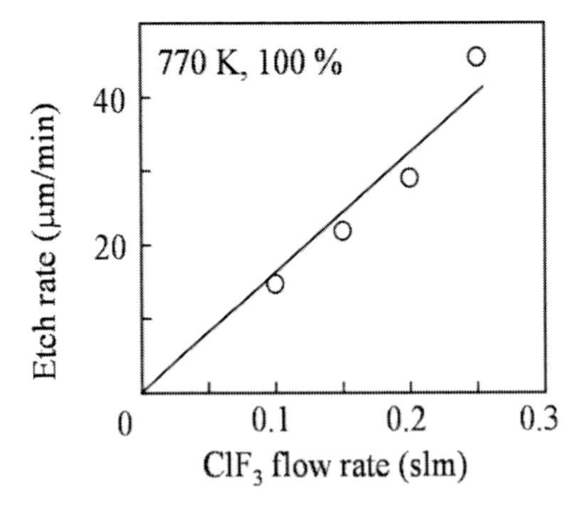

Figure 2.4. Etch rate of polycrystalline 3C-silicon carbide substrate surface using chlorine trifluoride at atmospheric pressure and at 770 K in the flow rate range between 0.1 and 0.25 slm.

In order to evaluate the influence of chlorine trifluoride gas concentration, the etch rate of the polycrystalline 3C-silicon carbide substrate surface using 10-100% chlorine trifluoride gas in ambient nitrogen was measured at the flow rate of 0.2 slm, atmospheric pressure and 670-970 K, as shown in Figure 2.5. In this figure, the square, reverse triangle, circle, diamond, and triangle show the substrate temperatures of 670, 730, 770, 870 and 970 K, respectively. Being consistent with Figure 2.3, the etch rate at the substrate temperature of 670K is very low. The etch rates at 730 K are significantly higher that those at 670 K.

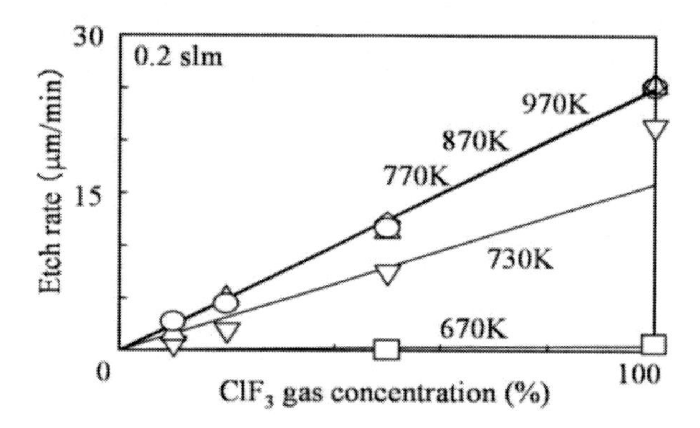

Figure 2.5. Etch rate of the polycrystalline 3C-silicon carbide substrate surface using chlorine trifluoride gas at 10-100%, 0.2 slm, atmospheric pressure, and 670-970K. Square: 670 K, reverse triangle: 730K, circle: 770K, diamond: 870 K, and triangle:970K.

At the substrate temperature of 770 K, the etch rate is proportional to the chlorine trifluoride gas concentration. At 870 and 970 K, the etch rate at each chlorine trifluoride gas concentration is the same as that at 770 K. Therefore, when the substrate temperature is higher than 770 K, the etch rate over a very wide chlorine trifluoride gas concentration range is not affected by the substrate temperature. This behavior is useful for obtaining an uniform etched depth, because a non-uniform distribution of the substrate temperature is allowed.

In order to quickly discuss the rate process of the etching in this study, the Arrhenius plot of the etch rate is shown in Figure 2.6. In the $1/T$ range lower than 0.0013, the etch rate is nearly constant; the etch rate decreases with increasing $1/T$. This behavior can be the change in the rate process from the transport limited regime to the reaction limited regime, with increasing $1/T$. The activation energy obtained near the $1/T$ of 0.0014 is 500 kJ/mol.

The mechanism to cause the temperature independent etch rate at higher temperatures can be the same as that for silicon etching [47] which is in the transport limited regime. In order to achieve the temperature-independent etch rate, the following mechanism of compensation is possible,

(i) The etch rate increase due to the increase in the diffusivity of chlorine trifluoride gas and the overall rate constant, and
(ii) The etch rate decrease due to the decrease in the chlorine trifluoride gas concentration by the gas volume expansion in the gas phase above the substrate, with the increasing substrate temperature.

The detail of these two transport factors has been discussed based on numerical calculation in our previous study [47] and in Section 3.5 for single-crystalline 4H-silicon carbide, which shows that the rate process of the etching is in the transport limited regime; an appropriate reactor height enables the compensation between (i) and (ii). However, because the surface reaction rate at the temperatures is sufficiently small, the etching process is in the reaction limited regime; the etch rate changes with the substrate temperature lower than 770K.

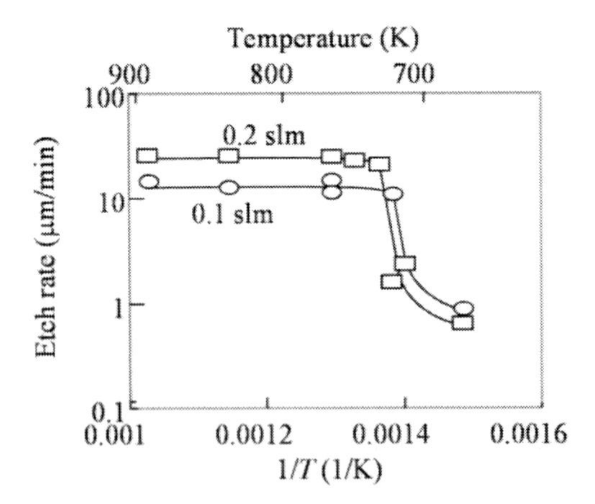

Figure 2.6. Arrhenius plot of etch rate of the polycrystalline 3C-silicon carbide substrate surface using chlorine trifluoride gas at 100 % in the temperature range between 670 and 970 K. Square: 0.2 slm, and circle: 0.1 slm.

Because the gaseous products measured using the QMS over the very wide chlorine trifluoride gas concentrations and temperatures are chlorine, silicon tetrafluoride and carbon tetrafluoride, the significant increase in the etch rate near 720K is simply due to the increase in the rate constant following the Arrhenuis law, not due to any change in the chemical reaction mechanism.

2.3. Surface Morphology and Roughness

The change in the surface morphology of the 3C-silicon carbide is explained. Figure 2.7 shows photographs of the silicon carbide surface etched using chlorine trifluoride gas at the flow rate of 0.1 slm and atmospheric pressure at 670, 720 and 770 K for 15 min. The values in parenthes are the etch depth. Figure 2.7 (a) shows the initial surface which has the periodically line-shaped hills and valleys. As shown in Figure 2.7 (b), although the surface still has the line-shaped morphology at 670 K after 15 minutes, its pattern remains but is unclear. At the higher temperatures of 720 and 770 K, the line-shaped appearance does not remain, as shown in Figures 2.7 (c) and (d). In these figures, there are very small and very shallow pits having a round edge. This shows that the etching using chlorine trifluoride can smooth the large hills and valleys which existed on the silicon carbide surface.

In order to show the detail of surface smoothing effect, the change in the surface appearance is shown, in Figure 2.8, versus etch period at a substrate temperature of 770 K and a flow rate of 0.1 slm of chlorine trifluoride. This figure shows photographs of the etched silicon carbide surface at (a) 0 min, (b) 5 min, (c) 10 min, (d) 15 min, and (e) 30 min. The values in parenthes are the etch depth. Figures 2.8 (a) and (d) are the same as Figures 2.7 (a) and (d), respectively. The relationship between the etch period in Figure 2.8 and the etch depth is shown in Figure 2.9 (a). The line-shaped pattern in Figure 2.8 (a) is slightly rounded after 5 minutes. At 10 minutes, there is only a trace of the line-shaped appearance, as shown in Figure 2.8 (c). The 3C-silicon carbide surface has a round-shaped morphology after 15 minutes as shown in Figure 2.8 (d), since the line-shaped pattern is removed during the etch period between 10 and 15 minutes. The surface morphology in Figure 2.8 (d) is maintained at 30 minutes in Figure 2.8 (e), and the rounded edges of the very shallow pits do not become sharp during the last 15 minutes. Thus, the round-shaped morphology of silicon carbide surface will be maintained during etching for longer than 30 minutes.

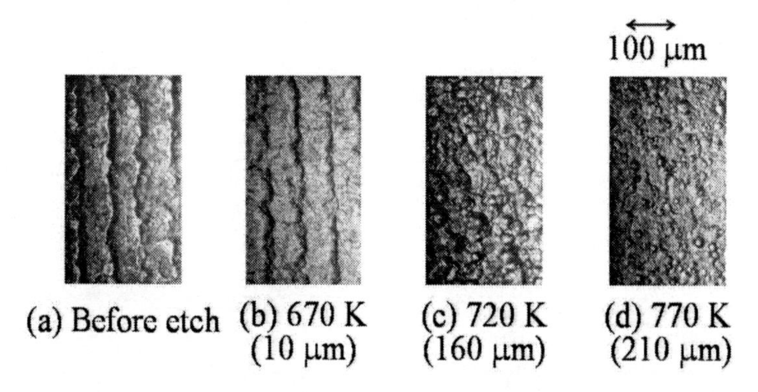

Figure 2.7. Photograph of the silicon carbide substrate surface etched using chlorine trifluoride at atmospheric pressure after 15 min at (b) 670 K, (c) 720 K, and (d) 770 K. (a) is the silicon carbide substrate surface before etching. The values in parenthes are the etch depth.

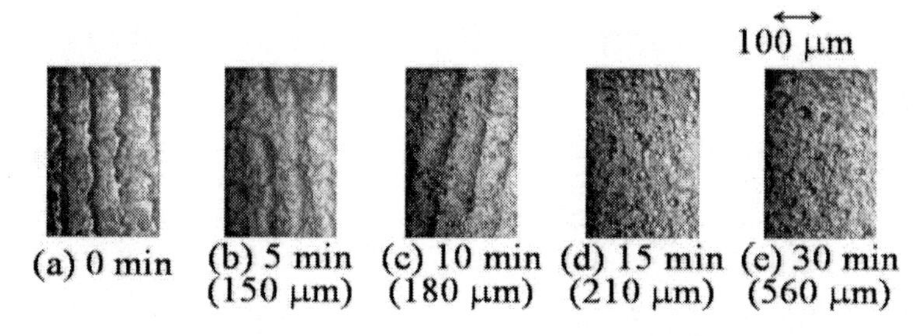

Figure 2.8. Photograph of the silicon carbide surface etched using chlorine trifluoride at atmospheric pressure and 770 K at (a) 0 min, (b) 5 min, (c) 10 min, (d) 15 min, and (e) 30 min. The values in parenthes are the etch depth.

In order to evaluate the surface smoothing effect of silicon carbide by chlorine trifluoride gas, the surface roughness is measured using the root-mean-square (RMS) roughness as shown in Figure 2.9 (b). Figure 2.9 (a) also shows the etch depth. The initial surface has an RMS roughness of 5 μm. The surface roughness decreases with increasing etch period. At 10 minutes, when 180 μm has been etched, the RMS roughness becomes a low value of 1 μm. Consistent with Figures 2.8 (d) and (e), the RMS roughness is maintained at nearly 1 μm at 30 minutes when the etch depth becomes greater than 500 μm. Solid line in Figure 2.9 (b) indicates the possibility that the chlorine trifluoride gas has a smoothing effect on the 3C-silicon carbide surface.

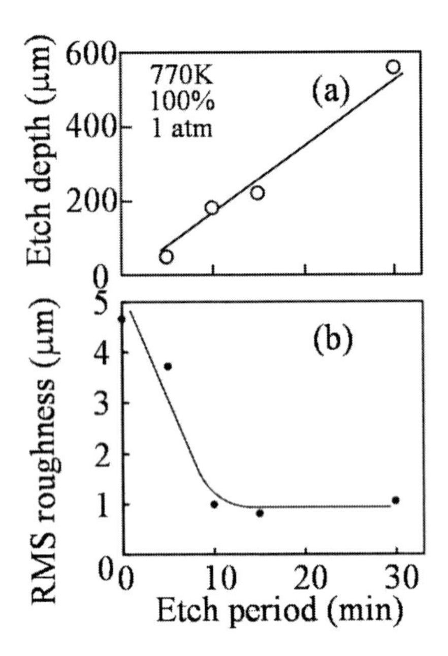

Figure 2.9. Etch depth and roughness of the silicon carbide surface etched by chlorine trifluoride gas at atmospheric pressure and 770 K within 30 min: (a) etch depth, and (b) root mean square (RMS) roughness.

Figure 2.10 shows photographs of the polycrystalline 3C-silicon carbide surface etched using a gas mixture of chlorine trifluoride and nitrogen at 670-870K for 15 min. The concentration and the flow rate of the gas mixture are 10-100% and 0.2 slm, respectively. Figure 2.11 shows the RMS roughness of the polycrystalline 3C-silicon carbide surface etched using chlorine trifluoride gas at atmospheric pressure and 670-870K for 15 min. The triangle, diamond and circle show the RMS roughness at the chlorine trifluoride gas concentrations of 10, 50 and 100 %, respectively.

The photograph indicated by 'Before etch' in Figure 2.10 shows the initial surface, which has very narrow, vague and shallow trenches formed by mechanical polishing. Using the chlorine trifluoride gas concentration of 10%, the change in the surface morphology is explained. The surface etched at 730K and 10% is recognized to have circular-shaped pits. Although the etch rate under this condition is very low, its surface shown in Figure 2.10 has an etched depth of 15μm. This surface shows many circle-like pits, the edge of which is clearly shown. These may be the grain boundary or some disordered region which can be etched at a slightly higher etch rate.

At 770K and 10 %, the shape of the circular-shaped pits still clearly exists, similar to that at 730K and 10 %. The photograph at 870K and 10% shows pits smaller than those at 770K and 10%. Simultaneously, the conical shape of the pits still exists. This shows that chlorine trifluoride gas at the low concentration of 10% has a quite small role of smoothing the surface, but rather tends to roughen it. This trend is measured using the RMS roughness, shown by the triangles in Figure 2.11. The RMS roughness before etching is nearly 0.5μm; it increases with the increasing substrate temperature at the chlorine trifluoride gas concentration of 10%.

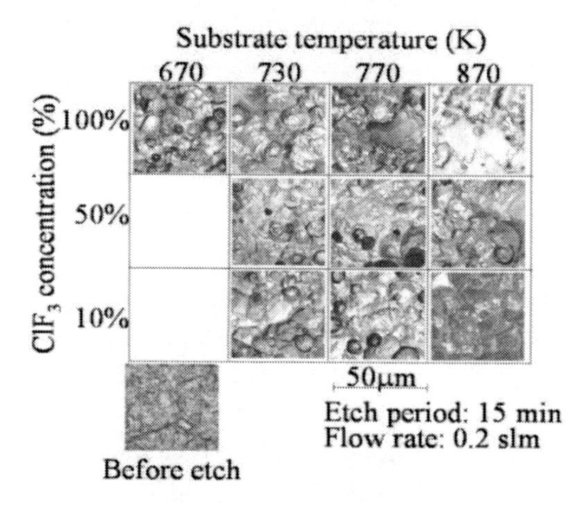

Figure 2.10. Photograph of the polycrystalline 3C-silicon carbide surface etched using chlorine trifluoride gas at atmospheric pressure for 15min at 670-870 K, 10-100% and 0.2 slm.

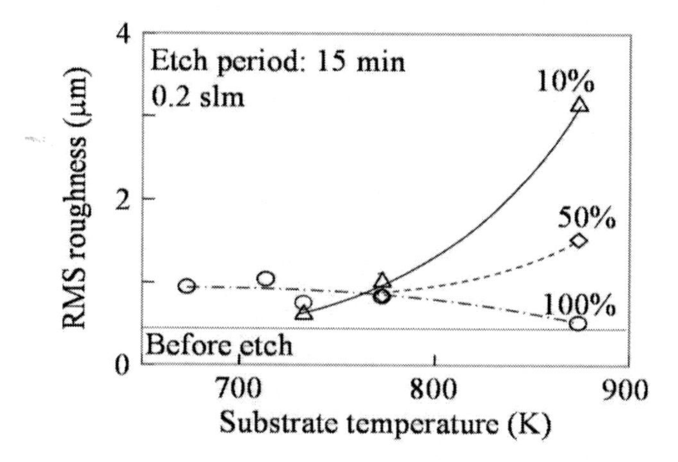

Figure 2.11. RMS roughness of the polycrystalline 3C-silicon carbide substrate surface etched using chlorine trifluoride gas at atmospheric pressure and 670-870 K for 15min. Triangle: 10 %, diamond: 50 %, and circle: 100%.

Next, at the fixed chlorine trifluoride gas concentration of 50 %, the change in the etched surface morphology is explained using Figure 2.10. The surface etched at 730K and 50% shows the clear edge shape of the pits. The surface etched at 770K and 50% still has a clear edge of the conical-shaped pits. Although some pits still have such the clearly observed edge shape, the rest of the surface has no clear edges. The surface morphology at 870K and 50% shows both clear and vague edges. Since semi-smoothed and clear pits coexist there, the RMS roughness, indicated by the diamonds in Figure 2.11, still slightly increases with the increasing substrate temperature.

The change in the surface morphology etched at 100% is also shown in Figure 2.10. The surface etched at 670K and 100% shows the clear edge of the pits. In contrast to this, the surface etched at 730K and 100% shows the slightly vague edge of the pits. For the surface etched at 770K and 100 %, the conical-shaped pits still remain, but are few. The edge of the conical-shaped pits disappears, when the surface is etched at 870 K. Since this trend in smoothing the surface appears in the RMS roughness behavior, the RMS roughness at 100%, indicated using circles in Figure 2.11, slightly decreases with the increasing substrate temperature.

From the view point of the effect of the etchant gas concentration at each temperature, the shape of the pits tends to become unclear with the increasing chlorine trifluoride gas concentration. Therefore, as the overall trend, the higher temperature and the higher chlorine trifluoride gas concentration produces a smoother surface of the polycrystalline 3C-silicon carbide.

The polycrystalline 3C-silicon carbide etch rate can be adjusted using the combination of gas flow rate, gas concentration and the substrate temperature, in order to obtain surfaces suitable for various purposes. This technique is expected to be used for various applications, such as the dry cleaning of the silicon carbide substrate surface instead of wet method, and the removal of the damaged layer formed during the chemical mechanical polishing using diamond slurry.

2.4. Surface Chemical Condition and Etch Rate

The fraction of silicon and carbon on the silicon carbide surface remaining after the etching is useful information for developing various processes which are performed after the etching. Thus, the chemical bonds at the silicon carbide surface are measured using X-ray photoelectron spectroscopy (XPS) before and after etching by chlorine trifluoride gas. Figure 2.12 shows the fraction of carbon, silicon, oxygen, chlorine and fluorine on the silicon carbide substrate before and after etching the depth greater than 150 μm using the chlorine trifluoride gas for 15 min at atmospheric pressure and at 720 K.

The silicon carbide surface before the etching has a carbon fraction nearly equal to that of silicon. However, the fraction of carbon significantly increases to 75 % after etching using chlorine trifluoride gas. This indicates that the production of volatile carbon compound is slower than that of silicon compound at this temperature.

In order to evaluate the state of silicon and carbon, the XPS spectra of Si 2p and C 1s are shown in Figures 2.13 and 2.14, respectively. The conditions of etching are the same as in Figure 2.12. The entire spectra of Si 2p and C 1s are decomposed and assigned to the chemical bonds indicated in these figures.

An almost amount of silicon at the silicon carbide surface has chemical bond with carbon before etching as shown in Figure 2.13 (a). However, after the etching, a significant amount of silicon oxides and oxidized or halogenated silicon carbide are present as shown in Figure 2.13 (b).

The chemical bonds of carbon simultaneously change, as same as those of silicon. Figure 2.14 (a) shows that chemical bonds of carbon with silicon dominate at the substrate surface before etching. After etching, an almost amount of carbon has chemical bonds with carbon and hydrogen as shown in Figure 2.14 (b), although the amount of hydrogen in the reactor was negligible at the etching step. The silicon carbide substrate surface after etching by chlorine trifluoride gas is covered with large amount of carbon having carbon-carbon bonds.

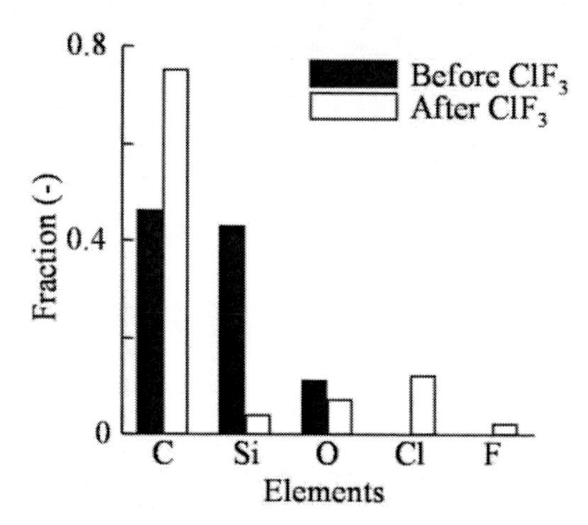

Figure 2.12. Fraction of carbon, silicon, oxygen, chlorine and fluorine on the silicon carbide substrate surface before and after etching using chlorine trifluoride for 15 min at atmospheric pressure in the reactor. The temperature of the silicon carbide substrate is 720 K.

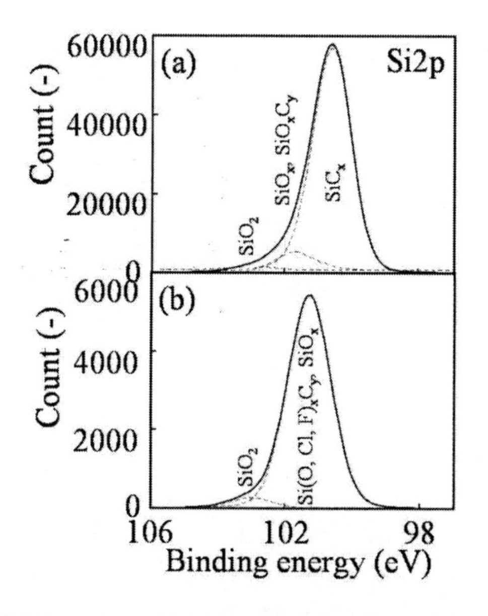

Figure 2.13. XPS spectra of Si 2p measured (a) before and (b) after etching using chlorine trifluoride gas for 15 min at atmospheric pressure. The temperature of the silicon carbide substrate is 720 K.

Although a large amount of carbon remains on the silicon carbide surface, the etch depth along with the etch period, shown in Figure 2.9 (a), indicates a linear increase in the etch depth proportional to the period of introducing chlorine trifluoride gas into the reactor. Therefore, the chemical reaction associated with silicon carbide etching occurs in a steady state, without any suppression due to the surface coverage of the carbon.

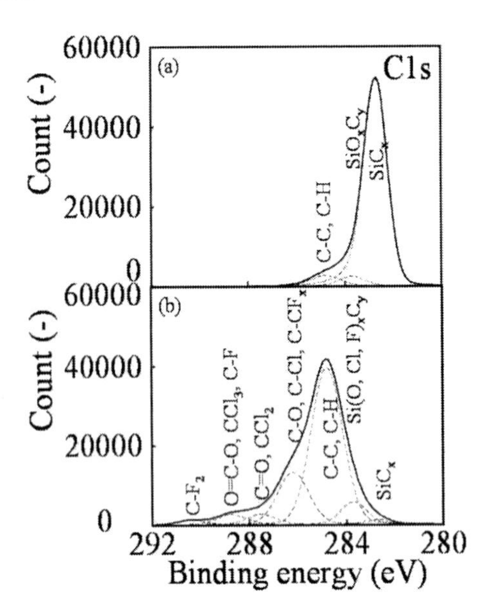

Figure 2.14. XPS spectra of C 1s measured (a) before and (b) after the etching using chlorine trifluoride gas for 15 min at atmospheric pressure. The temperature of the silicon carbide substrate is 720 K.

Very small amounts of chlorine and fluorine remaining on the silicon carbide surface are simultaneously detected, as shown in Figure 2.12. These halogen atoms may be removed by increasing the substrate temperature or baking in ambient hydrogen. The existence of oxygen may be due to the oxidation and the contamination of the silicon carbide surface by airborne organic compounds, that occurred during the storage of the substrate in air after etching.

2.5. Chemical Reactions

The gaseous products and the chemical reactions associated with silicon carbide etching are explained. Figure 2.15 shows the mass spectra of the gaseous species existing in the exhaust gas immediately after beginning silicon carbide substrate etching at 770 K using chlorine trifluoride gas at atmospheric pressure. The partial pressures are normalized using the pressure at the mass of 28 a.m.u., which is the largest partial pressure in this measurement and which can be assigned to silicon from the silicon tetrafluoride and nitrogen remaining in the QMS system.

Typical mass spectra, shown in Figure 2.15, are interpreted by taking into account the fragmentation in the QMS analyzer and the isotopic abundance of chlorine [48-50]. In this figure, the ion species at a mass of 14 a.m.u. is assigned to N^+, which is the fragment of nitrogen gas. The low partial pressures corresponding to chlorine trifluoride and its fragment, ClF_3^+, at masses of 92 and 94 a.m.u., and ClF^+ at 54 and 56 a.m.u., are detected. The partial pressures observed at masses of 70, 72 and 74 a.m.u. correspond to Cl_2^+, which is assigned to chlorine gas. Chlorine gas is produced due to the chemical reaction during silicon carbide etching, similar to that for the silicon etching [35, 51]. The partial pressures at masses of 35 and 37 a.m.u. can be assigned to Cl^+, which is a fragment from both chlorine trifluoride and chlorine (Cl_2). The partial pressures observed at masses of 19 and 20 a.m.u. correspond to F^+ and HF^+, respectively. F^+ and HF^+ are produced due to the fragmentation of chlorine trifluoride. Since the partial pressure of fluorine (F_2) at mass 38 a.m.u. did not appear in the mass spectra, the thermal dissociation of chlorine trifluoride gas [52] is negligible in the gas phase of the cold wall reactor used in this study.

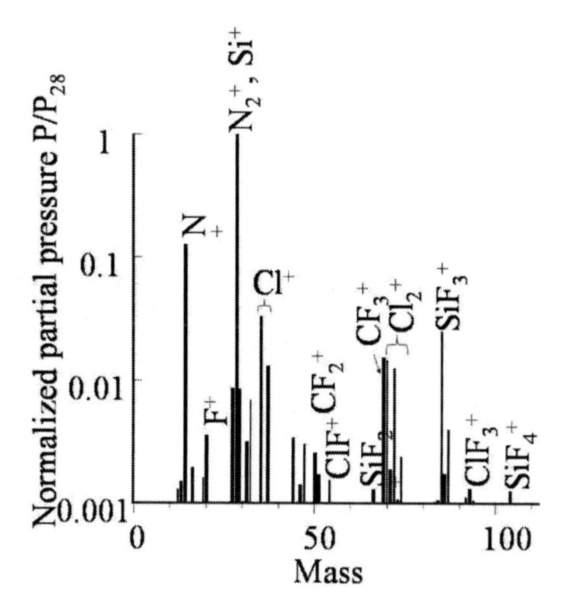

Figure 2.15. Mass spectra of gaseous species existing in the exhaust gas from the reactor during the etching of the silicon carbide surface using chlorine trifluoride gas at atmospheric pressure. The temperature of the silicon carbide substrate is 770 K. The ionization conditions are 70 eV and 1.73 mA.

The partial pressures at masses of 66, 85 and 104 a.m.u. can be assigned to SiF_2^+, SiF_3^+ and SiF_4^+, respectively, whose parent is silicon tetrafluoride, like silicon etching [51]. The gaseous carbon compound produced by etching is identified as carbon tetrafluoride (CF_4), because the partial pressures at masses of 50 and 69 a.m.u. correspond to CF_2^+ and CF_3^+, respectively, which can be assigned as fragments of carbon tetrafluoride.

Thus, silicon and carbon become the gaseous species of silicon tetrafluoride and carbon tetrafluoride, respectively. The overall chemical reaction between silicon carbide and chlorine trifluoride is as follows:

$$3SiC + 8ClF_3 \rightarrow 3SiF_4 \uparrow + 3CF_4 \uparrow + 4Cl_2 \uparrow \quad (2\text{-}1)$$

3. Single-Crystalline 4H-Silicon Carbide Etching Using Chlorine Trifluoride Gas

3.1. Substrate, Reactor and Process

The dry etching of single-crystalline 4H-silicon carbide using chlorine trifluoride gas over the wide temperature range of 570-1570 K is reviewed, particularly about the etch rate, surface chemical reaction rate constant, surface morphology and etch pits. Additionally, the etch rate and rate constant of fused silica (silicon dioxide, SiO_2) using chlorine trifluoride gas is reported, because its information is necessary for designing the safe process and reactor.

The substrate used is the n-type single-crystalline 4H-silicon carbide wafer having a (0001) surface, 8-degrees off-oriented to <11-20>. This substrate has nitrogen as the n-type dopant at the concentration of 3 - 5 x 10^{18} cm^{-3}. The 4H-silicon carbide substrate, having 5 mm wide x 5 mm long x 400 μm thick dimensions, is placed on the center of the polycrystalline 3C-silicon carbide susceptor, which has the dimension of 30 mm wide × 40 mm long × 0.2 mm thick produced by the chemical vapor deposition method (Admap Inc., Tokyo).

The reactor and process are the same as those shown in Figures 2.1 and 2.2 in the previous section. The Si-face (0001) and C-face (000-1) of the 4H-silicon carbide substrates are etched using chlorine trifluoride gas. The etching is performed at the temperatures between 570 – 1570 K at the chlorine trifluoride gas flow rate of 0.1 – 0.3 slm. The average etch rate of 4H-silicon carbide is determined by the decrease in the substrate weight. Because the 3C-silicon carbide susceptor and the 4H-silicon carbide substrate are simultaneously etched in the reactor, the etch rate obtained is comparable to the average value for the wide 4H-silicon carbide substrate. The etch rate of fused silica is obtained similar to 4H-silicon carbide.

The X-ray topograph of Si-face and C-face 4H-silicon carbide was taken at the beam-line BL15C of the Photon Factory of the High Energy Accelerator Research Organization (Proposal No. 2006G286), in order to evaluate the crystalline defects.

3.2. Numerical Calculation of Etch Rate

The etch rate of single-crystalline 4H-silicon carbide and fused silica is numerically calculated. The geometry of horizontal cold-wall reactor, shown in the previous section (Figure 2.1), is taken into account for a series of calculations. In order to evaluate the silicon carbide etch rate and the overall rate constant in steady state in non-uniformly distributed temperature and gas flow fields, the two-dimensional equations of mass, momentum, energy, species transport and surface chemical reaction, linked with the ideal gas law, are solved. The discretized equations are coupled and solved using the SIMPLE algorithm [53] on a CFD software package, Fluent version-6 (Fluent, Inc., Lebanon, NH, USA).

The silicon carbide etching is assumed to follow the overall reaction in Eq. (3-1) [36, 37],

$$3SiC + 8ClF_3 \rightarrow 3SiF_4 + 3CF_4 + 4Cl_2. \qquad (3\text{-}1)$$

The fused silica is assumed to be etched by the overall reaction as follows:

$$3SiO_2 + 4ClF_3 \rightarrow 3SiF_4 + 3O_2 + 2Cl_2. \qquad (3\text{-}2)$$

Mass changes due to the chemical reaction of Eqs. (3-1) and (3-2) are taken into account in the boundary conditions at the surface of silicon carbide and fused silica. The overall reactions shown in Eqs. (3-1) and (3-2) are assumed to be a first-order reaction.

$$\text{SiC etch rate} = 6 \times 10^{7}\, M_{SiC}\, k_{SiC}\, [ClF_3] / \rho_{SiC}\ (\mu m\ min^{-1}), \qquad (3\text{-}3)$$

$$\text{Fused silica etch rate} = 6 \times 10^7 M_{SiO_2} k_{SiO_2} [ClF_3] / \rho_{SiO_2} \; (\mu m \; min^{-1}), \qquad (3\text{-}4)$$

where ρ_{SiC} and ρ_{SiO_2} are the density of solid silicon carbide and solid fused silica (kg m^{-3}), respectively. M_{SiC} and M_{SiO_2} are the molecular weight of silicon carbide and silicon dioxide (kg mol^{-1}), respectively. The factor 6×10^7 is used for the unit conversion of m s^{-1} to μm min^{-1}. k_{SiC} and k_{SiO2} are the overall rate constant for the reaction of Eqs. (3-3) and (3-4) (m s^{-1}), respectively. $[ClF_3]$ is the concentration of chlorine trifluoride gas at the silicon carbide surface and at the fused silica surfece (mol m^{-3}). The concentration of each species at the surface is governed by a balance between the consumption due to the chemical reaction and the diffusion fluxes driven by the concentration.

The gas velocity and pressure at the inlet are 0.08 m s^{-1} and 1.0133×10^5 Pa, respectively. The heat capacities of chlorine trifluoride, nitrogen, tetrafluorosilane, tetrafluorocarbon and chlorine are taken from the literature [54]. The gas properties, such as the viscosity and the thermal conductivity of chlorine trifluoride, tetrafluorosilane, chlorine and nitrogen are estimated with the method described in the literature [55]. The Lennard-Jones parameters of σ and ε/k for chlorine trifluoride are 4.63 Angstroms and 355 K, respectively, which are obtained using a theoretical equation [55] taking the value of viscosity [56] into account. Each physical constant is expressed as a function of temperature. The properties of the mixed gas are estimated theoretically [57]. The binary diffusion coefficients of chlorine trifluoride, tetrafluorosilane and chlorine are estimated using the method described in the literature [55].

The overall rate constant, k_{SiC} and k_{SiO2} in Eqs. (3-3) and (3-4) are obtained so that the calculated etch rates agree with those measured at various conditions.

3.3. Etch Rate

Figure 3.1 shows the etch rate of the Si-face and C-face 4H-silicon carbide at the substrate temperatures between 570 K and 1570 K. The etch rate of the C-face 4H-silicon carbide is slightly higher than that of the Si-face 4H-silicon carbide. The etch rate of the Si-face and C-face 4H-silicon carbide is near 5 μm min^{-1} and it is still flat at the temperatures between 770 K and 1570 K. This flat etch rate behavior is similar to that of polycrystalline 3C-silicon carbide, shown in Figure 2.3 in the previous section. The various surface morphologies at higher temperatures shown in the latter part of this section are obtained at nearly the same etch rate.

Figure 3.2 shows the typical behavior of the etch rate changing with the chlorine trifluoride gas concentration, obtained at 1370 K. The etch rate is proportional to the chlorine trifluoride gas concentration, similar to that of polycrystalline 3C-silicon carbide, shown in Figure 2.5.

The fluctuation of the etch rate, shown in Figures 3.1 and 3.2, is entirely 20 %, which is due to the considerable distribution of the etch rate over the 3C-silicon carbide susceptor.

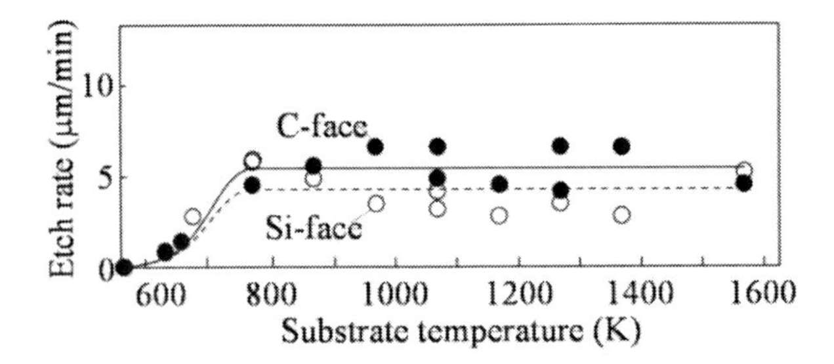

Figure 3.1. Etch rate of 4H-silicon carbide using chlorine trifluoride gas at 100 %, 0.1 slm and various temperatures. Dark circle: C-face, and white circle: Si-face.

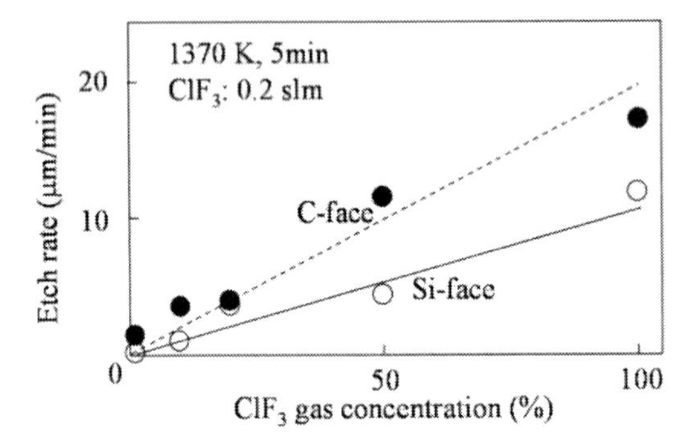

Figure 3.2. Etch rate of Si-face (white circle) and C-face (dark circle) 4H-silicon carbide changing with chlorine trifluoride gas concentration, at 1370 K and at the total gas flow rate of 0.2 slm for 5 min.

The relationship between the etch rate and the chlorine trifluoride gas flow rate is shown in Figures 3.3. Figures 3.3 (a) and 3.3 (b) show the etch rate of the Si-face and the C-face, respectively, of the 4H-silicon carbide substrate by chlorine trifluoride gas (100%) at atmospheric pressure, when changing with the chlorine trifluoride gas flow rate. The circle, square and triangle denote the etch rate at the substrate temperatures of 770, 870 and 970 K, respectively. As shown in these figures, the Si-face and C-face etch rates increase with the gas flow rate of the chlorine trifluoride gas. Additionally, the etch rate at 770, 870 and 970 K overlap each other for both the Si-face and the C-face, consistent with Figures 3.1. The etch rate of the Si-face is about 60 % of that of the C-face. The relationship between the Si-face etch rate and the C-face etch rate is similar to that of another empirically known etching technique, such as the potassium hydroxide method [58].

Next, the etch rate of fused silica is explained. The relationship between the SiO_2 etch rate and the concentration of chlorine trifluoride gas is shown in Figure 3.4. The total flow rate of the chlorine trifluoride gas and N_2 gas mixture is fixed at 0.2 slm. The SiO_2 substrate temperature is 770 K. The fused silica etch rate increases with the increasing concentration of chlorine trifluoride gas, as shown in Eq. (3-4). Therefore, SiO_2 etching is assumed to be a first-order chemical reaction.

The relationship between the fused silica etch rate and the flow rate of chlorine trifluoride gas is shown in Figure 3.5. The concentration of chlorine trifluoride gas is 100% without any dilution. The fused silica substrate temperature is 770 K. The fused silica etch rate increases with the increasing flow rate. From Figure 3.5, the relationship between the fused silica etch rate and the chlorine trifluoride gas flow rate is linear. This result shows that the inside wall of the fused silica chamber becomes thinner in a shorter period with higher chlorine trifluoride gas flow rate.

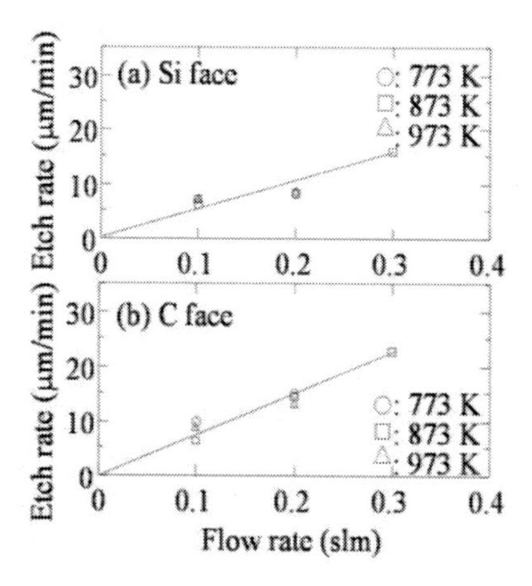

Figure 3.3. Etch rate of (a) Si-face and (b) C-face of 4H-silicon carbide substrate by chlorine trifluoride gas (100%) at atmospheric pressure in the flow rate range between 0.1 and 0.3 slm. Circle, square and triangle show the etch rates at the substrate temperatures of 770, 870 and 970 K, respectively.

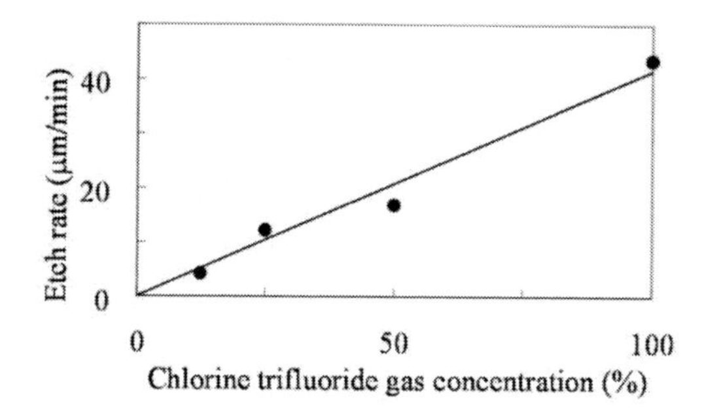

Figure 3.4. Etch rate of fused silica by chlorine trifluoride gas in ambient nitrogen, at the total gas flow rate of 0.2 slm and at 770 K.

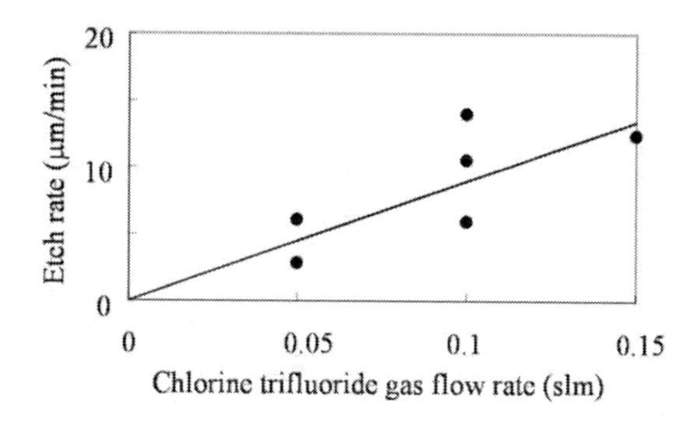

Figure 3.5. Etch rate of fused silica by chlorine trifluoride gas at 100 %, at atmospheric pressure, and at 770 K.

3.4. Surface Reaction Rate Constant

The rate constant of the surface chemical reaction is evaluated, using the numerical calculation taking into account the transport phenomena in the reactor. Figure 3.6 is the Arrhenius plot of the rate constants for etching of Si-face and C-face of 4H-silicon carbide. The rate constants is expressed in Eqs. (3-5) and (3-6).

k_{SiC}=4.1 exp(-6.6 x $10^4/RT$)(m/s) for Si-face. (3-5)

k_{SiC}=98 x exp(-8.3 x $10^4/RT$)(m/s) for C-face. (3-6)

R is the gas constant (J mol^{-1} K^{-1}).

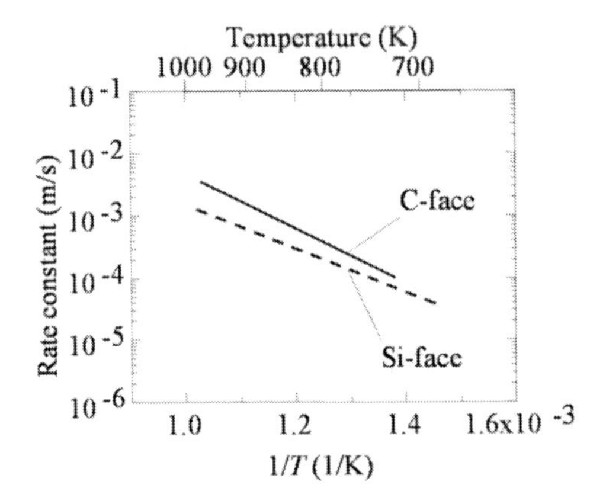

Figure 3.6. Rate constants for etching of Si-face (broken line) and C-face (solid line) of 4H-silicon carbide using chlorine trifluoride gas, obtained by numerical calculation.

In order to show that the rate constant of Eq. (3-5) can reproduce the measured etch rate behavior, the measured and the calculated etch rate values of Si-face are shown in Figure 3.7, as the Arrhenius plot. The calculation shows that the etch rate at the temperatures near 670K is near 1 μm/min, and it becomes near 10 μm/min at 1000K. The calculated etch rate tends to become flat at the higher temperatures at the chlorine trifluoride gas flow rate of 0.1 and 0.2 slm. Additionally, the etch rate obtained by the calculation increases with increasing the chlorine trifluoride gas flow rate. Because the great etchant flow rate can moderate the etchant depletion occurring in the downstream region, the average etch rate over the 4H-silicon carbide substrate and the 3C-silicon carbide susceptor can increase with the increasing flow rate of chlorine trifluoride gas. The etch rate for C-face calculated using Eq. (3-6) also shows the typical behavior of the measurement. Therefore, Eqs. (3-5) and (3-6) are applicable to reproduce the behavior of the 4H-silicon carbide (Si-face and C-face) etch rate.

The rate constant for the surface reaction of fused silica etching is shown in Figure 3.8.and Eq. (3-7).

k_{SiO2} = 0.11 exp(-4.6 x 10^4 / RT)(m s^{-1}), (3-7)

where T is the fused silica substrate temperature (K) and R is the gas constant (J mol^{-1} K^{-1}).

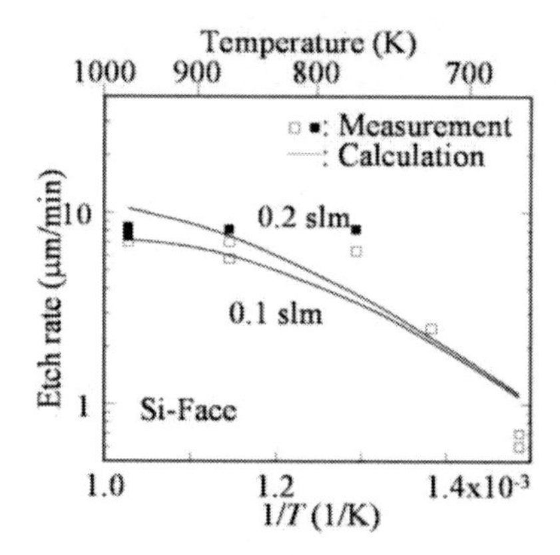

Figure 3.7. Arrhenius plot of 4H-silicon carbide Si-face etch rate at chlorine trifluoride flow rate of 0.1 and 0.2 slm. Square: measurement, solid line: calculation.

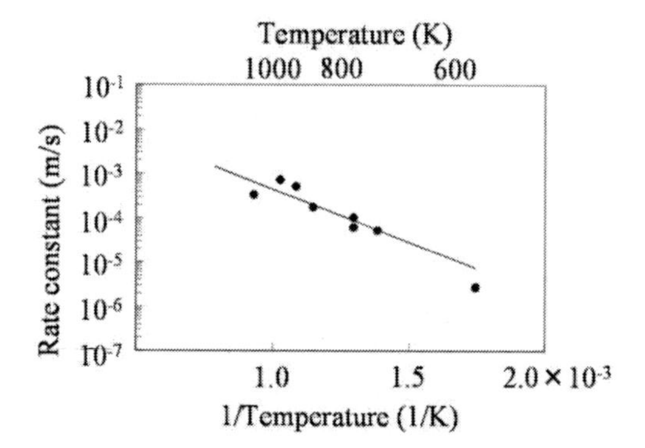

Figure 3.8. Arrhenius plot of rate constant for fused silica etching using chlorine trifluoride gas.

Figure 3.9 shows the measured and calculated etch rates. The circles denote the measured data and the solid line is obtained using Eq. (3-7). The flow rate of chlorine trifluoride is fixed at 0.1 slm.

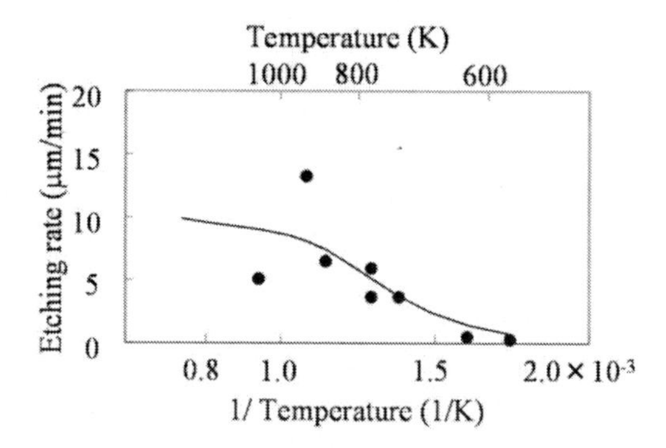

Figure 3.9. Arrhenius plot of fused silica etch rate at chlorine trifluoride gas flow rate of 0.1 slm. Circles denote measured data and the solid line is obtained by numerical calculation.

When the fused silica temperature is lower than 620 K, the etch rate increases moderately with increasing substrate temperature. In the temperature range between 620 and 1170 K, the etch rate increases significantly with temperature. At temperatures higher than 1170 K, the etch rate increases slowly with temperature: the etch rate tends to be flat.

3.5. Etch Rate Behavior

Using the surface reaction rate constant, the mechanism to cause the flat etch rate behavior for 4H-silicon carbide at higher temperatures is explained. Because the etch rate is expressed by the etchant transport rate and the surface chemical reaction rate, the mass flux for the etching is simply described using Eq. (3-8) [47], which is useful for the quick evaluation of the rate limiting process.

$$Mass\ flux\ for\ etching = \frac{C_g(T_g)}{\frac{h}{D(T_{ave})} + \frac{1}{k(T_s)}}, \qquad (3\text{-}8)$$

where C_g is the chlorine trifluoride concentration in the gas phase, h is the representative height for the diffusion transport of chlorine trifluoride gas from the gas phase to the substrate surface, D is the diffusivity of chlorine trifluoride gas, k is the surface chemical reaction rate constant, T_g is the gas phase temperature, T_s is the substrate surface temperature, and T_{ave} is the average of T_g and T_s.

In order to discuss the rate limiting process for the 4H-silicon carbide etching, the values of h/D and $1/k$ for 4H-silicon carbide etching are shown in Figure 3.10. At the low substrate temperatures, the $1/k$ value is significantly larger than the h/D value. Because the h/D value is negligible, the etch rate at low temperatures is dominated by the surface chemical reaction rate. However, the $1/k$ value decreases with the increasing substrate temperature. At the substrate temperatures near 1000K, $1/k$ becomes comparable to h/D; the etch rate is dominated by both the transport rate and the surface reaction rate. At the substrate temperatures higher than 1000K, $1/k$ becomes smaller than h/D; the etch rate is dominated by the transport rate. Thus, the change of the etch rate with the substrate temperature can be moderate.

The $1/k_{SiO2}$ value of fused silica etching is shown in Figure 3.10. The behavior of $1/k_{SiO2}$ is similar to that of $1/k_{SiC}$.

Figure 3.10 also shows the calculated $1/k$ value for silicon etching by chlorine trifluoride gas, because the silicon etch rate by chlorine trifluoride gas in the same reactor for obtaining silicon carbide etch rate is very flat over very wide temperature range. For silicon etching, the summation of $1/k$ value and h/D value moderately changes with the substrate temperature. Because the summation of $1/k$ value and h/D for 4H-silicon carbide etching at the substrate temperatures higher than 1000 K shows the moderate change similar to that for silicon etching, the 4H-silicon carbide etch rate is flat at high substrate temperatures.

Figure 3.11 shows the etch rate of silicon carbide, silicon and fused silica, which are calculated and are normalized using those at 1400 K. The Si-face 4H-silicon carbide etch rate, the C-face 4H-silicon carbide etch rate and the fused silica etch rate at low temperatures significantly increase with the increasing substrate temperatures; their values higher than 1000 K show the flat etch rate.

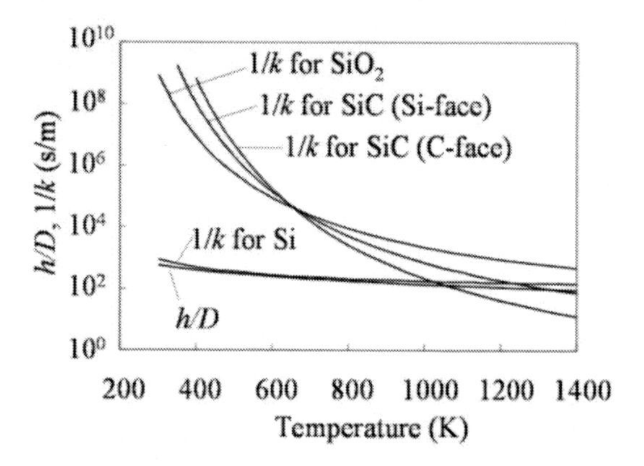

Figure 3.10. h/D values and 1/k values of silicon carbide (Si-face), silicon carbide (C-face), Si and fused silica, as a function of the substrate temperature.

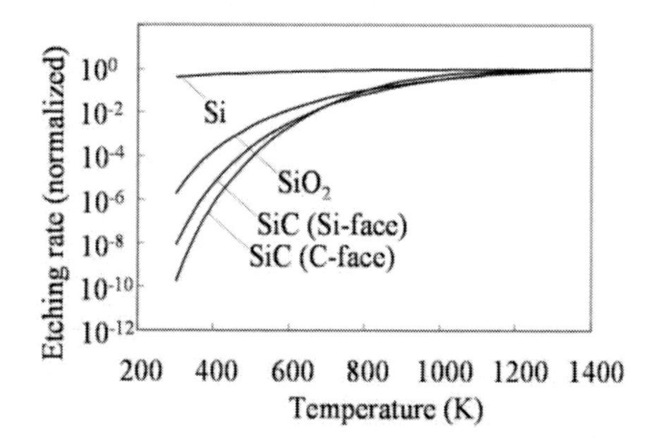

Figure 3.11. Silicon carbide (Si-face and C-face), silicon and fused silica etch rate, calculated and normalized using the etch rate at 1400 K.

3.6. Safe Design of Reactor System

The silicon carbide film is very often formed as a by-product on the susceptor surface and on the chamber wall surface, in addition to the substrate. Because such the film emits small particles to cause the surface deterioration of the silicon carbide epitaxial film, it must be removed using the chemical reaction with chlorine trifluoride gas. However, because the chlorine trifluoride gas can simultaneously etch the fused silica chamber, The etch rate of silicon carbide and fused silica should be compared.

Here, in the cold wall reactor using the infrared light heating unit, the silicon carbide film attached on the fused silica surface is heated and removed by chlorine trifluoride gas. After the silicon carbide film is removed, infrared light is not absorbed by the fused silica chamber wall. Then, the temperature of the chamber wall immediately decreases; the etching of the chamber wall surface by chlorine trifluoride gas is suppressed, following the etch rate shown in Figure 3.11.

When the temperature of the by-product, such as silicon or silicon carbide, film is 1370 K and the temperature of the transparent fused silica chamber wall is 770 K, the etch rate ratios, R, between SiO_2 and Si, and between SiO_2 and SiC are as follows:

$$R_{SiO_2-Si} = \frac{Fused\ silica\ etch\ rate\ at\ 773\,K}{Silicon\ etch\ rate\ at\ 1373\,K} = 0.087\,. \tag{3-9}$$

$$R_{SiO_2-SiC} = \frac{Fused\ silica\ etch\ rate\ at\ 773\,K}{Siliconcarbide\ etch\ rate\ at\ 1373\,K} = 0.089\,. \tag{3-10}$$

From Figures (3.9) and (3.10), the etch rate of the fused silica chamber wall surface by chlorine trifluoride gas is quite low, when the by-products (silicon and silicon carbide) film is removed.

As described above, when transparent fused silica is used for the chamber, a reactor system with a sufficiently low etch rate of the fused silica chamber by chlorine trifluoride gas can be achieved using the IR lamp heating unit and the chamber cooling unit, which maintain the transparent chamber wall temperature sufficiently below 570 K.

3.7. Surface Morphology

Figure 3.12 shows the surface morphology of the Si-face 4H-silicon carbide before and after the etching at the chlorine trifluoride gas concentration of 100 % and at the flow rate of 0.1 slm. The etching was performed at the substrate temperatures of (b) 570, (c) 620, (d) 770, (e) 970, (f) 1270, (g) 1370 and (h) 1570 K. The etched depth was 5-18 μm.

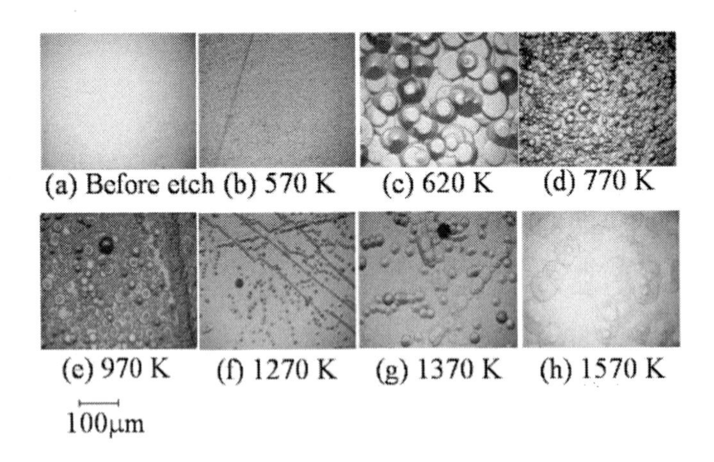

Figure 3.12. Surface morphology of Si-face 4H-silicon carbide (a) before and after the etching using chlorine trifluoride gas at the concentration of 100 %, at the substrate temperature of (b) 570, (c) 620, (d) 770, (e) 970, (f) 1270, (g) 1370 and (h) 1570 K and at the flow rate of 0.1 slm. The etched depth is 5-18 μm.

Figure 3.12 (a) shows the surface morphology of the Si-face 4H-silicon carbide substrate before the etching. Figure 3.12 (b) shows that there are many small pits after the etching at 570 K. At 620 K, the pits are very large, nearly 50 μm in diameter, as shown in Figure 3.12 (c). With the increasing temperature, the pits tends to become small and shallow, as shown in Figures 3.12 (d) - (h). The pit diameter after etching at 770 K, shown in Figure 3.12 (d), is nearly 10 % of that at 620 K, shown in Figure 3.12 (c). Figure 3.12 (e) shows that the pit diameter becomes even smaller and the pit density decreases at 970 K. This trend is very clear at the temperatures higher than 1270 K, as shown in Figs 3.12 (f), (g) and (h). Figure 3.12 (g) shows that the pit density significantly decreases at 1370 K. The sharp-shaped pits, presented at the lower temperatures, are not there, on the silicon carbide surface after the etching at 1570 K, as shown in Figure 3.12 (h).

Next, the influence of chlorine trifluoride gas concentration on the surface morphology is explained. Figure 3.13 shows the surface morphology of the Si-face 4H-silicon carbide surface taken by the optical microscope before and after the etching for 5 min at the total flow rate of 0.2 slm, at the substrate temperature of 1370 K and at various chlorine trifluoride gas concentrations. Figure 3.13 (a) shows the surface morphology of the Si-face 4H-silicon carbide before the etching.

Although there are many large pits at the chlorine trifluoride gas concentration of 100%, as shown in Figure 3.13 (b), they become significantly small and less at 20 %, as shown in Figure 3.13 (d). Figure 3.13 (e) shows that the surface etched at 10% is flat with only a small number of pits. At 1 %, most of the surface is flat, except for scratches, as shown in Figure 3.13 (f).

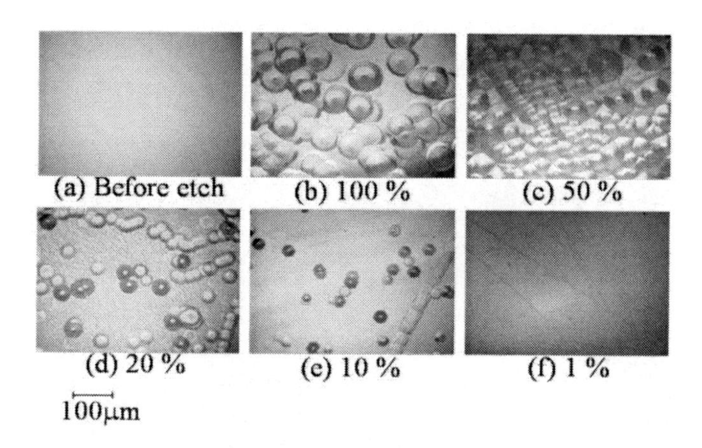

Figure 3.13. Surface morphology of Si-face 4H-silicon carbide surface (a) before and after etching at the total gas flow rate of 0.2 slm, 1370 K and various chlorine trifluoride gas concentrations of (b) 100, (c) 50, (d) 20, (e) 10 and (f) 1 %, for 5 min.

From Figures 3.12 and 3.13, the Si-face 4H-silicon carbide surface after etching tends to be flat with the increasing temperature and decreasing chlorine trifluoride gas concentration. Following this trend, the Si-face 4H-silicon carbide surface is etched at 1570 K at the chlorine trifluoride gas concentration of 1 % for 0.5 min. Figure 3.14 shows the AFM photograph of the etched surface. Figures 3.14 (a), (b) and (c) are the plan view, A-A' cross section and B-B' cross section, respectively. Although the etched depth is only about 0.03 μm, it can reveal the trend of the surface, causing pit or not. As shown in Figure 3.14 (a), this surface does not show any etch pit; Figures 3.14 (b) and (c) showes no periodical shape reflecting the 4H-silicon carbide crystal step [59]. The root-mean-square (RMS) roughness are 0.1 and 0.2 nm on A-A' line and B-B' line, respectively, which are comparable to that of the polished 4H-silicon carbide substrate surface. Thus, the shallow etching for removing thin layer, such as damaged layer, is possible with maintaining the specular surface of the Si-face 4H-silicon carbide.

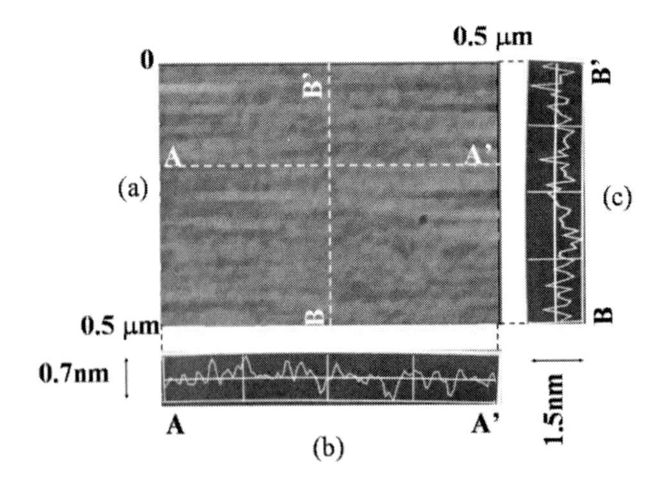

Figure 3.14. AFM photograph of Si-face of 4H-silicon carbide, etched for 0.5 min using the chlorine trifluoride gas concentration of 1 % diluted in ambient nitrogen at the substrate temperature of 1570 K and at the total flow rate of 4 slm. (a): plan view, (b): A-A' cross section, and (c): B-B' cross section.

Figure 3.15 shows the surface morphology of the C-face 4H-silicon carbide before and after the etching at the chlorine trifluoride gas concentration of 100 %, at (b) 570, (c) 620, (d) 770, (e) 1070, (f) 1270, (g) 1370 and (h) 1570 K and at the flow rate of 0.1 slm. The etched depth is 10-30 μm. Figure 3.15 (a) is the C-face 4H-silicon carbide surface before the etching.

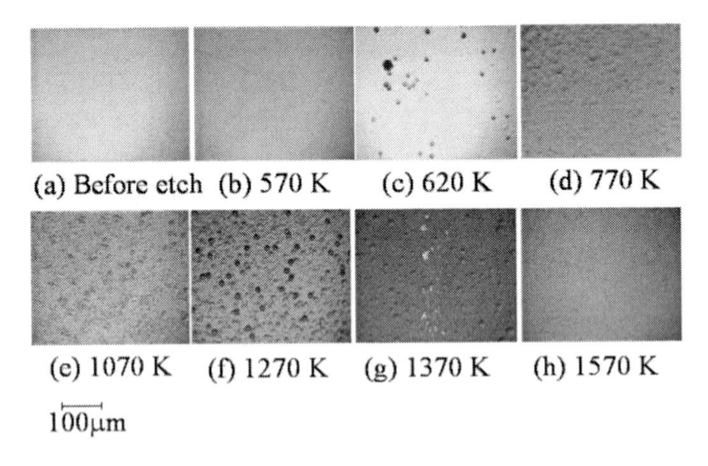

Figure 3.15. Surface morphology of C-face 4H-silicon carbide (a) before and after the etching using chlorine trifluoride gas at the concentration of 100 %, at the substrate temperature of (b) 570, (c) 620, (d) 770, (e) 1070, (f) 1270, (g) 1370 and (h) 1570 K and at the flow rate of 0.1 slm. Etched depth is 10-30 μm.

There is the flat surface after the etching at 570 K, as shown in Figure 3.15 (b), because of significantly small etch rate. However, Figure 3.15 (c) shows that pits are produced at 620 K. The surface etched at the temperatures between 770 K and 1270 K have many small pits, as shown in Figures 3.15 (d), (e) and (f) . Figures 3.15 (g) and (h) show that the pit diameter decreases at the temperatures higher than 1370 K. The surface etched at 1570 K shows a flat surface as shown in Figure 3.15 (h).

Next, the influence of the chlorine trifluoride gas concentration on the surface morphology of the C-face 4H-silicon carbide is explained. Figure 3.16 shows the morphology of the C-face 4H-silicon carbide surface before and after the etching for 5 min at the various chlorine trifluoride gas concentrations of (b) 100, (c) 50, (d) 20, (e) 10 and (f) 1 % at the total flow rate of 0.2 slm. The substrate temperature is fixed at 1370 K. The etched depth is 3-84 μm.

Figure 3.16 (b) shows that the C-face 4H-silicon carbide surface shows large and shallow pits after etching at the chlorine trifluoride gas concentration of 100 %. As shown in Figure 3.16 (c), the etch pits become shallow at the chlorine trifluoride gas concentration of 50 %. Figures 3.16 (d), (e) and (f) show that the etched surface is entirely flat at the chlorine trifluoride gas concentrations less than 20 %. Particularly, the surface etched at 1% is flat, as shown in Figure 3.16 (f). Overall, the trend in the etched surface morphology of the C-face 4H-silicon carbide is similar to that of the Si-face 4H-silicon carbide, although the pit size of the C-face is smaller than that of Si-face.

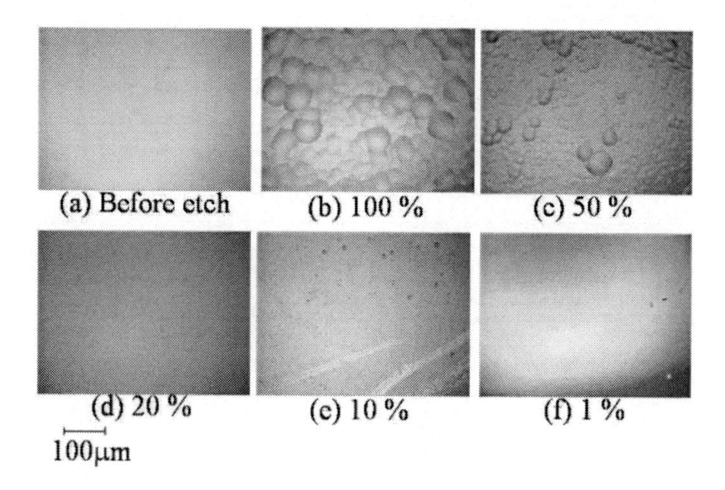

Figure 3.16. Surface morphology of C-face 4H-silicon carbide surface before and after etching at the total gas flow rate of 0.2 slm, 1370 K and various chlorine trifluoride gas concentrations for 5 min. (a) before etching, (b) 100, (c) 50, (d) 20, (e) 10 and (f) 1 %. Etched depth is 3-84 μm.

Similar to the Si-face 4H-silicon carbide surface, the C-face 4H-silicon carbide surface is etched at 1570 K at the chlorine trifluoride gas concentration of 1 % for 0.5 min. Figures 3.17 (a), (b) and (c) are the AFM photographs of the plan view, A-A' cross section and B-B' cross section, respectively. The etched depth is near 0.03 μm. Figure 3.17 (a) does not show any shape like an etch pit; Figs 3.17 (b) and (c) show no periodical shape reflecting the 4H-silicon carbide crystal step [59]. The RMS roughness is 0.4 and 0.3 nm on A-A’ line and B-B’ line, respectively, which are comparable to that of the polished 4H-silicon carbide substrate surface. Thus, the shallow etching without producing any trace of pit shape is possible, for the C-face 4H-silicon carbide.

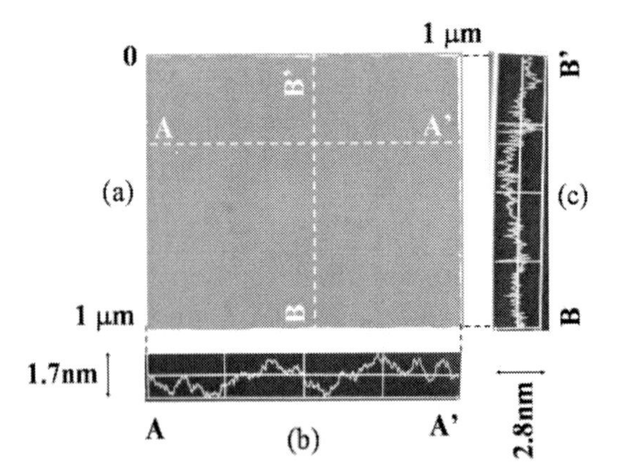

Figure 3.17. AFM photograph of C-face 4H-silicon carbide, etched for 0.5 min using the chlorine trifluoride gas concentration of 1 % diluted in ambient nitrogen at the substrate temperature of 1570 K and at the total flow rate of 4 slm. (a): plan view, (b): A-A' cross section, and (c): B-B' cross section.

3.8. Surface Morphology Behavior and Its Rate Process

Entire surface morphology behavior changing with the substrate temperature for Si-face and C-face 4H silicon carbide is summarized as follows:

(i) Very small change at very low temperatures (lower than 570 K),

(ii) Significant pit formation between 570 K and 1270 K, and

(iii) Pit formation reduced at high temperatures (higher than 1370 K).

The process of pit formation is described following the rate theory, assuming that the etch pit is formed due to the difference of the etch rate between the perfect crystal region and the weak spot having any kinds of damage and crystalline defect [10].

The rate constant of the etching in the perfect crystal region, k_P, is assumed to be expressed in Eq. (3-11).

$$k_P = A\,exp\left(-\frac{E}{RT}\right), \tag{3-11}$$

where A is the pre-exponential factor, E is the activation energy, R is the gas constant, and T is the substrate temperature. In contrast to this, the weak spot, which has larger etch rate to cause pit, is assumed to have the slightly smaller activation energy than that in the perfect region. The rate constant at the weak spot, k_W, is assumed to be expressed in Eq. (3-12), using the difference of the activation energy from that in the perfect region, ΔE.

$$k_W = A\,exp\left(-\frac{E-\Delta E}{RT}\right). \tag{3-12}$$

Assuming that the etchant gas concentration is the same in the perfect region and at the weak spot, the pit depth is expressed in Eq. (3-13), taking into account that *ΔE/RT* is very small.

$$Pit\,depth = V_{\mathrm{E}}\left(\frac{k_{\mathrm{W}} - k_{\mathrm{P}}}{k_{\mathrm{P}}}\right) = V_{\mathrm{E}}\left(exp\left(\frac{\Delta E}{RT}\right) - 1\right) \cong V_{\mathrm{E}}\frac{\Delta E}{RT}, \tag{3-13}$$

where V_{E} is the etch rate in the perfect region.

Here, assuming that V_{E} shown in Figure 3.1 is the etch rate in the perfect region, the normalized pit depth, *h*, is evaluated and shown in Figure 3.18. The *h* value is defined using the maximum value of the pit depth, in Eq. (3-14).

$$h = \frac{Pit\,depth}{Pit\,depth_{\mathrm{MAX}}} = \frac{\frac{V_{\mathrm{E}}}{T}}{\left(\frac{V_{\mathrm{E},}}{T}\right)_{Pit\,depth_{\mathrm{MAX}}}}, \tag{3-14}$$

In Figure 3.18, the *h* value at the temperatures lower than 500 K is very small; it significantly increases near 700 K. After showing its maximum, the *h* value gradually decreases with the increasing substrate temperature. Near 1600 K, the *h* value is significantly smaller than the maximum value. This trend qualitatively agrees with that of the 4H-silicon carbide surface etched using chlorine trifluoride gas. Thus, the surface morphology trend over wide temperature range can be understood mainly by the rate process.

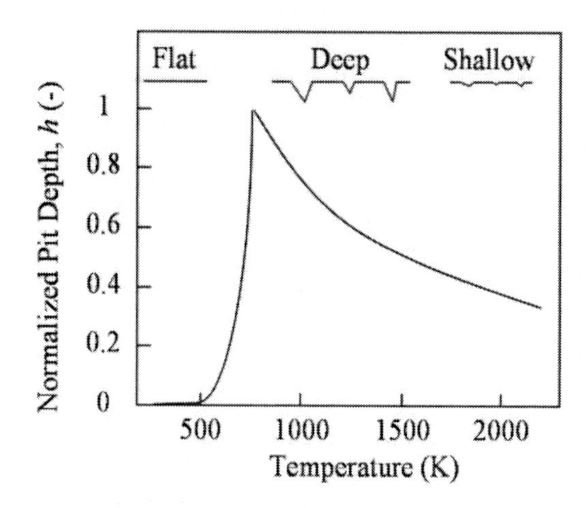

Figure 3.18. Normalized pit depth and temperature-dependent surface morphology behavior following the rate theory.

3.9. Etch Pits and Crystalline Defect

AFM photographs of the pit shape formed by the chlorine trifluoride gas are shown in Figure 3.19. Figures 3.19 (a) and (b) show the pits on the Si-face and C-face, respectively, after the etching using the chlorine trifluoride gas (100%) at atmospheric pressure for 3 min at 870 K and 0.3 slm. Figure 3.19 (c) is the etch pit of Figure 3.19 (a), the edge of which is traced using dotted line.

Figures 3.19 (a) and (c) reveals a nearly hexagonal edge shape having a flat-shaped bottom. Its diameter and depth are about 0.03 mm and 2000-3000 nm, respectively. The many pits that exist over the entire surface are shown to have the same edge and bottom shape as that shown in Figure 3.19 (a). Figure 3.19 (b) shows the pit shape on the C-face. This surface is very flat and smooth having only a very small number of circular shaped pits, which are very shallow with a diameter of 0.01 mm and a depth of about 200 nm.

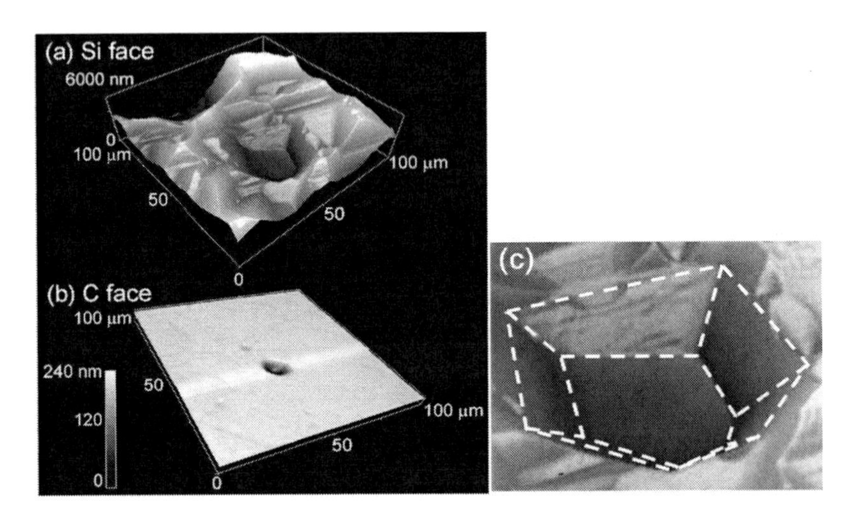

Figure 3.19. Etch pits on (a) Si-face and (b) C-face (870 K, chlorine trifluoride: 100 %, 0.3 slm). (c) is the magnification of the pit in (a), the edge of which is traced using dotted line.

Many etch pits at the 4H-silicon carbide surface, produced by the etching using chlorine trifluoride gas, are expected to show a relationship with various crystalline defects, when the etching condition is appropriate. Thus, the Si-face and C-face 4H-silicon carbide surface are etched using chlorine trifluoride gas at 100 % and at 700 K so that the pit depth become maximum value, as predicted by Figure 3.18. Additionally, the etch pits are compared with the X-ray topograph, because the X-ray topograph is suitable in order to evaluate an origin of etch pit [59, 60].

Figures 3.20 (a) and 3.21 (a) are the X-ray topograph of the Si-face and C-face 4H-silicon carbide surface, respectively. Figures 3.20 (b) and 3.21 (b) are the photograph of Si-face and C-face 4H-silicon carbide surface, respectively, etched using the chlorine trifluoride gas at 100 % and at 700 K for 60 min. White arrows in these figures indicate the position of the spots in the X-ray topograph and the etch pits at the etched surface.

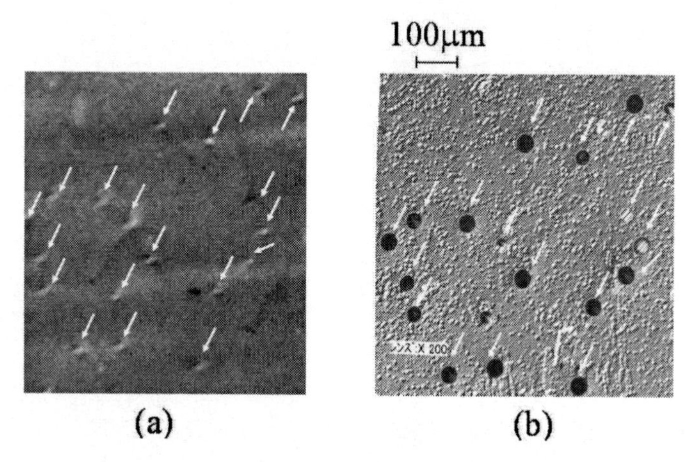

Figure 3.20. Comparison between the X-ray topograph and the etched Si-face 4H-silicon carbide surface. (a) X-ray topograph of the Si-face 4H silicon carbide surface, and (b) the Si-face 4H-silicon carbide surface etched using chlorine trifluoride gas at 100 % and at 700 K for 60 min. Arrows in this figure indicate the spot in the X-ray topograph and the etch pit at the etched surface.

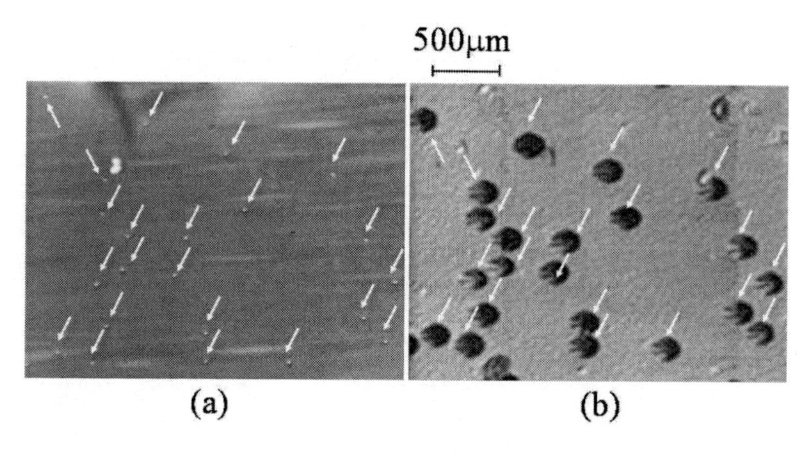

Figure 3.21. Comparison between the X-ray topograph and the etched C-face 4H-silicon carbide surface. (a) X-ray topograph of the Si-face 4H silicon carbide surface, and (b) the C-face 4H-silicon carbide surface etched using chlorine trifluoride gas at 100 % and at 700 K for 60 min. Arrows in this figure indicate the spot in the X-ray topograph and the etch pit at the etched surface.

As indicated using white arrows, there are many etch pits, the positions of which correspond to those of the spots in the X-ray topograph. The dimension of spot in Figure 3.21 (a) is larger than that in Figure 3.20 (a); the diameter of etch pits in Figure 3.21 (b) is about 250 μm which is similarly larger than that in Figure 3.20 (b), about 40 μm. Taking into account the report [61] about the dimension of etch pits formed by KOH, Figures 3.21 (a) and (b) may show the screw dislocation; Figures 3.20 (a) and (b) may show the threading edge dislocation. Because the etching technique using the chlorine trifluoride gas may reveal the crystalline defects, like the KOH technique [61], the relationship between etch pits and various crystalline defects should be further studied in future.

Because of the functions to produce the specular surface and to reveal the crystalline defects, chlorine trifluoride gas is expected to be more useful than the other wet and dry techniques [58], for silicon carbide industrial process.

4. POLYCRYSTALLINE SILICON CARBIDE FILM DEPOSITION USING MONOMETHYLSILANE GAS AT LOW TEMPERATURES

4.1. Reactor and Process

In order to obtain a silicon carbide film by chemical vapor deposition (CVD), the horizontal cold-wall reactor shown in Figure 4.1 is used. This reactor consists of a gas supply system, a quartz chamber and infrared lamps. A 30-mm-wide x 40-mm-long (100) silicon substrate manufactured by the Czochralski method is horizontally placed on the bottom wall of the quartz chamber. The silicon substrate is heated by infrared rays from halogen lamps through the quartz chamber walls. Because the quartz chamber absorbs only a slight amount of heat from the lamps, the temperature of the quartz chamber wall is kept low to achieve a cold-wall thermal condition. The height and the width of the quartz chamber are 10 mm and 40 mm, respectively.

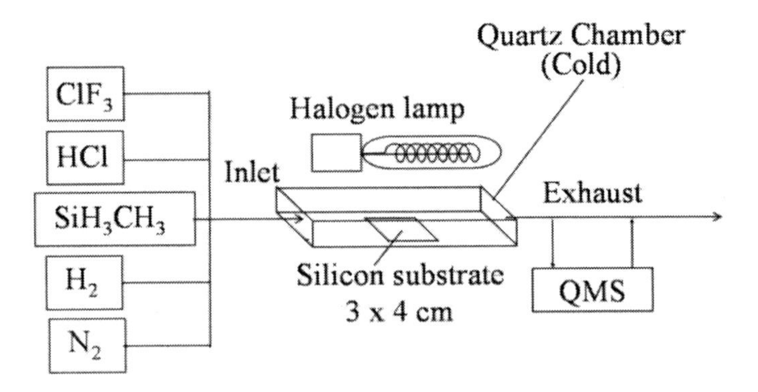

Figure 4.1. Horizontal cold-wall chemical vapor deposition reactor used for silicon carbide film deposition.

The gas supply system has the function of introducing gases, such as hydrogen, nitrogen, monomethylsilane, hydrogen chloride and chlorine trifluoride. Hydrogen gas is used as the carrier gas. This gas is also used to remove the silicon oxide film and organic contamination that exist on the silicon substrate surface [35] immediately before the film deposition. Chlorine trifluoride gas is used in order to remove the silicon carbide film deposited on the inner wall of the quartz chamber.

The process for film deposition is shown in Figures 4.1, 4.2 and 4.3. The silicon carbide film deposition is performed by the following steps.

- Step (A): cleaning of the silicon surface at 1370 K for 10 minutes in ambient hydrogen.
- Step (B): silicon carbide film deposition using monomethylsilane or a gas mixture of monomethylsilane and hydrogen chloride at 870-1220 K.
- Step (C): annealing of the silicon carbide film in ambient hydrogen at 1270 K for 10 minutes.

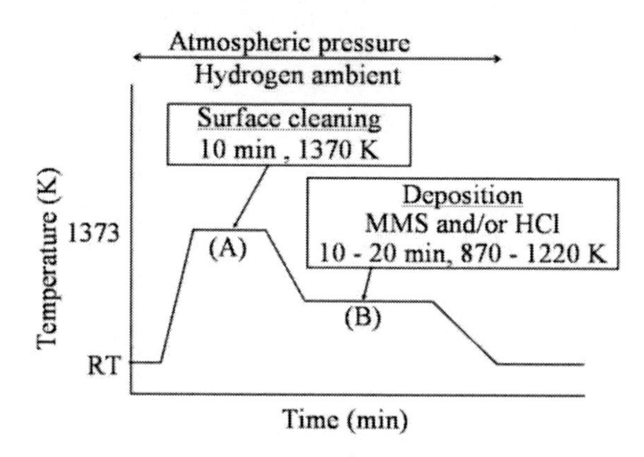

Figure 4.2. Process for silicon carbide film deposition, using gases of monomethylsilane, hydrogen chloride and hydrogen. Step (A): cleaning surface at 1370 K and Step (B): film deposition.

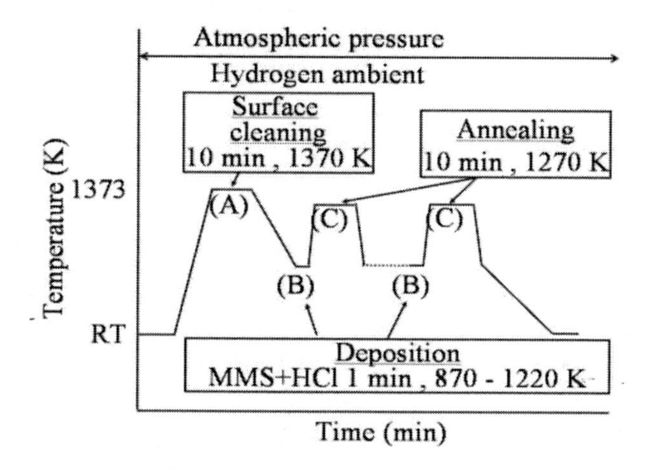

Figure 4.3. Process for silicon carbide film deposition, using gases of monomethylsilane, hydrogen chloride and hydrogen. Step (a): baking at 1370 K, Step (b): film deposition and Step (c): annealing at 1270 K.

For the process shown in Figure 4.2, Step (B) is performed after Step (A). In contrast to this, the process shown in Figure 4.3 involves first Step (A) and then the repetition of Steps (B) and (C).

The typical process for evaluating the robustness of the coating film is Steps (A), (D) and (E) shown in Figure 4.4.

- Step (A): cleaning of the silicon surface at 1370 K for 10 minutes in ambient hydrogen.
- Step (D): silicon carbide film formation using a gas mixture of monomethylsilane and hydrogen chloride, at room temperature - 1070 K after Step (A).
- Step (E): etching by hydrogen chloride gas at 1070 K for 10 minutes.

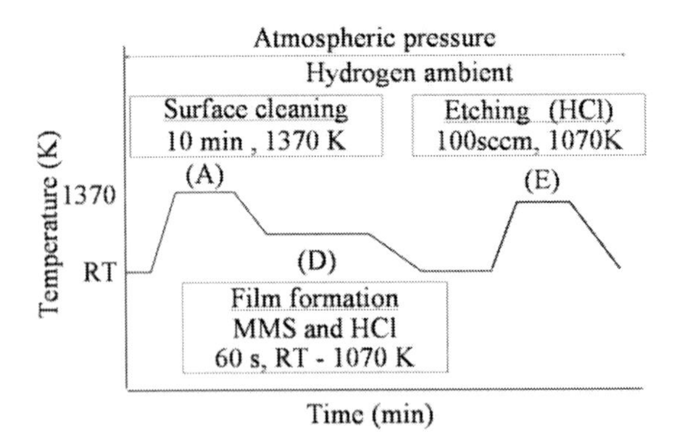

Figure 4.4. Process for silicon carbide film deposition and etching using gases of monomethylsilane, hydrogen chloride and hydrogen. Step (A): cleaning silicon surface at 1370 K, Step (D) film formation after Step (A), and Step (E): etching of film surface using hydrogen chloride gas at 1070 K.

Silicon carbide film cannot be etched by the hydrogen chloride gas; silicon surface suffers significant etching by hydrogen chloride gas at 1070 K [62]. Thus, in order to quickly determine the formation of silicon carbide, some of the films obtained by Steps (A) and (D) are exposed to the gas mixture of hydrogen chloride and hydrogen at the flow rate of 0.1 slm and 2 slm, respectively, at 1070 K, as Step (E) in Figure 4.4.

Throughout the process, hydrogen gas is introduced into the reactor at atmospheric pressure at a flow rate of 2 slm. Water vapor in the hydrogen gas is removed by passing the gas through a liquid nitrogen trap (77 K) at the entrance of the reactor. After removing the substrate from the reactor, the quartz chamber is cleaned using 10% chlorine trifluoride gas in ambient nitrogen at 670-770 K for 1 min at atmospheric pressure, which is an application of the silicon carbide etching technique using chlorine trifluoride gas described in the previous sections. The average thickness of the silicon carbide film is evaluated from the increase in the substrate weight.

In order to evaluate the gaseous products produced during the film deposition in the quartz chamber, a part of the exhaust gas from the reactor is fed to a quadrupole mass spectra (QMS) analyzer (Microvision, Spectra International LLC), as shown in Figure 4.1. For the measurement of the mass spectra, the ionization energy and current are 40 eV and 1.73 mA, respectively.

In order to evaluate the condition of the chemical bonds at the center of the silicon carbide film, the X-ray photoelectron spectra (XPS) is *ex situ* measured at the Foundation of Promotion of Material Science and Technology of Japan (Tokyo). Additionally, the infrared absorption spectra through the obtained film are measured.

The surface morphology is observed using an optical microscope, a scanning electron microscope (SEM) and an atomic force microscope (AFM) . Surface microroughness is evaluated by AFM. In order to observe the surface morphology and the film thickness, a transmission electron microscope (TEM) is used.

The cleaning technique is an application of the silicon carbide etching technique using chlorine trifluoride gas [36, 37]. After finishing the film deposition, the quartz chamber is cleaned, as shown in Figure 4.5, using chlorine trifluoride gas (Kanto Denka Kogyo Co., Ltd., Tokyo, Japan) at the concentration of 10 % in ambient nitrogen at 670-770 K for 1 minute at atmospheric pressure. Figure 4.5 (a) shows the quartz chamber which has thick silicon carbide film formed at its inner surface. Most of the deposited film is removed by chlorine trifluoride gas at 670 K, as shown in Figure 4.5 (b). Although very small amount of silicon carbide film remains, it can be removed using chlorine trifluoride gas at 10 % and 770 K, as shown in Figure 5.4 (c).

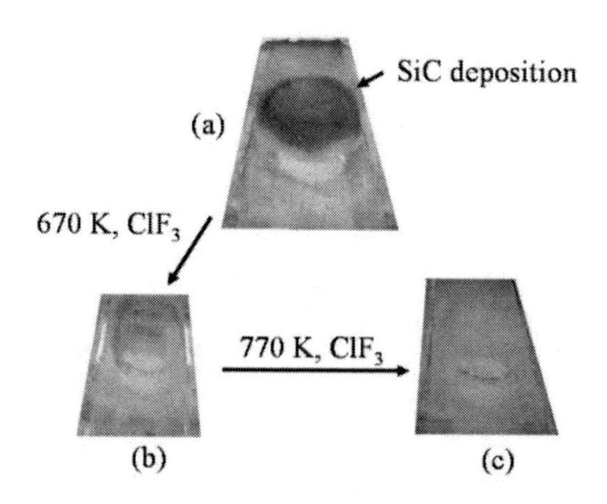

Figure 4.5. Photograph of quartz chamber (a) after silicon carbide film deposition using monomethylsilane gas at high temperatures, (b) after cleaning using chlorine trifluoride gas at 10% in ambient nitrogen and at 670 K, and (c) after cleaning using chlorine trifluoride gas at 10% and 770 K.

4.2. Thermal Decomposition of Monomethylsilane

First, the thermal decomposition behavior of the monomethylsilane gas is shown in order to evaluate the temperature suitable for silicon carbide film deposition, while maintaining the silicon-carbon bond present in the molecular structure.

Figure 4.6 shows the quadrupole mass spectra at the substrate temperatures of (a) 300 K, (b) 970 K, and (c) 1170 K. The concentration of monomethylsilane gas is 5% in ambient hydrogen at atmospheric pressure. The measured partial pressure is normalized using that of hydrogen molecule.

Figure 4.6(a) shows the three major groups at masses greater than 12, 28 and 40 a. m. u., corresponding to CH_x^+, SiH_x^+ and $SiH_xCH_y^+$, respectively. Because no chemical reaction occurs at room temperature, CH_x^+ and SiH_x^+ are produced by fragmentation in the mass analyzer. In this figure, the pressure of SiH_x^+ is higher than that of CH_x^+, because ionization of species including silicon is easier than that of species with carbon. [49] Because a small amount of chlorine from the chlorine trifluoride used for the *in situ* cleaning remains in the reactor, Cl^+ is detected, as shown in Figure 4.6(a). Figure 4.6(b) also shows that the three major groups of CH_x^+, SiH_x^+ and $SiH_xCH_y^+$ exist at 970 K without any significant change in their peak height compared with the spectrum in Figure 4.6(a). Therefore, Figure 4.6(b) indicates that the amount of monomethylsilane thermally decomposed at 970 K is relatively small. However, at 1170 K, the partial pressure of the CH_x^+ group increases and that of the $SiH_xCH_y^+$ group decreases, as shown in Figure 4.6 (c). Simultaneously, the $Si_2H_x^+$ group appears at a mass greater than 56. The appearance of $Si_2H_x^+$ is due to the formation of the silicon-silicon bond among SiH_x produced in the gas phase due to the thermal decomposition of the monomethylsilane.

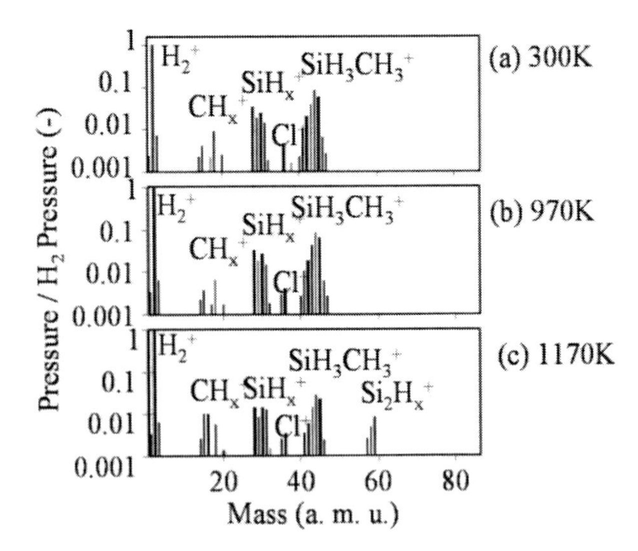

Figure 4.6. Quadrupole mass spectra measured during silicon carbide film deposition by the process in Figure 4.2. The substrate temperatures are (a) 300 K, (b) 970 K, and (c) 1170 K. The monomethylsilane concentration is 5%.

4.3. Film Deposition from Monomethylsilane

From Figure 4.6, a substrate temperature lower than 970 K is considered to be suitable for suppressing the thermal decomposition of monomethylsilane. Therefore, the silicon carbide film deposition is performed at 950K following the process shown in Figure 4.2. Here, the monomethylsilane concentration is 5% in ambient hydrogen at the total flow rate of 2 slm. After the deposition, the chemical bond and the composition of the obtained film are evaluated using the XPS.

Figures 4.7(a) and (b) show the XPS spectra of C 1s and Si 2p, respectively, of the film obtained using monomethylsilane. Because very large peaks due to the Si-C bond exist near 282 eV and near 100 eV, most of the deposited film is silicon carbide. This coincides with the fact that the infrared absorption spectrum of this film showed a peak near 793 cm^{-1}, which corresponds to the silicon-carbon bond [21].

However, the existence of an XPS peak below 100 eV indicates that this film includes a considerable amount of silicon-silicon bonds. The formation of the silicon-silicon bond is considered to be due to the silicon deposition from the SiH_x produced in the gas phase. This shows that the film obtained from the monomethylsilane includes a considerable amount of excess silicon. Figure 4.6 indicates that the substrate temperature of 950 K is sufficiently low to suppress the thermal decomposition of monomethylsilane gas in the gas phase. Therefore, a method of reducing the excess silicon is necessary.

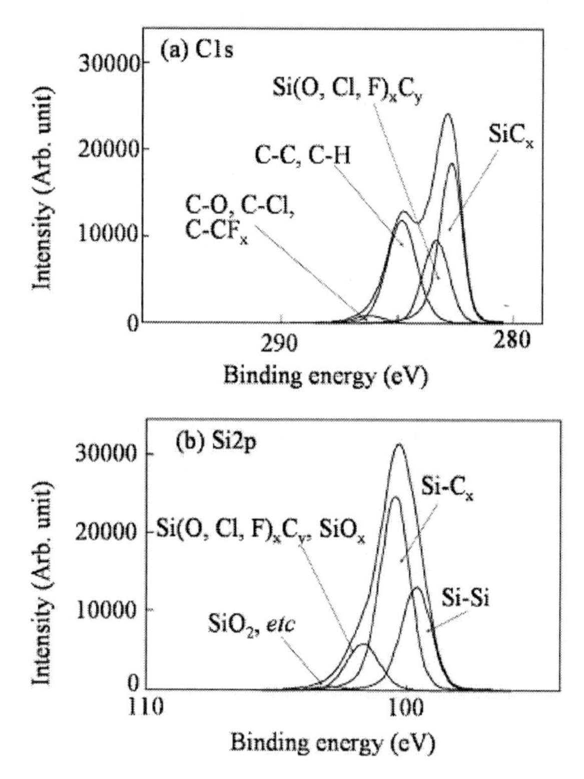

Figure 4.7. XPS spectra of (a) C 1s and (b) Si 2p of silicon carbide film deposited at the monomethylsilane concentration of 5%, and at the substrate temperature of 950K.

In Figure 4.7, there is the peak corresponding to Si(O, Cl, $F)_xC_y$, SiO_x. As the gas mixture used did not include considerable amount of chlorine and fluorine and since the XPS measurements were performed *ex-situ*, the film surface oxidization occurs during its storage in air. This oxidation is attributed to monomethylsilane species remaining at the growth surface. The other peaks related to carbon are considered to be organic contamination on the film surface [63], which accounts for nearly 30% of the carbon at the film surface.

4.4. Film Deposition from Monomethylsilane and Hydrogen Chloride

Here, the method of reducing the excess silicon in the film is explained, adopting the process using hydrogen chloride gas shown in Figure 4.2.

Figure 4.8 shows the quadrupole mass spectrum measured during the silicon carbide film deposition using monomethylsilane gas and hydrogen chloride gas. The gas concentrations of monomethylsilane and hydrogen chloride are 2.5% and 5%, respectively, in hydrogen gas at the flow rate of 2 slm. In Figure 4.8, the partial pressure of the various species is normalized using that of hydrogen (H_2). The substrate temperature is 1090K. This temperature is set to be higher than 970 K, because the higher temperature increases all the chemical reaction rates, and the changes due to the addition of the hydrogen chloride gas can be clearly seen. At this temperature, a considerable number of silicon-carbon bonds can be maintained, according to

Figure 4.6(c). Additionally, this temperature is near the optimum temperature for silicon carbide film growth using monomethylsilane gas, as reported by Liu and Sturm [22].

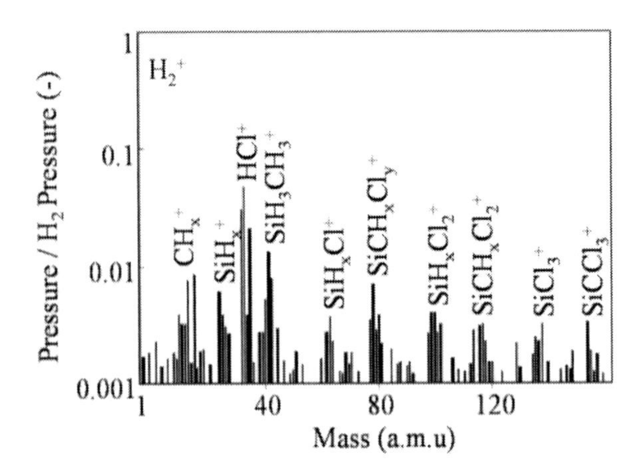

Figure 4.8. Quadrupole mass spectra measured during silicon carbide film deposition by the process in Figure 4.2. The substrate temperature is 1090K. The monomethylsilane gas concentration is 2.3%. The hydrogen chloride gas concentration is 4.7%.

Figure 4.8 shows the CH_x^+, SiH_x^+, $SiH_xCH_y^+$ and HCl^+ groups, which are assigned to the monomethylsilane gas, its fragments and hydrogen chloride gas. In this figure, the $Si_2H_x^+$ group was not observed, unlike Figure 4.6. In addition to these, there are the chlorosilane groups (SiH_xCl_y) at masses over 63 (y=1), 98 (y=2) and 133 (y=3) and the chloromethylsilane group ($SiH_xCl_yCH_z$) at masses over 75 (y=1), 110 (y=2) and 145 (y=3). Therefore, the chlorination of monomethylsilane and silanes occur in a monomethylsilane-hydrogen chloride system.

Figure 4.9 (a) shows the XPS spectra of C 1s of the obtained film. The carbon-silicon bond is observed at 283 eV; its oxidized or chlorinated state, Si(O, Cl, $F)_xC_y$, also exists, as shown in this figure. The other peaks related to the organic contamination on the film surface [64] accounts for nearly 40% of the carbon at the film surface. Figure 4.9 (b) shows the XPS spectra of Si 2p of the film obtained under the same conditions as those in Figure 4.9 (a). Consistent with Figure 4.9 (a), Figure 4.9 (b) shows that the silicon-carbon bond and Si(O, Cl, $F)_xC_y$ bond exist at the film surface. Because the infrared absorption spectra through the obtained film showed a peak near 793 cm^{-1}, which corresponded to the silicon-carbon bond [21], most of this film is silicon carbide.

The most important information obtained from Figures 4.9 (a) and (b) is the reduction of the number of silicon-silicon bonds, in spite of the higher substrate temperature than that in the case of Figure 4.7 (a); many carbon-carbon bonds are shown to exist at the film surface. Therefore, the hydrogen chloride plays a significant role in reducing the amount of excess silicon.

Because a small number of Si-O bonds are detected in Figure 4.9 (b), some of the silicon-carbon bonds in the remaining intermediate species is oxidized during storage in air.

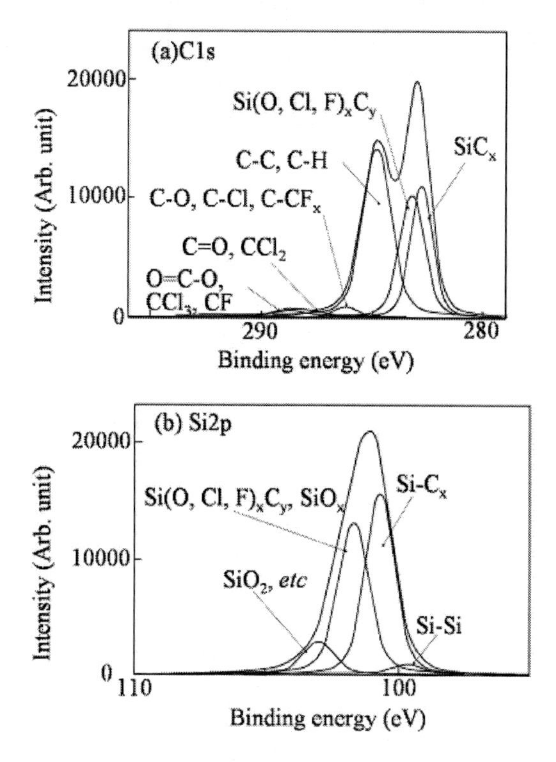

Figure 4.9. XPS spectra of (a) C 1s and (b) Si 2p of silicon carbide film. The substrate temperature is 1090K. The monomethylsilane gas concentration is 2.3%. The hydrogen chloride gas concentration is 4.7%.

4.5. Chemical Reaction in Monomethylsilane and Hydrogen Chloride System

On the basis of the information obtained from Figures 4.6 – 4.9, the chemical reactions in the gas phase and at the substrate surface can be described as shown in Figure 4.10 and as follows:

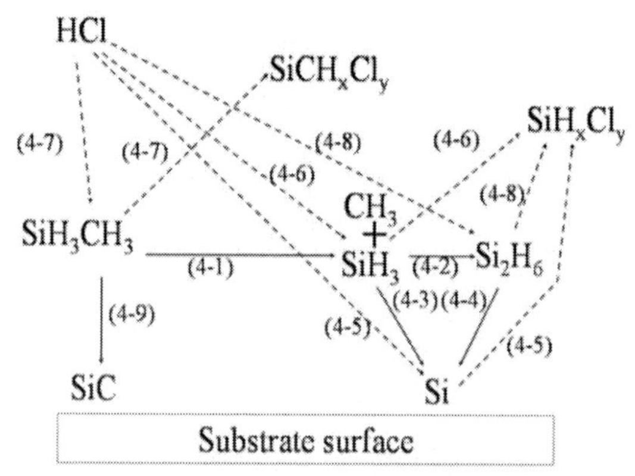

Figure 4.10. Chemical process of silicon carbide film deposition in a monomethylsilane and hydrogen chloride system. (4-*i*) is the equation number.

Thermal decomposition of SiH_3CH_3: $SiH_3CH_3 \rightarrow SiH_3+CH_3$ (4-1)

Si_2H_6 production: $2SiH_3 \rightarrow Si_2H_6$ (4-2)

Si production: $SiH_3 \rightarrow Si\downarrow + \frac{3}{2}H_2$ (4-3)

Si production: $Si_2H_6 \rightarrow 2Si\downarrow +3H_2$ (4-4)

Si etching [22]: $Si+3HCl \rightarrow SiHCl_3\uparrow +H_2$ (4-5)

Chlorination of SiH_3: $SiH_3+3HCl \rightarrow SiHCl_3+\frac{5}{2}H_2$ (4-6)

Chlorination of SiH_3CH_3: $SiH_3CH_3+3HCl \rightarrow SiCl_3CH_3 +\frac{3}{2}H_2$ (4-7)

Chlorination of Si_2H_6: $Si_2H_6+6HCl \rightarrow 2SiHCl_3 +5H_2$ (4-8)

silicon carbide production: $SiH_3CH_3 \rightarrow SiC\downarrow +3H_2$ (4-9)

In these chemical reactions, a small amount of monomethylsilane gas is thermally decomposed to form SiH_3, as shown by Eq. (4-1). SiH_3 forms silicon-silicon chemical bonds with each other to produce Si_2H_6 following Eq. (4-2). Both SiH_3 and Si_2H_6 can produce silicon in the gas phase and at the substrate surface, following Eqs. (4-3) and (4-4), respectively.

One of the possible origins of chlorosilanes, as shown in Figure 4.8, is the etching of silicon at the substrate surface, as described in Eq. (4-5), because the silicon etch rate using hydrogen chloride is considerably high, more than 5 μm/min, at the substrate temperature of 1070 K and at the hydrogen chloride gas concentration of 5% [65]. Another reason for the production of chlorosilanes is the chemical reaction in the gas phase between SiH_3 and hydrogen chloride and between Si_2H_6 and hydrogen chloride, as described in Eq. (4-6) and Eq. (4-8), respectively. Chloromethylsilanes are simultaneously produced, because monomethylsilane reacts with hydrogen chloride, as shown in Eq. (4-7). In addition to these reactions, silicon carbide is produced by the chemical reaction in Eq. (4-9).

The chemical reactions, Eqs. (4-1)-(4-8), are expected to affect the film composition. Si_2H_x is very easily decomposed to produce silicon clusters in the gas phase and on the substrate surface, in Eq. (4-4). However, the formation of Si_2H_6 is suppressed because of the production of $SiHCl_3$ from SiH_3, in Eq. (4-6), immediately after the SiH_3 formation. Therefore, the number of silicon clusters produced in the gas phase is reduced by adding the hydrogen chloride gas; this change can affect the composition of the film. Here, the composition of the film measured by XPS shows that the film surface formed without using hydrogen chloride gas, shown in Figures 4.7 (a) and (b), has silicon and carbon contents of

46% and 40%, respectively. This film is indicated to have an excess amount of silicon. In contrast, the film surface obtained using hydrogen chloride gas has a silicon content of 29% and a carbon content of 45%. This result shows that hydrogen chloride gas can reduce the excess silicon on the film surface. The film composition is expected to be adjusted by optimizing the ratio of hydrogen chloride gas to monomethylsilane gas.

4.6. Film Deposition Rate

Figure 4.11 shows the relationship between the silicon carbide film thickness and the deposition time using the process shown in Figure 4.2. The substrate temperature is maintained at 1070 K; the flow rates of the monomethylsilane gas and hydrogen chloride gas are 0.05 slm and 0.2 slm, respectively, in hydrogen gas of 2 slm. As shown in this figure, the film thickness is maintained at around 0.14 μm from 1 min to 30 min. Therefore, the film deposition stops within 1 min. This result coincides with those obtained by Ikoma et al. [18] and Boo et al. [20] using monomethylsilane gas. Therefore, the hydrogen chloride gas has no effect on the saturation of film deposition.

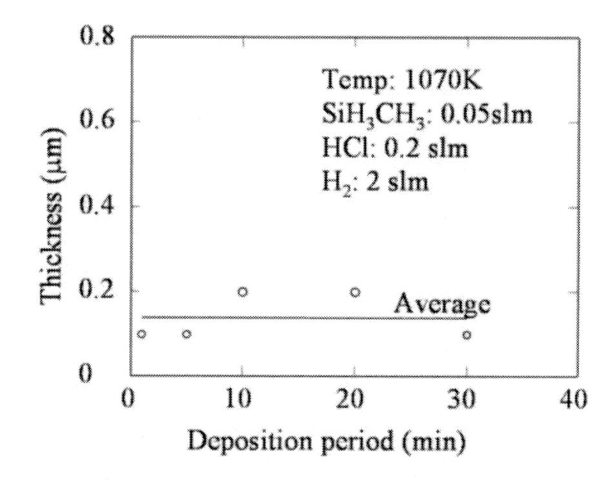

Figure 4.11. Relationship between silicon carbide film thickness and deposition period. The substrate temperature was 1070 K; the flow rates of monomethylsilane and hydrogen chloride were 0.05 slm and 0.2 slm, respectively, in hydrogen gas of 2 slm.

Because the carbon-hydrogen bond is very strong [66, 67], and because the abundance of carbon at the surface of film obtained using hydrogen chloride gas is much higher than that of silicon, the surface on which the deposition stopped is considered to be carbon rich; carbon at the surface is assumed to be terminated with hydrogen. This assumption is consistent with the following issues:

(1) The bonding energy between carbon and hydrogen is much higher than that of other chemical bonds among silicon, hydrogen and chlorine [67].
(2) The silicon-hydrogen and silicon-chlorine chemical bonds cannot perfectly terminate the surface to stop the film deposition, because the silicon epitaxial film can continue to grow in a chlorosilane-hydrogen system at 1070 K [68].

(3) Hydrogen bonded with carbon remains at temperatures less than 1270 K [15].

Here, in order to remove the hydrogen bonded with carbon at the surface, high-temperature annealing is useful. Using the process shown in Figure 4.3, the substrate is heated at 1270 K for 10 min before the next film deposition at 1070 K. Here, the film deposition period in each step is 1 minute.

Figure 4.12 shows the thickness of silicon carbide film obtained by the process employing the high-temperature annealing, Step (C), between the film deposition steps, as shown in Figure 4.3. The flow rates of hydrogen gas and hydrogen chloride gas are fixed to 2 slm and 0.2 slm, respectively. The flow rate of monomethylsilane gas is 0.05 and 0.1 slm. The film deposition period at each step is 1 minute. The film thickness increases with the increasing flow rate of monomethylsilane gas. Simultaneously, the film thickness is increased by repeating the deposition and annealing. The thickness of the obtained film is greater than 2 μm with the total deposition period of 4 min.

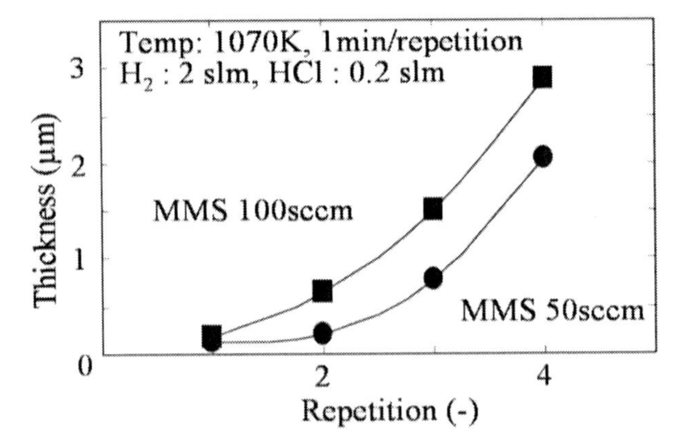

Figure 4.12. silicon carbide film thickness increasing with the repetition of the supply of gas mixture of monomethylsilane and hydrogen chloride (Step (D)) and the annealing at 1270 K (Step (D)). At the deposition, Substrate temperature is 1070 K; the flow rate of monomethylsilane gas is 0.05 slm and 0.1 slm. The flow rate of hydrogen chloride and hydrogen is 0.2 slm and 2 slm, respectively.

Figure 4.13 shows the infrared spectra of the films corresponding to those at the monomethylsilane flow rate of 0.05 slm in Figure 4.12. The numbers in this figure indicate the number of repetitions of Steps (B) and (C) in Figure 4.3. Although these spectra are very noisy, a change in the transmittance clearly appears at the silicon carbide reststrahl band (700-900 cm^{-1}) [69]. With the increasing number of repetitions of Steps (B) and (C), the transmittance near 793 cm^{-1} of 3C-silicon carbide [21] significantly decreases while maintaining the wave-number having a very wide absorption bandwidth. Therefore, the thick film obtained by the process shown in Figure 4.3 is polycrystalline 3C-silicon carbide.

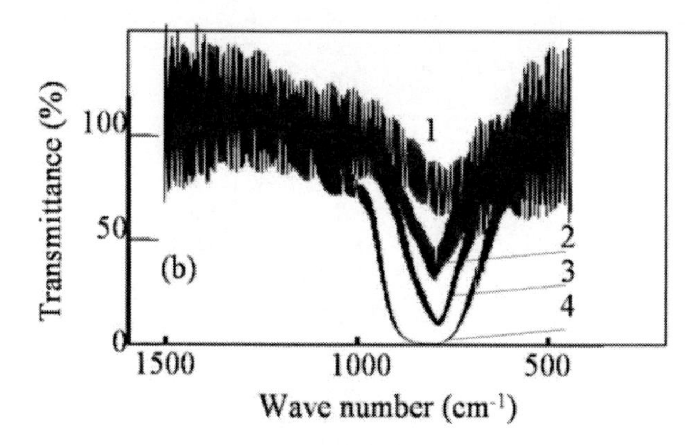

Figure 4.13. Infrared absorption spectra of silicon carbide film after repeatedly supplying gas mixture of monomethylsilane and hydrogen chloride for 1 min (Step (B)) and annealing at 1270 K for 10 min (Step (C)). In Step (B), the substrate temperature is 1070 K; the flow rates of monomethylsilane and hydrogen chloride are 0.05 slm and 0.2 slm, respectively, in hydrogen gas of 2 slm.

Figures 4.14 and 4.15 show the surface morphology of the film obtained at 1070K, corresponding to 4 repetitions in Figures 4,12 and 4.13. The substrate surface is covered with the film having small grains. The film surface has neither porous nor needle-like appearance. Therefore, the smooth film is expected to be obtained optimizing the condition of the film deposition. Additionally, the roughening of silicon substrate surface due to etching by hydrogen chloride is not significant.

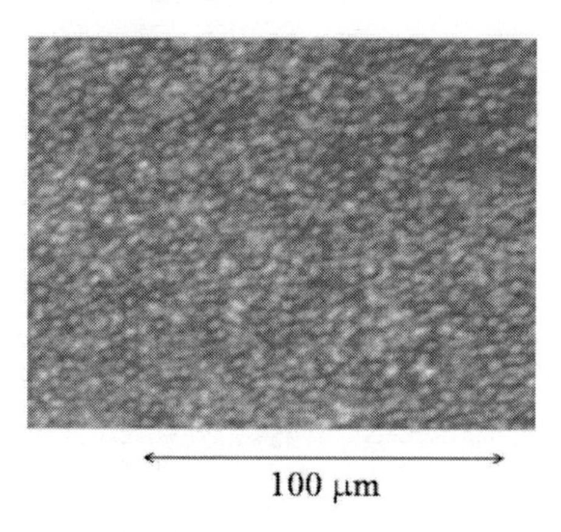

Figure 4.14. Surface morphology of the silicon carbide film after four repetitions of Steps (B) and (C), observed using optical microscope. The condition of silicon carbide film is the same as that in Figure 4.13.

Figure 4.16. shows the morphology of the film surface which is obtained after (R1) one, (R2) two, (R3) three and (R4) four repetitions of Steps (B) and (C). At the deposition, substrate temperature is 1070 K; the flow rate of monomethylsilane gas is 0.05 slm. The flow rate of hydrogen chloride and hydrogen is 0.2 slm and 2 slm, respectively. With increasing the repetitions, the film surface tends to be slightly rough, and shows very small grains. However, no significant roughening occurs.

Figure 4.15. SEM image of the film surface after four repetitions of Steps (B) and (C). The condition of silicon carbide film is the same as that in Figure 4.13.

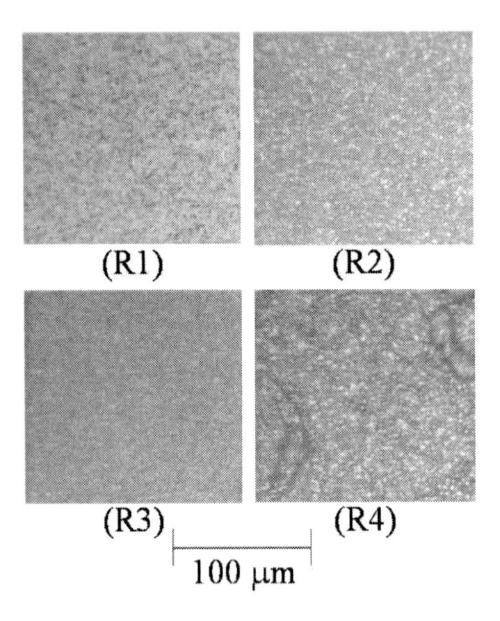

Figure 4.16. Photograph of the film surface. At the deposition, substrate temperature is 1070 K; the flow rate of monomethylsilane gas is 0.05 slm. The flow rate of hydrogen chloride and hydrogen is 0.2 slm and 2 slm, respectively. (R1), (R2), (R3) and (R4) are obtained after one, two, three and four repetitions, respectively, of Steps (B) and (C) in Figure 4.3.

When the film formation occurs due to the deposition of particles formed in the gas phase, the film deposition can continue for the period that the monomethylsilane gas is supplied. However, the film deposition saturated [43]. Therefore, the film showing the small grain appearance is formed dominantly by the surface process. Additionally, it is noted here that the roughening of silicon substrate surface due to etching by hydrogen chloride is not significant, because the film surface can be covered with silicon carbide, immediately after initiating the film deposition.

4.7. Surface Chemical Process

The role of the high temperature annealing and the surface chemical process for restarting the silicon carbide film growth is explained using Figure 4.17.

The silicon carbide film deposition starts at the silicon surface, as shown in Figure 4.17 (i). During Step (B), silicon carbide film is formed as shown in Figure 4.17 (ii). However, because the carbon-hydrogen bond could remain at high temperatures [15, 70], carbon at the surface is terminated with hydrogen, as shown in Figure 4.17 (iii).

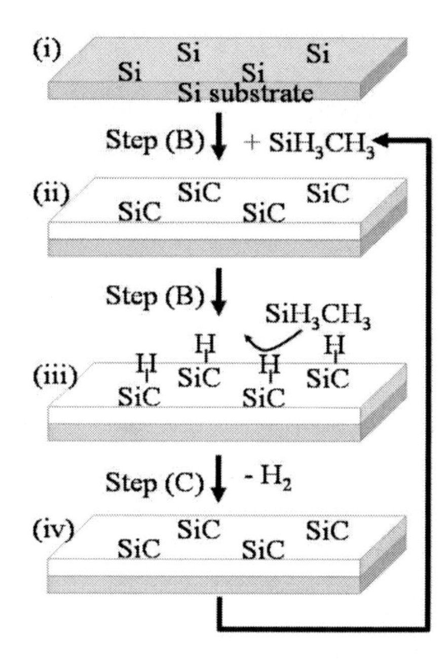

Figure 4.17. Chemical process for silicon carbide film formation from monomethylsilane gas. (i): silicon substrate, (ii): silicon carbide deposition using monomethylsilane gas, (iii): surface termination by hydrogen, and (iv) desorption of hydrogen.

The hydrogen, which terminates at the silicon carbide film surface, is removed using high temperature annealing, as shown in Figure 4.17 (iv). Here, because the bare silicon carbide surface can be formed, the process can return to the surface shown in Figure 4.17 (ii), which can perform Step (B).

Methane is thermally decomposed at atmospheric pressure to cause hydrogen and carbon, as described in Eq. (4-10).

$$CH_4 \rightarrow C + H_2 . \tag{4-10}$$

Because this reaction is endothermic (ΔH_{298K} of 74.9 kJ/mol) [54], the thermal decomposition at atmospheric pressure becomes significant at higher temperatures [71]. Thus, only trace amount of methane remains above 1200 K. Therefore, the hydrogen ambient at atmospheric pressure has very small influence to suppress the hydrogen desorption from the silicon carbide surface as shown in Figure 4.17 (iii).

The effective method to increase the film thickness, other than the repetition of Steps (B) and (C), is to increase the growth rate during Step (B), while the hydrogen-terminated surface is built.

Figure 4.12 shows that the obtained film thickness at the monomethylsilane flow rate of 0.1 slm is greater than that at 0.05 slm. Thus, the silicon carbide growth rate increases with the monomethylsilane gas concentration. This indicates that the silicon carbide film deposition continues, till the hydrogen termination at the silicon carbide surface is completed.

4.8. Hydrogen Chloride Gas Flow Rate and Silicon Carbide Film Thickness

The silicon carbide film thickness at various gas compositions of monomethylsilane and hydrogen chloride for 5 minutes at 1070 K is shown in Figure 4.18. The hydrogen gas flow rate is 2 slm; the hydrogen chloride gas flow rate is 0.1 slm (circle), 0.15 slm (square) and 0.2 slm (triangle).

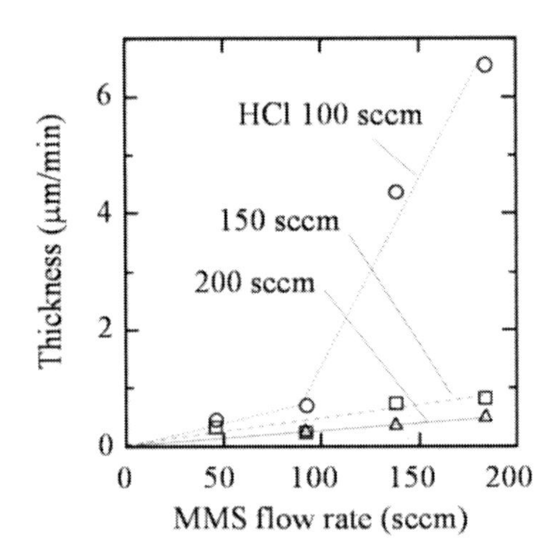

Figure 4.18. silicon carbide film thickness produced for 60 seconds at 1070 K at various gas flow rates of monomethylsilane and hydrogen chloride. Hydrogen chloride gas flow rate is 0.1 slm (circle), 0.15 slm (square) and 0.2 slm (triangle). Hydrogen gas flow rate is 2 slm.

In Figure 4.18, the film thickness entirely decreases with the increasing hydrogen chloride gas flow rate. The square and triangle show that the silicon carbide film thickness is very small but it gradually increases with the increasing monomethylsilane gas flow rate between 0.05 and 0.2 slm. In contrast to this, the silicon carbide film thickness obtained at the hydrogen chloride gas flow rate of 0.1 slm, indicated by the circle, shows a significant increase at the monomethylsilane gas flow rate greater than 0.1 slm. Here, the surface appearance of the film showing such a significant thickness increase is dark and very rough.

Here, it should be noted that the silicon substrate surface was significantly etched by hydrogen chloride gas at its flow rate of 0.1 slm for 60 s, without monomethylsilane gas. This indicates that the silicon-silicon bond present at the film surface can be removed by hydrogen chloride gas. Thus, based on these results, the amount of excess silicon in the silicon carbide film is decreased, however, the insufficient amount of hydrogen chloride gas can not sufficiently suppress the incorporation of excess silicon. From Figure 4.18, the amount of hydrogen chloride gas, comparable to or greater than that of the monomethylsilane gas, is necessary for the effective removal of the excess silicon. Because the hydrogen chloride gas flow rate larger than 0.15 slm is sufficient for the film formation at the monomethylsilane gas flow rate between 0.05 and 0.2 slm, the film thickness could linearly increase with the increasing monomethylsilane gas flow rate, as indicated using square and triangle in Figure 4.18.

4.9. Surface Morphology

The surface morphology of the silicon carbide film is evaluated by the AFM, because some of the silicon carbide films obtained from monomethylsilane gas shows a very smooth appearance by visual inspection. Figure 4.19 shows the AFM photograph of (a) silicon surface before the film formation, and (b) silicon carbide film surface with a thickness of 0.2 μm obtained at 1070 K for 5 min at the monomethylsilane gas flow rate of 0.092 slm and hydrogen chloride gas flow rate of 0.15 slm. The measured area was 0.2 x 0.2 μm.

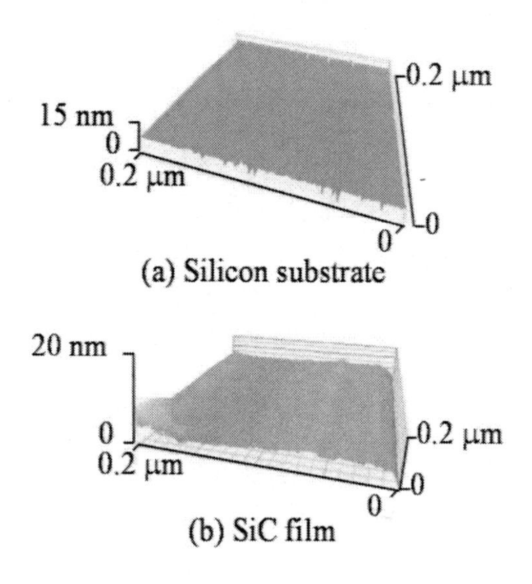

Figure 4.19. AFM photograph of (a) silicon surface before the silicon carbide film deposition and (b) silicon carbide film surface with the thickness of 0.2 μm obtained at 1070 K for 5 minutes at the monomethylsilane gas flow rate of 0.092 slm and hydrogen chloride gas flow rate of 0.15 slm. The Ra and RMS microroughness are 0.6 nm and 0.7 nm, respectively.

Figure 4.19 (a) shows that the silicon substrate surface before the film formation is very smooth with the average roughness (Ra) and the root-mean-square roughness (RMS) of 0.2 nm and 0.3 nm, respectively. After the silicon carbide film formation, the surface roughness slightly increases due to the formation of short hillocks, as shown in Figure 4.19 (b). However, its surface appearance is still specular by visual inspection. The Ra and RMS microroughness are 0.6 nm and 0.7 nm, respectively.

Some of the silicon carbide films obtained from monomethylsilane gas at 1070 K show a small grain-like surface [43, 44], but the other films show a very smooth surface. Because the specular surface is expected to have a higher coating quality than that of a grain-like surface, the condition for obtaining the smooth surface with a high reproducibility should be studied in future.

4.10. Room Temperature Deposition

In this section, the low temperature silicon carbide film formation is explained. For maintaining the gas condition in a series of film deposition, hydrogen chloride gas is introduced with monomethylsilane gas, even at room temperature, at which hydrogen chloride gas hardly reacts with silicon. In the silicon carbide film formation for 60 seconds at various temperatures between 1070 K and room temperature following Steps (A) and (D) in Figure 4.4, the obtained film thickness is very small, around 0.1 μm, and their surface often has a grain-like morphology, as shown in Figure 4.20 and a yellowish appearance indicating the existence of the silicon carbide film. Thus, the film formation at the lowest temperature, that is, at room temperature, is further explained.

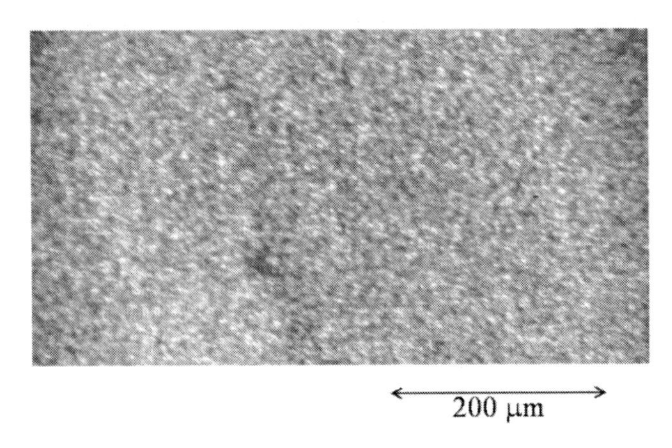

Figure 4.20.Surface morphology of the film formed at room temperature for 60 seconds using monomethylsilane gas (0.092 slm) and hydrogen chloride gas (0.15 slm), immediately after the surface cleaning in ambient hydrogen at 1370 K for 10 min.

The average film thickness obtained at room temperature, following Steps (A) and (D) in Figure 4.4, is 0.1 μm, which is comparable to the thickness obtained at 1070 K. In order to quickly evaluate the coating quality of the silicon carbide film, the film surface is further exposed to hydrogen chloride gas at 1070 K, following Step (E) in Figure 4.4. Because the film shows no decrease in weight and because its surface appearance is maintained, the film formed at room temperature is expected to be silicon carbide.

In order to show the necessary condition for the film formation at room temperature, monomethylsilane gas is supplied to silicon substrate skipping its surface cleaning (Step (A)) of the process shown in Figure 4.4. This resulted in no weight increase to indicate no film formation; its surface was significantly etched by hydrogen chloride gas at 1070 K, by Step (E) in Figure 4.4. This shows that the surface cleaning in ambient hydrogen (Step (A)) takes an important role for the silicon carbide film formation from monomethylsilane gas at low temperatures.

In order to verify the silicon carbide film formation, Figure 4.21 shows the XPS spectra of C 1s of the 0.1 μm-thick deposited film which is obtained from monomethylsilane gas at room temperature, and further etched by hydrogen chloride gas at 1070 K for 10 minutes, following Steps (A), (D) and (E) shown in Figure 4.4.

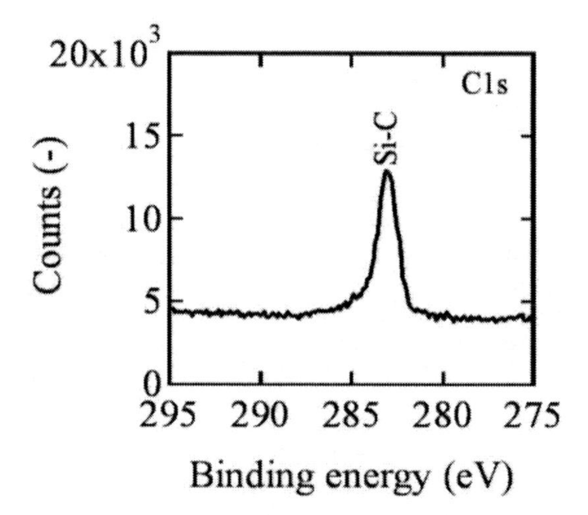

Figure 4.21. XPS spectra of C1s of the silicon carbide film, obtained from monomethylsilane gas and hydrogen chloride gas on silicon surface at room temperature after annealing in hydrogen ambient. This film was further exposed to hydrogen chloride gas at 1070 K for 10 min before the XPS measurement.

Figure 4.21 clearly shows the existence of the Si-C bond at 283 eV. Because silicon atom bonding with the carbon atom is consistently detected at 101 eV, the obtained film contains the Si-C bond.

In order to show the influence of hydrogen chloride gas at room temperature, the silicon carbide film formation is performed only with monomethylsilane gas. This resulted in the same film formation when using monomethylsilane and hydrogen chloride gases. Thus, hydrogen chloride gas reasonably has less significant influence at room temperature.

4.11. Robustness of Film Formed at Room Temperature

In order to evaluate the robustness of the silicon carbide film formed at room temperature, the deposited film is exposed to hydrogen chloride gas at 1070 K, following Step (C) in Figure 4.4; its surface is compared with that of a silicon substrate after exposed to hydrogen chloride gas.

Figure 4.22 shows SEM photograph of the silicon substrate surface and silicon carbide film. Figure 4.22 (a) is the silicon substrate surface after etching using hydrogen chloride gas at the flow rate of 0.1 slm diluted by hydrogen gas of 2 slm, at the substrate temperature of 1070 K for 10 min, without silicon carbide film formation. This figure shows the existence of many pits indicating the occurrence of etching by hydrogen chloride gas.

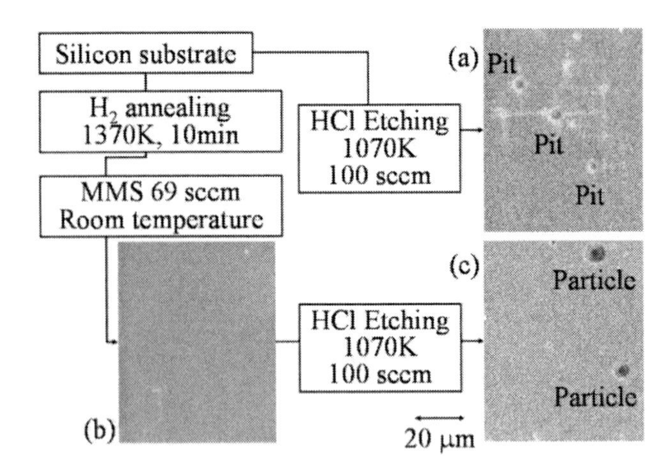

Figure 4.22. Surface of (a) silicon substrate after etching using hydrogen chloride gas at 1070 K for 10 min, (b) deposited film using monomethylsilane gas of 0.069 slm at room temperature, and (c) the film of (b) further etched using hydrogen chloride gas at 1070 K for 10 min.

Figure 4.22 (b) shows the silicon carbide film surface formed using monomethylsilane gas of 0.069 slm at room temperature for 1 min. This figure shows that there is no large pit at the deposited film surface. Next, this surface is exposed to hydrogen chloride gas at the flow rate of 0.1 slm diluted in hydrogen gas of 2 slm, at 1070 K for 10 min. This condition is exactly the same as that performed for the silicon surface, shown in Fig, 4.22 (a). As shown in Figure 4.22 (c), a considerable morphology change is not observed at the deposited film surface, except of particles intentionally taken in order to clearly focus the surface for SEM observation.

Figure 4.23 is the TEM micrograph of the cross section of the silicon carbide film, produced from monomethylsilane gas and hydrogen chloride gas on silicon surface at room temperature after annealing at 1370 K in hydrogen ambient. This film was further exposed to hydrogen chloride gas at 1070 K for 10 min, before the TEM measurement.

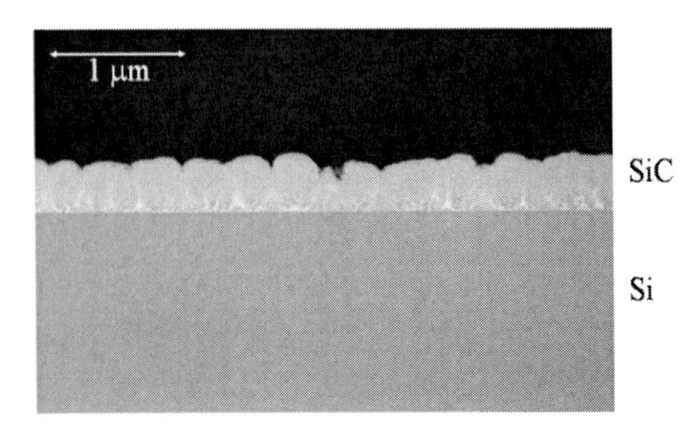

Figure 4.23. TEM micrograph of the cross section of the silicon carbide film, shown in Figure 4.22 (c).

Figure 4.23 shows that the entire silicon substrate surface is sufficiently covered with the silicon carbide film consisted of arranged many grains, diameter of which is about 0.2 – 0.3 μm. The average film thickness in the observed area is about 0.3 μm. Additionally, there are no etch pit and pin-hole caused due to etching by hydrogen chloride gas at the silicon carbide-silicon interface. Thus, the silicon carbide film deposited at room temperature is robust in a hazardous ambient including hydrogen chloride gas.

4.12. Film Formation Mechanism at Room Temperature

Based on the result that the surface cleaning in ambient hydrogen is necessary for producing the silicon carbide film, the surface chemical process for the low temperature silicon carbide formation using monomethylsilane gas is schematically shown Figure 4.24.

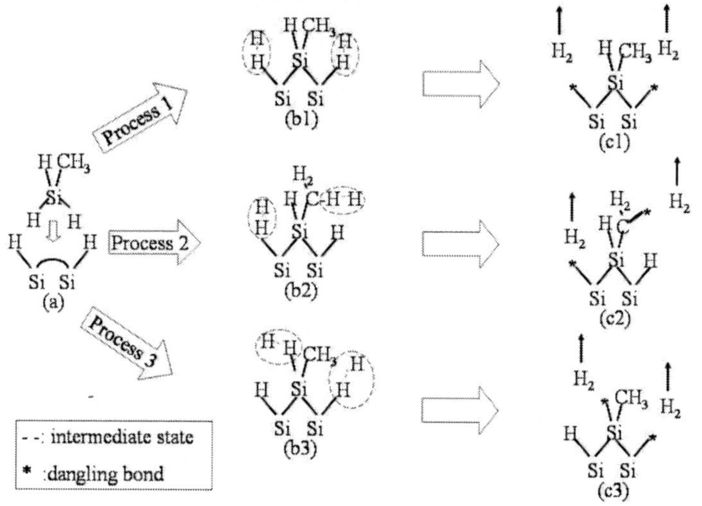

Figure 4.24. Surface processes for low temperature silicon carbide film growth. (a) approach of monomethylsilane to silicon dimer at hydrogen-terminated silicon surface. Process 1: (b1) chemisorption of monomethylsilane and production of hydrogen radicals, and (c1) production of hydrogen molecules, and silicon dangling bonds at the neighboring silicon. Process 2: (b2) chemisorption of monomethylsilane and production of hydrogen radicals, and (c2) production of hydrogen molecules, silicon dangling bond at the neighboring silicon, and carbon dangling bond in the chemisorbed monomethylsilane. Process 3: (b3) chemisorption of monomethylsilane and production of hydrogen radicals, and (c3) production of hydrogen molecules, and silicon dangling bonds at the neighboring silicon and in the chemisorbed monomethylsilane.

The silicon carbide film formation is initiated by Step (a), as shown in Figure 4.24. At Step (a), monomethylsilane molecule approaches to silicon dimer present at hydrogen-terminated silicon surface. The silicon dimer is assumed to be broken in order to accept monomethylsilane molecule. Here, Step (a) is for an initiation of the surface chemical reaction; Steps (b) and (c) are for a repetition of the surface chemical reaction to produce multilayer film. After Step (a), Processes 1, 2 and 3, are expected to occur.

At Step (b1) in Process 1, silicon atom in monomethylsilane forms covalent bonds with silicon atom of the substrate. Here, hydrogen radicals are produced. The hydrogen radicals bond to the neighboring silicon atoms. At Step (c1) in Process 1, two hydrogen atoms are produced; dangling bonds remain at the neighboring silicon atoms.

At Step (b2) in Process 2, silicon atom in monomethylsilane forms covalent bonds with silicon atom of the substrate, similar to Process 1. Next, one of the hydrogen radicals produced can approach the hydrogen atom bonding with the carbon atoms in the chemisorbed monomethylsilane molecule. At Step (c2) in Process 2, two hydrogen atoms are produced; dangling bonds remain at the neighboring silicon atom and at the carbon atom in the monomethylsilane.

At Step (b3) in Process 3, silicon atom in monomethylsilane forms covalent bonds with silicon atom of the substrate, similar to Processes 1 and 2; one of the hydrogen radical produced can approach the hydrogen atom bonding with the silicon atom in the chemisorbed monomethylsilane molecule. At Step (c3) in Process 3, two hydrogen atoms are produced; dangling bonds remain at the neighboring silicon atom and at the silicon atom in the monomethylsilane. Because the dangling bonds formed after Steps (c1), (c2) and (c3) can accept more monomethylsilane molecules, chemisorption of monomethylsilane is expected to be spread and repeated over the substrate surface.

When Process 2 is slower than Process 3, a larger amount of C-H bond can remain at the film surface. Because this induces the C-H termination over the entire surface, the silicon carbide film formation finally stops.

Here, silicon dimer was reported to be very weak [72]; many research groups [15, 73] reported the occurrence of the dissociative adsorption of organosilane on silicon dimer at room temperature. Additionally, Silvestrelli et al. [74] reported that SiH_2CH_3 can bond to silicon dimer, when monomethylsilane molecule vertically approached the surface. Taking into account these previous studies, the surface process, shown in Figure 4.24, is consistent with the results of the low-temperature silicon carbide film formation and its saturation, using monomethylsilane gas.

5. SUMMARY

Etching and film formation of silicon carbide performed using highly reactive gases of chlorine trifluoride (ClF_3) and monomethylsilane (MMS, SiH_3CH_3), respectively, are reviewed.

Silicon carbide etching using chlorine trifluoride gas with high etch rate occurs at the temperatures higher than 770 K. Its chemical reaction is as follows:

$$3SiC + 8ClF_3 \rightarrow 3SiF_4 \uparrow + 3CF_4 \uparrow + 4Cl_2 \uparrow$$

The etch rate is 10-20 μm/min and 5 μm/min, for polycrystalline 3C-silicon carbide and single-crystalline 4H-silicon carbide, respectively. The etch rate of Si-face of 4H-silicon carbide is slightly smaller than that of C-face. The etched surface tends to be carbon rich. The etched surface of Si-face of 4H-silicon carbide shows various kinds of morphology: pitted at low temperatures of 570 – 1270 K, and smooth at 1570 K. The C-face of 4H-silicon carbide shows the similar trend, and is entirely very smoother than that of Si-face. Most of the etch pits formed near 700 K at the Si-face and C-face 4H-silicon carbide show relationship between dislocations revealed by the X-ray topograph.

The 3C-silicon carbide thin film is formed on silicon surface using monomethylsilane gas at the temperatures between room temperature and 1270 K. Although silicon, produced by thermal decomposition in gas phase and substrate surface, is incorporated into the silicon carbide film, it can be significantly reduced by means of the addition of hydrogen chloride gas. Although the silicon carbide film formation saturates within 1 minute due to the surface termination by C-H bonds, it can start again after annealing at 1270 K in order to remove hydrogen atoms. In order to develop the low-temperature silicon carbide film formation process, monomethylsilane gas is introduced to silicon substrate at room temperature. After the silicon surface is cleaned at 1370 K and cooled down in hydrogen ambient, monomethylsilane molecule can adsorb on the silicon surface to produce silicon carbide film, even at room temperature. Such the low temperature film formation is possible, because the hydrogen terminated silicon surface has a reactive structure, such as silicon dimer. The silicon carbide film formed at room temperature is robust, because it can maintain after the etching using hydrogen chloride gas at 1070 K.

ACKNOWLEDGMENTS

Various studies reviewed in this Chapter were performed with Ms. Satoko Oda, Mr. Yusuke Katsumi, Mr. Yu Kasahara, Ms. Keiko Tanaka, Mr. Kazuchika Furukawa, Mr. Takashi Sekiguchi, Mr. Mikiya Nishida, Ms. Mayuka Watanabe, Mr. Hiroshi Ohmori, Mr. Yusuke Ando, Dr. Yutaka Miura, Mr. Yoichi Negishi, Dr. Takashi Takeuchi, Prof. Masahiko Aihara, Prof. Minoru Takeda, Prof. Hironobu Kunieda and Prof. Kenji Aramaki of Yokohama National University. The silicon carbide etching was collaborated with Mr. Yasushi Fukai, Mr. Katsuya Fukae, Mr. Naoto Takechi, Dr. Yuan Gao, and Mr. Shinji Iizuka, of Kanto Denka Kogyo Co., Ltd., and Dr. Tomohisa Kato, Dr. Hajime Okumura and Dr. Kazuo Arai of National Institutes of Advanced Science and Technology.

Mr. N. Okumura of Keyence Co., Ltd., is very much appreciated for the surface roughness evaluation. The author thanks Prof. Maki Suemitsu of Tohoku University for fruitful discussion on chemical vapor deposition using monomethylsilane gas.

X-ray topography experiment has been performed under the approval of the Photon Factory Program Advisory Committee (Proposal No. 2006G286).

REFERENCES

[1] Y. Gogotsi, S. Welz, D. A. Ersoy and M. J. McNallan, *Nature*, 411(6835) (2001) 283.

[2] K. Vyshnyakova, G. Yushin, L. Pereselentseva, Y. Gogotsi, *Int. J. Appl. Ceramic Tech.*, 3 (2006) 485.

[3] A.V. Zinovev, J.F. Moore, J. Hryn and M.J. Pellin, *Surf. Sci.*, 600 (2006) 2242.

[4] M. Cooke, *III-Vs Review*, 18 (Dec. 2005) 40.

[5] C. Chai, Y. Yang, Y. Li and H. Jia: *Optical Materials*, 23 (2003) 103.

[6] M. Mehregany, C. A. Zorman, S. Roy, A. J. Fleischman, C. H. Wu and N. Rajan: *International Materials Reviews*, 45 (2000) 85.

[7] C. R. Stoldt, C. Carraro, W. R. Ashurst, D. Gao, R. T. Howe and R. Maboudian: *Sensors and Actuators a-Physical*, 97-98 (2002) 410.

[8] N. Rajan, M. Mehregany, C. A. Zorman, S. Stefanescu and T. P. Kicher: *Journal of Microelectromechanical Systems*, 8 (1999) 251.
[9] W. R. Ashurst, M. B. J. Wijesundara, C. Carraro and R. Maboudian: *Tribology Letters*, 17 (2004) 195-198.
[10] F. Shimura, Semiconductor Silicon Crystal Technology, p.244, Academic Press (San Diego, USA, 1989).
[11] N. N. Greenwood and A. Earnshaw: Chemistry of the Elements, (Butterworth and Heinemann, Oxford, 1997).
[12] T. Kimoto and H. Matsunami, *J. Appl. Phys*., 75 (1994) 850-859.
[13] R. L. Myers, Y. Shishkin, O. Kordina and S. E. Saddow, *J. Cryst. Growth*, 285 (2005) 486.
[14] E. R. Sanchez and S. Sibener, *J. Phys. Chem.* B, 106 (2002) 8019-8028.
[15] H. Nakazawa and M. Suemitsu, *Appl. Surf. Sci.* 162–163 (2000) 139-145.
[16] K. Yasui, K. Asada, T. Maeda, T. Akahane, *Appl. Surf. Sci.* 175–176 (2001) 495-498.
[17] T. Kaneko, N. Miyakawa, H. Yamazaki and Y. Iikawa, *J. Cryst. Growth*, 237-239 (2002) 1260-1263.
[18] Y. Ikoma, T. Endo, F. Watanabe and T. Motooka, *Jpn. J. Appl. Phys*., 38 (1999) L301-303.
[19] J. H. Boo, S. A. Ustin and W. Ho, *Thin Solid Films,* 343-344 (1999) 650-655.
[20] S. J. An and G. C. Yi, *Jpn. J. Appl. Phys*., 40 (2001) 1379-1383.
[21] S. Madapura, A. J. Steckl and M. Loboda, *J. Electrochem. Soc*., 146 (1999) 1197-1202.
[22] C. W. Liu and J. C. Sturm, *J. Appl. Phys*., 82 (1997) 4558-4567.
[23] B. Kim, S. Kim, S. Ann and B. Lee: *Thin Solid Film*, 434 (2003) 276.
[24] U. Schmid, M. Eickoff, C. Richter, G. Kroetz and D. Schmitt-Landsiedel: *Sensors and Actuators* A, 94, (2001) 87.
[25] R. Wolf and R. Helbig: *J. Electrochem. Soc*., 143 (1996) 1037.
[26] P. H. Yih and A. J. Steckl: *J. Electrochem. Soc.,* 142 (1995) 312.
[27] Z. Y. Xie, C. H. Wei, L. Y. Li, Q. M. Yu and J. H. Edgar: *J. Cryst. Growth*, 217 (2000) 115.
[28] E. K. Sanchez, S. Ha, J. Grim, M. Skowronski, W. M. Vetter, M. Dudley, R. Bertke and W. C. Mitchel: *J. Electrochem. Soc*., 149 (2002) G 131.
[29] M. Syvajarvi, R. Yakimova and E. Janzen: *J. Electrochem. Soc*., 147 (2000) 3519.
[30] J. S. Shor, X. G. Zhang and R. M. Osgood: *J. Electrochem. Soc*., 139 (1992) 1213.
[31] P. Chabert, *J. Vac. Sci. Technol*. B, 19 (2001) 1339.
[32] P. A. Khan, B. Roof, L. Zhou and I. Asesida: *J. Electron. Mater.,* 30 (2001) 212.
[33] L. Jiang, R. Cheung, R. Brown and A. Mount: *J. Appl. Phys*., 93 (2003) 1376.
[34] L. Jiang, N. O. V. Plamk, M. A. Blauw, R. Cheung and E. van der Drift: *J. Phys. D: Appl. Phys*., 37 (2004) 1809.
[35] H. Habuka, H. Koda, D. Saito, T. Suzuki, A. Nakamura, T. Takeuchi and M. Aihara, *J. Electrochem. Soc*., 150 (2003) G461-464.
[36] H. Habuka, S. Oda, Y. Fukai, K. Fukae, T. Sekiguchi, T. Takeuchi, and M. Aihara, *Jpn. J. Appl. Phys*. 44 (2005) 1376-1381.
[37] H. Habuka, S. Oda, Y. Fukai, K. Fukae, T. Takeuchi and M. Aihara, *Thin Solid Films*, 514 (2006) 193-197.
[38] Y. Miura, Y. Katsumi, S. Oda, H. Habuka, Y. Fukai, K. Fukae, T. Kato, H. Okumura and K. Arai, *Jpn. J. Appl. Phys*., 46 (2007) 7875.

[39] H. Habuka, Y. Katsumi, Y. Miura, K. Tanaka, Y. Fukai, T. Fukae, Y. Gao, T. Kato, H. Okumura and K. Arai, *Materials Science Forum*, 600-603 (2008) 655-658.
[40] H. Habuka, K. Tanaka, Y. Katsumi, N. Takechi, K. Fukae and T. Kato, *Materials Science Forum*, 645-648 (2010) 787-790.
[41] H. Habuka, K. Tanaka, Y. Katsumi, N. Takechi, K. Fukae, and T. Kato, *J. Electrochem. Soc.,* 156 (2009) H971-H975.
[42] Y. Miura, Y. Kasahara, H. Habuka, N. Takechi, and K. Fukae, Jpn. J. Appl. Phys., 48 (2009) 026504.
[43] [43] H. Habuka, M. Watanabe, Y. Miura, M. Nishida, T. Sekiguchi, *J. Crystal Growth,* 300 (2007) 374-381.
[44] H. Habuka, M. Watanabe, M. Nishida and T. Sekiguchi, *Surf. Coat. Tech.*, 201 (2007) 8961-8965.
[45] H. Habuka H. Ohmori and Y. Anndo, *Surf. Coat. Tech.* 204 (2010) 1432-1437.
[46] Y. Chinone, S. Esaki and F. Fujita: *Shirikon no Kagaku*, (Realize, Tokyo, 1996) p. 890.
[47] H. Habuka, T. Sukenobu, H. Koda, T. Takeuchi and M. Aihara, J. *Electrochem. Soc.,* 151 (2004) G783-787.
[48] CRC Handbook of Chemistry and Physics, ed.: D. R. Lide (CRC Press, Boca Raton, FL, 1997) 78th ed., p. 1.
[49] T. E. Lee, A Beginner's Guide to Mass Spectral Interpretation (Wiley, Chichester, 1998).
[50] J. R. Chapman, Practical Organic Mass Spectrometry (Wiley, Chichester, 1993).
[51] H. Habuka, T. Otsuka and W. F. Qu: *Jpn. J. Appl. Phys*., 38 (1999) 6466.
[52] Y. Saito, M. Hirabaru and A. Yoshida: *J. Vac. Sci. Technol*. B, 10 (1992) 175.
[53] S. V. Patankar, Numerical Heat Transfer and Fluid Flow, McGraw-Hill, New York, USA (1980)
[54] http://webbook.nist.gov/chemistry
[55] R. C. Reid, J. M. Prausnitz, and B. E. Poling: The Properties of Gases and Liquids (McGraw-Hill, New York, 1987) 4th ed.
[56] Nihon Gakujutsushinkokai, Fussokagaku Nyumon, p. 86 Nikan Kogyo Shinbunsha, Tokyo, Japan (1997) [in Japanese]
[57] FLUENT User's Manual, Ver. 5.5 (Fluent, Inc., Hanover, 2001)
[58] H. Matsunami, *Technology of Semiconductor SiC and Its Application,* Nikkan Kogyo, (Tokyo, Japan, 2003).
[59] T. Ohno, H. Yamaguchi, S. Kuroda, K. Kojima, T. Suzuki, and K. Arai, *J. Cryst. Growth,* 260 (2004) 209–216.
[60] X. Ma, M. Dudley, W. Vetter and T. Sudarshan, *Jpn. J. Appl. Phys*., 42 (2003) L1077.
[61] J. Takanashi, M. Kanaya and Y. Fujiwara, *J. Cryst. Growth*, 135 (1994) 61,
[62] H. Habuka, T. Suzuki, S. Yamamoto, A. Nakamura, T. Takeuchi and M. Aihara, *Thin Solid Films,* 489 (2005) 104-110.
[63] K. Teker, *J. Cryst. Growth*, 257 (2003) 245.
[64] S. Ishiwari, H. Kato and H. Habuka, *J. Electrochem. Soc*., 148 (2001) G644.
[65] H. Habuka, T. Suzuki, S. Yamamoto, A. Nakamura, T. Takeuchi and M. Aihara, *Thin Solid Films,* 489 (2005) 104.
[66] H. G. Yoon, J. H. Boo, W. L. Liu, S. B. Lee, S. C. Park and H. Kang, *J. Vac. Sci. Technol.* A, 18 (2000) 1464.
[67] Kagaku Binran (Iwanami, Tokyo, 1984) 3rd ed. [in Japanese].

[68] H. Habuka, T. Nagoya, M. Mayusumi, M. Katayama, M. Shimada and K. Okuyama, *J. Cryst. Growth,* 169 (1996) 61.

[69] M. F. MacMillan, R. P. Devaty, W. J. Choyke, D. R. Goldstein, J. E. Spanier and A. D. Kurtz, *J. Appl. Phys*., 80 (1996) 2412.

[70] H. G. Yoon, J. H. Boo, W. L. Liu, S. B. Lee, S. C. Park and H. Kang, *J. Vac. Sci. Technol.* A18 (2000) 1464.

[71] A. W. Weimer, J. Dahl, J. Tamburini, A. Lewandowski, R. Pitts and C. Bingham, Proc. 2000 DOE Hydrogen Program Review, NREL/CP-570-28890.

[72] A. Redondo and W. A. Goddard III, *J. Vac. Sci. Technol*., 21 (1982) 344.

[73] D. G. J. Sutherland, L. J. Terminello, J. A. Carlisle, I. Jimenez, F. J. Himpsel, K. M. Baines, D. K. Shuh and W. M. Tong, *J. Appl. Phys*., 82 (7) (1997) 3567-3571.

[74] P. L. Silvestrelli, C. Sbraccia, A. H. Romero, and F. Ancilotto, *Surf. Sci*., 532 (2003) 957.

In: Silicon Carbide: New Materials, Production ... ISBN: 978-1-61122-312-5
Editor: Sofia H. Vanger

Chapter 2

SILICON CARBIDE PARTICULATE REINFORCED ALUMINUM ALLOYS MATRIX COMPOSITES FABRICATED BY SQUEEZE CASTING METHOD

Adem Onat
Sakarya University, Vocational School of Sakarya,
54188 SAKARYA - TURKEY

ABSTRACT

In this paper, production and characterization of SiC particles reinforced Al alloy matrix composites produced by squeeze casting has been investigated. Al-4.5Cu-3Mg alloy was used for matrix alloy and 5, 10, 15 %vol. SiC particles reinforced composites have been produced at 100 MPa pressure and 500 °C die temperature applied at casting operations. The effects of volume fraction of SiC on microstructure, mechanical properties have been investigated. Also gravity cast matrix alloy has been produced to determine effect of pressure on microstructure and properties.

The results showed that, the composites have homogeneously distributed porosity free SiC particles. Squeeze casting also resulted in refining microstructure, porosity and other casting defects such as shrinkage cavities were minimized due to pressure application during solidification. The porosity of unreinforced alloys decreased from 6.63% to 1.08%. In composites, increasing volume fraction of particles caused to increase porosity of composite. Hardness of matrix alloy was increased from 94 BHN to 100 BHN by squeeze casting method. In composites, hardness increased with increasing volume fraction of SiC. The highest value was recorded at 15 %vol. SiCp, 139 BHN.

Squeeze casting method, also, improved tensile properties of matrix alloy, the UTS values increased from 195 MPa to 247 MPa and elongation from 1.78% to 3.30%. The elongation of composites decreased from 1.53% to 0.45% with increasing volume fraction of SiC from 5 to 15 percent.

Fracture surface analysis showed that, the failure in composites occurs in both matrix and particles simultaneously implying good bonding between matrix and SiC particles and the fracture initiation did not occur at matrix particle interface.

Wear tests were carried out at 0.5, 1 and 2.0 m/s sliding speeds under the loads of 5, 10, and 15 N for a 1000 m sliding distance versus AISI D2 steel disc. Test results proved that friction coefficient of the composites decreased with an increase in the applied load and the sliding speed. In addition, the higher the applied load and the faster the sliding speed are, the higher the wear rate is. SEM analysis indicated that worn surfaces consisted of plastically deformed and oxidized particles removed by the micro-machining effects of the reinforcement phase.

Keywords: Production of metal matrix composites; Squeeze casting; Al/SiC particulate composite; Microstructure of Al/SiC_p Composites, Characterization of Al/SiCp composites, Mechanical properties, Fracture, Wear.

1. INTRODUCTION

Metal Matrix Composites (MMCs) are now attracting enormous interest, because of their superiority in strength, stiffness, wear resistance, elevated temperature strength or other engineering properties [1-4]. Parallel to commercialization, there are research centers throughout the world that are actively researching further development and exploitation of net or near net shape fabrication process. This is evidenced by the publication of more than thousands papers in various engineering and scientific journals. These are mainly related to aluminum and magnesium-based alloys with special emphasis on metal matrix composites MMCs [5]. This is because of aluminum matrix composites have many advantages: high strength, high modulus, high wear resistance, and low density and coefficient of thermal expansion [6].

The composites contain hard refractory ceramic particles such as SiC, TiC and Al_2O_3 in a soft aluminum matrix, the improved properties resulting from the hard particles. The particulate reinforced aluminum matrix composites not only have good mechanical and wear properties, but are also economically viable [7-10]. Therefore, SiC-particulate-reinforced aluminum composites have found many applications in the aerospace and automotive industry [9].

In recent years, discontinuous MMCs have attracted considerable attention on account of

a) availability of spectrum of reinforcements at competitive costs,
b) successful development of manufacturing process to produce MMCs with reproducible microstructures and properties,
c) availability of standard and near net standard metal working methods that can be utilized to form these materials.

Furthermore, use of discontinuous reinforcements minimizes problems associated with fabrication of continuously reinforced MMCs such as fiber damage, microstructural heterogeneity, fiber mismatch and interfacial reactions. For applications subjected to severe loads of extreme thermal fluctuations such as in automotive components, discontinuously reinforced MMCs have been shown to offer near isotropic properties with substantial improvements in strength and stiffness, relative to those available in monolithic materials.

Properties of discontinuously reinforced MMCs are critically dependent on the nature of the reinforcing particle-matrix interface, morphology, physical properties, distribution and volume fraction of the reinforcement phase, any intermetallics associated with the reinforcement, and the overall mechanical properties of resultant composite. These factors can be controlled by optimizing of manufacturing process.

Overall, strength of such particle-reinforcement MMCs depends on:

a) the diameter of the reinforcing particles,
b) the interparticle spacing,
c) the volume fraction of the reinforcement, and
d) condition at matrix-reinforcement interface.

In order to improve the mechanical properties, it is very important to understand to strengthening mechanism, including the study of interfacial structure, dislocation generation and precipitation during ageing process. Matrix properties, including the work hardening coefficient that improves the effectiveness of the reinforcement constraint, are also important [11].

2. MATRIX MATERIALS AND REINFORCEMENTS

2.1. Matrix Materials

The matrix is the body constituent, serving to enclose the composite and gives it its bulk form [12]. In general, pure metals are not used as matrix materials. The metal matrix is an alloy that can be relatively simple but, in general it is multiple element alloys. There are a number of phases that can exist within the alloy.

Numerous metals have been used as matrix, for example Al, Cu, Fe, Mg, Ti, Zn and Pb. However, almost all of the structural alloy system has been considered as matrix materials in MMCs.

2.2. Reinforcements

The reinforcements such as fibers, particles, laminates, flakes, and fillers are the structural constituents of the composite. They determine the internal structure of the composite. Generally, but not always, they are the additive phase [12]. The reinforcement used in MMCs can be divided into two groups:

- Continuous
- Discontinuous

The most important discontinuous reinforcements have been SiC and Al_2O_3 in both whisker and particulate form. In terms of continuous reinforcements most are non-metals. However continuous metal filaments are employed, the most important being W, but stainless steel is also being used as a reinforcement.

The choice of reinforcement is not arbitrary. Several factors must be account on determine reinforcement [13]:

- The application
- The method of composite manufacture
- Cost

The recently developed some matrix and reinforcement material and their composite applications are given in Table 1:

Table 1. Representative MMC materials [14, 15]

MATRIX	REINFORCEMENT	POTENTIAL APPLICATIONS
Aluminum	Graphite	Satellite, missile, and helicopter structures
Magnesium	Graphite	Space and satellite structures
Lead	Graphite	Storage battery plates
Copper	Graphite	Electrical contacts and bearings
Aluminum	Boron	Compressor blades and structural supports
Magnesium	Boron	Antenna structures
Titanium	Boron	Jet-engine fan blades
Aluminum	Borsic	Jet-engine fan blades
Titanium	Borsic	High-temperature structures and fan blades
Aluminum	Alumina	Superconductor restraints in fusion power reactors
Lead	Alumina	Storage battery plates
Magnesium	Alumina	Helicopter transmission structures
Aluminum	Silicon Carbide	High-temperature structures
Titanium	Silicon Carbide	High-temperature structures
Superalloy[1]	Silicon Carbide	High-temperature engine components
Superalloy	Molybdenum	High-temperature engine components
Superalloy	Tungsten	High-temperature engine components

[1] cobalt base.

3. Production Method of MMCs

Several processing techniques have been evolved in an attempt to optimize the microstructure and mechanical properties of MMCs. The processing methods utilized to manufacture MMCs can be grouped according to the temperature of metal matrix during processing. Accordingly, the processes can be classified into three categories [11]:

1. Solid-phase processes
2. Liquid-phase processes
3. Two-phase (solid/liquid) processes

3.1. Solid-Phase Processes

The fabrication of particulate-reinforced MMCs from blended elemental powders involves number of stages prior to final consolidation. The important features of two of the processes, namely, powder metallurgy and high energy rate processing rate will be described below.

3.1.1. Powder Metallurgy

Because of difficulty in wetting ceramic particles with molten metal, the powder metallurgy route was the first method developed. Solid-phase processes involve the blending of rapidly solidified powders with particulates, plates or whiskers, through a series of steps as summarized in Figure 1.

The sequence of steps includes:

i. sieving of the rapidly solidified particles,
ii. blending of the particles with the reinforcement phase or phases,
iii. compressing the reinforcement and particle mixture to approximately 75% density,
iv. degassing and final consolidation by extrusion, forging, rolling or any other hot working method.

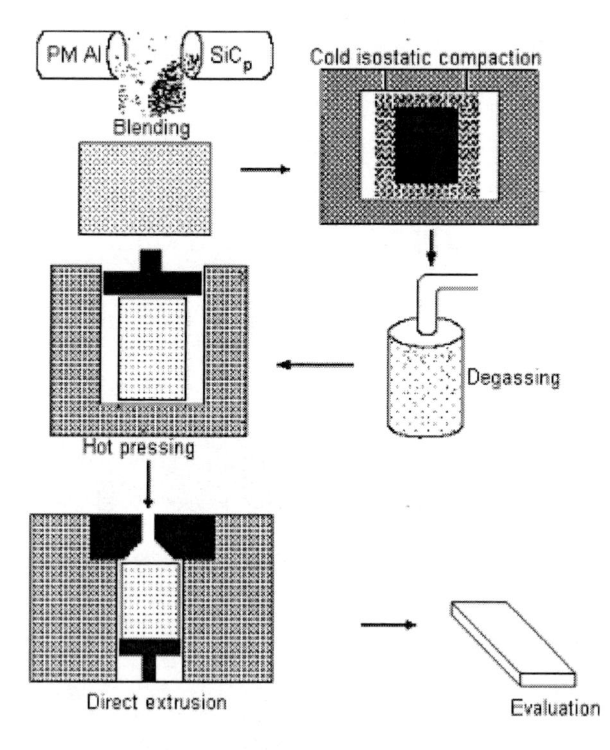

Figure 1. Schematic interpretation of the processing route for Powder Metallurgy (PM) Al-SiC_p composites [11].

This technology has been developed to various degrees of success by various commercial manufacturers. Powder metallurgy processed Aluminum-SiC MMCs posses higher overall strength levels compared to equivalent material processed by liquid-phase process, moreover exhibit a slight improvement in tensile ductility. In terms of final microstructural refinement, the powder metallurgy approach is superior and preferred in view of rapid solidification experienced by powders. This permits the developments of novel matrix materials outside the compositional limits dictated by equilibrium thermodynamics in conventional solidification processes.

The powder metallurgy route has also several attractive features [13]:

1. It allows essentially any alloy to be used as matrix.
2. It also any type of reinforcement to be used because reaction between the matrix and reinforcement can be minimized by using solid state processing.
3. Non-equilibrium alloys can be used for the matrix by using rapidly solidified material. This is particularly important where the composite is to be used for high temperature applications, and the rapidly solidified alloys have better elevated temperature strength than conventional alloys.
4. High volume fractions of reinforcement are possible, thus maximizing the modulus and minimizing the coefficient of thermal expansion of composite.

The powder metallurgy route also has some major disadvantages:

1. It involves handling large quantities of highly reactive, potentially explosive powders.
2. Manufacturing route is relatively complex.
3. It can produce the initial product forms limitedly.
4. The most important disadvantage is that the product is expensive in comparison with conventional material.

3.1.2. High Energy-High-Rate Process

An approach that has been successfully utilized to consolidate rapidly quenched powders containing a fine distribution of ceramic particulates is known as the high energy-high rate processing. In this approach the consolidation of metal-ceramic mixture is achieved through the application of a high energy in a short period of time. An examination of the open literature reveals that both mechanical energy and electrical energy sources can be utilized to consolidate MMCs. The high energy-high rate pulse (1 MJs^{-1}) facilitates rapid heating of conducting powder in die with cold walls. The short time at temperature approach offers an opportunity to control:

- phase transformations, and
- the degree of microstructural coarsening not readily possible trough standard powder processing techniques.

The process has been successfully used for the manufacturing of Aluminum/SiC and (titanium aluminide+niobium)/SiC composites. Although the results were encouraging, extensive work still remains in order to access the potential application of this approach.

3.2. Two Phase Processes

Two phase processes involve the mixing of ceramics and matrix in a regime of the phase diagram where the matrix contains both liquid and solid phases. Applicable two phase processes include the spray deposition, rheocasting and variable co-deposition of multiphase materials (VCM).

3.2.1. Spray Deposition

The spray deposition process for unreinforced alloys was developed Singer, and put into commercial use by Osprey Metals. In the osprey process, the reinforcement particulates are introduced into a stream of molten alloy that is subsequently atomized by jets of inert gas. The sprayed mixture is collected on substrate in the form of reinforced metal matrix billet [13]. This approach was introduced by ALCAN as a modification of the Osprey process [11]. Schematic diagram of the modified Osprey technique can be seen in Figure 2.2.

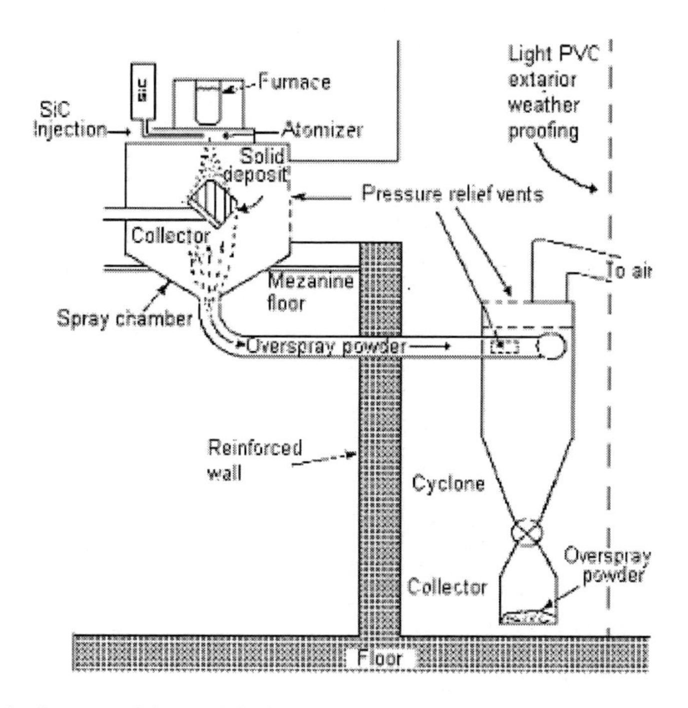

Figure 2. Schematic diagram of the modified Osprey technique [11, 13].

The process combines the blending and consolidation steps of powder metallurgy process and promises major savings in the production of MMCs [11]. Also the process has advantage that the contact time between the melt and the reinforcing particles is brief, so reaction between the two is limited and wider ranges of reinforcements are possible provided as the sprayed billets are not remelted. The initial billets are typically 95-98% dense, and required secondary fabrication step to achieve full density. As long as the alloy has a sufficient freezing range to achieve atomization at moderate superheats, any matrix composition can be used, including the advanced aerospace alloys, such as the Al-Li 8090 alloy. The cost of composites produced by spray deposition should be intermediate between powder processed composites and material made by mixing method [12].

3.2.2. Rheocasting

In the rheocasting process, fine ceramic particulates are added to a metallic alloy matrix at a temperature within the solid-liquid range of the alloy. This is followed by agitation of mixture to form low viscosity slurry. This approach takes advantage of fact that many metallic alloys behave like low viscosity slurry, when subjected to agitation during solidification. This behavior, which has been observed for fraction solids as high as 0.5, occurs during stirring and results in breaking of the solid dendrites into spherodial solid particles that are suspended in the liquid as fine-grained particulates. This unique characteristic of numerous alloys, known as thixotrophy, can be regained even after complete solidification by raising the temperature. This approach has been successfully utilized in die casting of Al and Cu base alloys.

The slurry characteristic of matrix during stirring permits the addition of reinforcements during solidification. The ceramic particulates are mechanically entrapped initially, and are prevented from agglomeration by presence of primary alloy solid particles. Subsequently, the ceramic particulates interact with the liquid matrix to affect bonding. Furthermore, continuous deformation and breakdown of solid phases during agitation prevent particulate agglomeration and settling. This method has been successfully utilized by several researchers to produce particulate MMCs.

3.2.3. Variable Co-Deposition of Multiphase Materials (VCM)

In the variable co-deposition of materials process, the matrix material is disintegrated into fine dispersion of droplets using high velocity inert gas jets. Simultaneously, one or more jets of strengthening phases are injected into atomized spray at a prescribed spatial location. Interfacial control is achieved by injection of reinforcing particulates at a spatial location where the atomized matrix spray contains a limited amount of volume fraction of liquid. Hence, contact time and thermal exposure of the ceramic particulates with the partially solidified matrix are minimized, and interfacial reactions are closely controlled. In addition, tight control of the environment during processing minimizes oxidation and other environmental effects. For situations where reactivity between the matrix material and reinforcement is negligible, the reinforcing phases are introduced into the liquid alloy matrix prior to spray deposition.

3.3. Liquid-Phase Processes

In liquid-phase processes, the ceramic particulates are incorporated into a molten metallic matrix using various proprietary techniques. This is followed by mixing and eventual casting of resulting composite mixture into either shaped components or billets for further fabrication. The process involves a careful selection of ceramic reinforcement depending on the matrix alloy. In addition to compability with the matrix, selection criteria for ceramic reinforcement include following factors:

a) Elastic modulus
b) Tensile strength
c) Density

d) Melting temperature
e) Thermal stability
f) Size and shape of reinforcement
g) Coefficient of thermal expansion
h) Cost

The most commercial discontinuous ceramic reinforcements using production of MMCs are SiC and Al_2O_3.

Silicon Carbide, SiC

Silicon Carbide is the only chemical compound of carbon and silicon. It was originally produced by a high temperature electro-chemical reaction of sand and carbon. Silicon carbide is composed of tetrahedral of carbon and silicon atoms with strong bonds in the crystal lattice. This produces a very hard and strong material. Silicon carbide is an excellent abrasive and has been produced and made into grinding wheels and other abrasive products for over one hundred years. Today the material has been developed into a high quality technical grade ceramic with very good mechanical properties. It is used in abrasives, refractories, ceramics, and numerous high-performance applications.

Silicon carbide is not attacked by any acids or alkalis or molten salts up to 800°C. In air, SiC forms a protective silicon oxide coating at 1200°C and is able to be used up to 1600°C. The high thermal conductivity coupled with low thermal expansion and high strength gives this material exceptional thermal shock resistant qualities. Silicon carbide ceramics with little or no grain boundary impurities maintain their strength to very high temperatures, approaching 1600°C with no strength loss. Chemical purity, resistance to chemical attack at temperature, and strength retention at high temperatures has made this material very popular as wafer tray supports and paddles in semiconductor furnaces. The electrical conduction of the material has lead to its use in resistance heating elements for electric furnaces, and as a key component in thermistors (temperature variable resistors) and in varistors (voltage variable resistors).

The key properties of SiC are [16];

- Low density
- High elastic modulus
- High strength
- High hardness
- Low thermal expansion
- High thermal conductivity
- Excellent thermal shock resistance
- Superior chemical inertness

Aluminum Oxide, Al_2O_3

Aluminum oxide commonly referred to as alumina. Alumina is the most cost effective and widely used material in the family of engineering ceramics. The raw materials from which this high performance technical grade ceramic is made are readily available and reasonably priced, resulting in good value for the cost in fabricated alumina shapes. With an excellent combination of properties and an attractive price, it is no surprise that fine grain technical grade alumina has a very wide range of applications.

Alumina possesses strong ionic inter atomic bonding giving rise to its desirable material characteristics. It can exist in several crystalline phases which all revert to the most stable hexagonal alpha phase at elevated temperatures. This is the phase of particular interest for structural applications.

Alpha phase alumina is the strongest and stiffest of the oxide ceramics. Its high hardness, excellent dielectric properties, refractoriness and good thermal properties make it the material of choice for a wide range of applications.

High purity alumina is usable in both oxidizing and reducing atmospheres to 1925°C. Weight loss in vacuum ranges from 10^{-7} to 10^{-6} g/cm^2.sec over a temperature range of 1700° to 2000°C. It resists attack by all gases except wet fluorine and is resistant to all common reagents except hydrofluoric acid and phosphoric acid. Elevated temperature attack occurs in the presence of alkali metal vapors particularly at lower purity levels.

The composition of the ceramic body can be changed to enhance particular desirable material characteristics. An example would be additions of chrome oxide or manganese oxide to improve hardness and change color. Other additions can be made to improve the ease and consistency of metal films fired to the ceramic for subsequent brazed and soldered assembly.

The key properties of Al_2O_3 are given below [17]:

- High strength
- High stiffness
- High hardness
- High wear resistance
- Excellent size and shape capability
- Good thermal conductivity
- Available in purity ranges from 94%, an easily metallizable composition, to 99.5% for the most demanding high temperature applications
- Resists strong acid and alkali attack at elevated temperatures
- Excellent dielectric properties from DC to GHz frequencies

Engineering properties of SiC and Al_2O_3 reinforcements are given Table 2. A basic requirement for liquid phase processes (casting techniques) of MMCs is initial intimate contact and intimate bonding between the ceramic phase and the molten alloy. This is achieved either by mixing the ceramic dispersoids into molten alloys, in fully or partially molten states or by pressure infiltration of preforms of ceramic phase by molten alloys.

Because of the poor wettability of most ceramics with molten metals, intimate contact between reinforcement and alloy can be promoted only by artificially inducing wettability or by using external forces to overcome the thermodynamic surface energy barrier and viscous drag [18]. Mixing techniques generally used for introducing and homogeneously dispersing a discontinuous phase in a melt are [19]:

1. Addition of particles to a vigorously agitated fully or partially molten alloy.
2. Injection of discontinuous phase in the melt with the help of an injection gun.
3. Dispersing pellets or briquettes, formed by compressing powders of base alloys and ceramic phase, in a mildly agitated melt.

4. Addition of the powder to an ultrasonically irradiated melt. The pressure gradients resulting from the cavitations phenomenon promote homogeneous mixing of ceramics in metallic melts.
5. Addition of powders to an electromagnetically stirred melt. Turbulent flow conditions achieved trough electromagnetic stirrings are used to obtain a uniform suspension.
6. Centrifugal dispersion of particles in a melt.

Table 2. Properties of particulate SiC and Al_2O_3 reinforcement *[16, 17]

Properties	SI/Metric (Imperial)	SiC		Al_2O_3 (99,5 % Purity)	
		SI/Metric	(Imperial)	SI/Metric	(Imperial)
Density	gm/cc (lb/ft^3)	3.1	(193.5)	3.89	(242.8)
Porosity	% (%)	0	(0)	0	(0)
Color	—	black	—	ivory	—
Flexural Strength	MPa (lb/in^{2}x10^3)	550	(80)	379	(55)
Elastic Modulus	GPa (lb/in^{2}x10^6)	410	(59.5)	375	(54.4)
Shear Modulus	GPa (lb/in^{2}x10^6)	—	—	152	(22)
Bulk Modulus	GPa (lb/in^{2}x10^6)	—	—	228	(33)
Poisson's Ratio	—	0.14	(0.14)	0.22	(0.22)
Compressive Strength	MPa (lb/in^{2}x10^3)	3900	(566)	2600	(377)
Hardness	Kg/mm^2	2800	—	1440	—
Fracture Toughness K_{IC}	MPa•m$^{1/2}$	4.6	—	4	—
Maximum Use Temperature (no load)	°C (°F)	1650	(3000)	1750	(3180)
Thermal Conductivity	W/m•°K (BTU•in/ft^2•hr•°F)	120	(830)	35	(243)
Coefficient of Thermal Expansion	10^{-6}/°C (10^{-6}/°F)	4.0	(2.2)	8.4	(4.7)
Specific Heat	J/Kg•°K (Btu/lb•°F)	750	(0.18)	880	(0.21)

*All properties are room temperature values.

In all these techniques, external force is used to

i. transfer a non-wettable ceramic phase into a melt,

ii. create a homogeneous suspension in the melt.

The melt-particle suspension thus created can be cast either by conventional foundry techniques or centrifugal casting, or by novel techniques such as squeeze casting. The various techniques used to solidify melt-particle slurries are discussed below:

3.3.1. Sand and Die Casting

The slow freezing rates obtained in sand moulds leads to preferential concentration of particles lighter than Aluminum alloys (e.g., mica, graphite, porous alumina) near the top surface of sand casting and segregation of heavier particles (sand, zircon, glass, SiC, etc.) near bottom part of castings. Depending upon the intrinsic hardness of dispersed particles, these high-particle-volume fraction surfaces serve as selectively reinforced surfaces, for instance, tailor-made lubricating or abrasion-resistant containing surfaces for various tribological applications.

The relatively rapid freezing rates in metallic moulds generally give rise to more homogeneous distribution of particles in cast matrix. Further improvements in particle distribution can be achieved by use of water-cooled moulds, Cu chills, and other agitation techniques.

3.3.2. Centrifugal Casting Method

Centrifuging during solidification of an axisymmetric part can be used effectively to segregate heavy or light non-metallic additions to the outside and inside extremities of the part prior to solidification [1].

Solidification in rotating moulds of composite melts containing particle dispersion of mica, graphite, porous alumina, zircon, and carbon microballons exhibits two distinct zones-a particle rich zone near the inner circumference for lower density particles and a particle poor zone near the outer circumference. The outer zone is particle-rich if the particles are denser than the melt, as is the case with zircon and SiC particles in Aluminum [18].

3.3.3. Compocasting

Compocasting is a promising method for preparing MMCs reinforced by discontinuous fibers, whiskers or particles. Mehribian et. al. used the term Compocasting to describe an application of the rheocasting which is directly termed by M.C. Flemings [1].

In the compocasting process particulates and discontinuous fibers of SiC, TiC, alumina, silicon nitride, graphite mica, glass, slag, MgO, and boron carbide have been incorporated into vigorously agitated partially solidified Aluminum alloy slurries. An outstanding advantage of compocasting is the ability to incorporate material which is not wetted by the alloy. The non-wetting discontinuous ceramic phase is mechanically entrapped between the proeutectic phase present in the alloy slurry, which is held between its liquidus and solidus temperatures. Settling, flotation or agglomerations are also prevented by the presence of solid alloy particles. This semifusion process allows near-net shape fabrication by extrusion or forging, since deformation resistance is considerably reduced as a result of the semifused state of the composite slurry [18].

Several schematics of compocasting equipment have been reported in literature. A schematic of the compocasting equipment can be seen in Figure 3. A hollow rotor is incorporated in the design of the equipment to facilitate. The addition of the particulate materials (graphite powder in this case). The required quantity of graphite particles is stored in the hollow rotor so that their temperature equilibrates with that of the alloy slurry. This ensures that addition of graphite particles to the alloy slurry does not result in a reduction in temperature or a consequent reduction in viscosity. The rotor prevents surface agitation of the melt and consequent air. It also permits graphite particles to be introduced below the surface of the melt at the shearing zone.

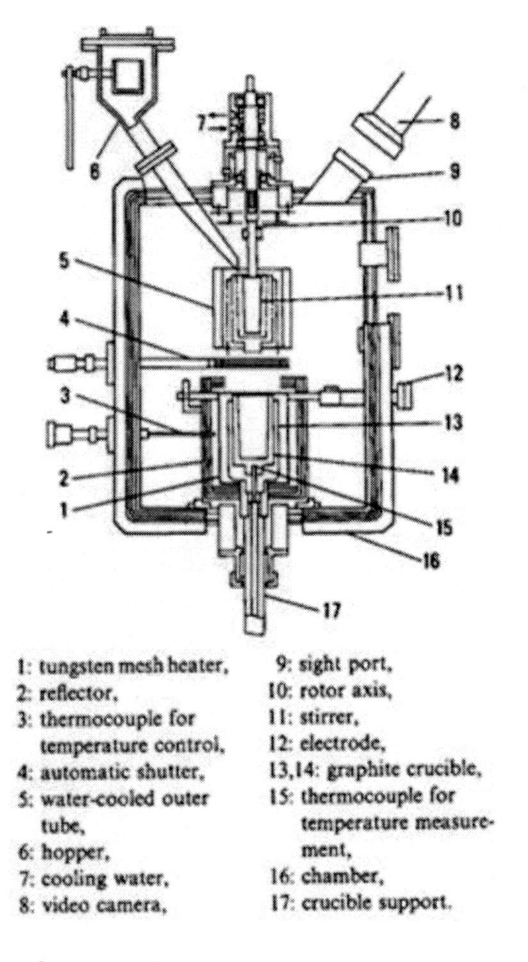

Figure 3. Schematic diagram of compocasting equipment [20].

The compocasting procedure is as follows: The melt is poured in to the preheated compocasting unit and stirring is initiated at the rotor speed required to provide the desired shear rate. Then, the temperature is gradually lowered until the alloy is 30-50% solid. At this temperature the non metallic particulates are added by turning the injector down. The powering is controlled such that the total percentage of solid, non-metals and solid spheroids of the alloy, does not exceed about 50%. Stirring is continued until interface interactions between the particulates and the matrix promote wetting. After mixing the composed slurry is bottom poured into at preheated crucible for die casting or squeeze casting.

3.3.4. Pressure Casting

Several pressure casting methods have been used for preparing MMCs. Pressure casting of composites allows larger-sized and more complex component shapes to be rapidly produced at relatively low pressures (<15 MPa) for an equivalent capital expenditure.

Pressurized gas and hydraulic ram in a die-casting machine have been employed to synthesize porosity-free fiber and particle composites. It has been reported that high pressures, short infiltration paths, and columnar solidification toward the gate produced void-free composite castings. The pressure die-casting particle composites exhibit lower bulk and interfacial porosities, more uniform particle distribution, less agglomeration of particles, and occasional exfoliation/ fragmentation of soft particles (e.g., graphite in Aluminum alloys), with melt penetrating into fine exfoliation cleavages.

Pressure die casting can be used to increase the fiber concentration in the compocasting and for simultaneous two-dimensional alignment of the fibers following compocasting. This process also can be used for producing high-volume fraction and dual-layered composites of particle/metal mixed melt after compocasting.

Another pressure casting technique is unidirectional pressure infiltration into a fibrous preform under controlled processing conditions. Figure 4 shows a device developed for this technique.

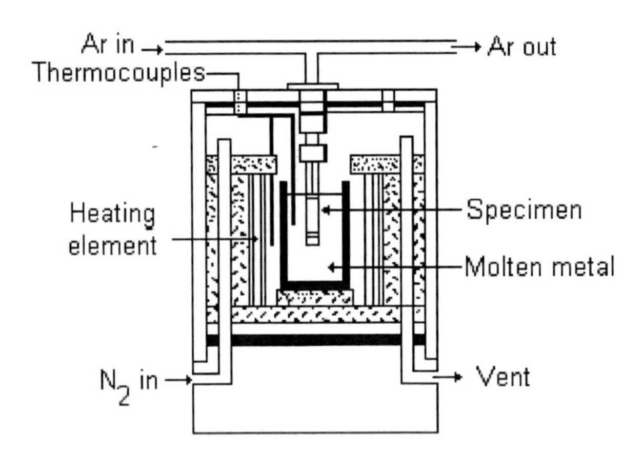

Figure 4. Schematic of pressure infiltration device [21].

The operating principle of the device is to use pressurized gas to force molten metal into an evacuated die. The device consists of two main components: the pressure vessel, containing the melt furnace and crucible, and the cap, attached to which are the control valves and the die. The die is attached to the centre of the cap can be evacuated. The vessel is pressurized with nitrogen about 3 MPa pressure forcing the metal up through the die and into the fibers, forming composite.

3.3.5. Investment Casting

In investment casting of MMCs, filament winding or pregreg handling procedures developed for fiber reinforced plastics (FRPs) are used to position and orient the proper volume fraction of continuous fibers within the casting. The layers of reinforcing fibers are stacked in the proper sequence and orientation, and the fiber preform thus produced is infiltrated either under pressure or by creating a vacuum in the permeable preform.

Continuous graphite fiber reinforced Mg has been produced by this method by MCI in Cleveland in collaboration with Martin Marietta.

3.3.6. Vacuum Infiltration Process

Several fiber-reinforced metals (FRMs) are prepared by the vacuum infiltration process. Making MMCs under vacuum, better wettability between non-metallic reinforcements and metal melts can be achieved with less oxidation occurring [1]. In the first step, the fiber yarn is made into a handleable tape with fugitive binder in a manner similar to processing resin matrix composite pregreg. Fiber tapes are then laid out in the desired orientation, fiber volume fraction, and shape. They are inserted into suitable casting mould. The fugitive organic binder is burned away, and the mould is infiltrated with molten matrix metal. The best casting in terms of soundness (voidage, porosity, and distribution of fibers) and mechanical properties are obtained if the mould assembly is preheated and pre-evacuated before the liquid metal is introduced.

3.3.7. Vortex Method

The vortex method employing an impeller mixer is a technique for producing particle/metal composites. The impeller mixer stirs the melt to create a vortex into which particles can be placed and incorporated in the melt. The same impeller mixer, at reduced speeds, is an effective means of controlling particle dispersion after introduction. The graphite particle coated by nickel or copper has been tried successfully in Aluminum alloys. Later investigations showed that it is possible to disperse less than 3 wt. % of uncoated but pre-heat-treated (heated to 400°C in air for 1 h) graphite in Aluminum alloy melts by vortex method.

3.3.8. Squeeze Casting

Squeeze casting is a recently commercialized casting process in which metal is solidified under direct action of pressure sufficient to prevent the appearance of either porosity or shrinkage porosity. Squeeze casting is unique in this respect: all other casting processes leave some residual porosity. The process also known variously as Liquid Metal Forging, Squeeze Forming, Extrusion Casting and Pressure Crystallization are essentially a combination of Die-casting and Closed Die Forging.

These processes have their individual merits and demerits, and one may satisfy certain requirements of which the other is incapable. For instance, more complex shapes can be obtained in a single operation in casting than in die forging. But defects in the form of gas and shrinkage porosity are ubiquitously produced in castings, together with both microsegregation and macrosegregation effects. These features are difficult to eliminate in shaped castings and they lead to reduced mechanical properties. Additionally, it is often difficult to obtain a uniform cast microstructure in section of varying thickness. Die forging, on the other hand, produces full-density material that contains a degree of anisotropy resulting from the directionality of microstructure caused by non-hydrostatic forming stresses. Also, several processing stages are usually required in die forging to produce articles of complex shapes [22].

The squeeze casting process is shown basically in Figure 5. In this process, preform used to selectively reinforce a casting, containing the required composition of fibers at a chosen volume fraction, with desired shape, is first preheated to proper temperature. The heated preform is put into the die cavity just prior to the addition of the molten alloy. A measured quantity of modified and filtered metal is poured into a preheated die cavity located on the bed of hydraulic press. The press is moved to close off cavity and pressurize the liquid metal. Pressure is maintained until solidification is complete. Then the press is opened and component is ejected.

The key of the process is that solidification is finished under high pressure, which is several orders of magnitude greater than the melt pressure developed in conventional foundry practice. Pressures applied to the system are up to 150 MPa, these are necessary to overcome capillary effects and frictional forces experienced by liquid flowing through the interfiber channels. Once pressure is applied, the melt front passes quickly through the preform at a velocity of about 30 mm per second. During the squeeze casting the pressure can amount to a few thousand atmospheres, with a consequent increasing in the freezing temperature. For aluminum-based alloys the freezing temperature may increase by 10-25°C, depending on pressure, during squeeze casting.

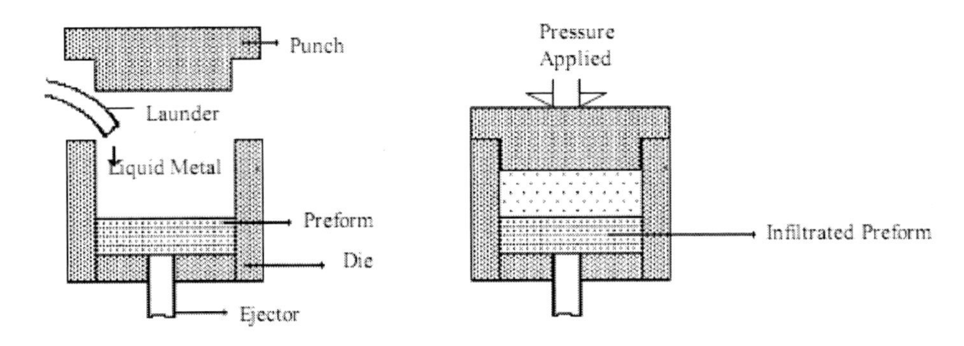

Figure 5. Schematic of squeeze casting of composite [1].

Furthermore, the high rate of heat removal achieved in squeeze casting is also helpful to rate of solidification. This is because the applied pressure tends to force the solidification metal against the die and eliminate the air gaps between the casting surface and die wall. The combination of undercooling with high heat transfer rates ensures a fine-grained equiaxed structure. The high pressures employed during solidification result in the formation of a porosity-free, high-integrity near net shape casting with smooth surfaces and excellent microstructure. Strength and fatigue life of this casting are much longer than those of conventional gravity casting. The high pressure employed promotes wetting of most ceramic fibers. Selective reinforcement is realized by placing preforms of ceramic fibers in appropriate locations within casting die. A major limitation is the size of parts that may be cast because of the high pressure requirements.

The major advantages of squeeze casting over other casting processes are as follows [23]:

1. Material having neither gas porosity nor shrinkage porosity is produced.
2. No feeders or risers are required and therefore no metal wastage occurs.
3. The inherent cast ability of the alloy is of little or no concern since the applied pressures deviate the need for customary high fluidity. Both common casting alloys and apparently wrought alloys can be squeeze cast to finished shape.

4. Control of microstructure is possible only by control of dominant process parameters such as pouring temperature and mould temperature. Nucleating agents can be used but they are not normally necessary.
5. Because there are no internal or external defects on properly produced squeeze cast component, costly post-solidification examination by non-destructive testing techniques is of very limited value.
6. Squeeze casting can have mechanical properties as good as, and in some cases even better than, wrought products of the same composition: their casting factor should therefore be set at unity.
7. Squeeze casting provides the most effective and efficient route to produce near-net shape composite components for engineering applications.
8. Besides the densification achieved, the moderate applied pressure causes intimate contact between the solidifying casting and die. As a result, relatively fine grains and large number of nuclei formed due to the low casting temperature and elevated pressure.

The excellent mechanical properties, microstructural refinement and integrity of squeeze cast products are used to advantage in many critical applications. Parts made recently include Al-alloy truck hubs, barrel heads, and hubbed flanges, automotive wheel and pistons, etc.

In any squeeze casting operation, certain practical problems have to be solved with regard to liquid metal engineering, including clean metal melting, metal metering, metal movement into the die coat; magnitude of pressure is applied (the dwell time). Two forms of squeeze casting machines are available, based upon different approaches to metal metering and movement: direct and indirect.

In the direct squeeze casting method, Figure 6, the pressure is applied to entire surface of the liquid metal during freezing, producing castings of full density. This technique consequently gives the most rapid heat transfer, yielding the finest grain structure. The principal stages involved in component manufacture by direct squeeze casting are:

1. A metered amount of molten metal is transferred into a female die cavity or mould.
2. The male die, or punch, of the appropriate shape is driven into the die cavity to exert a pressure on molten metal while it is solidifying and form the required shape.
3. After the casting has solidified under the applied pressure the punch is withdrawn.
4. The shaped solid casting is ejected from the die.

Although flat-topped ingots can be produced by way of direct process, it is customary to manufacture more complex casting to finished shape by means of shaped punch. The insertion of punch into the liquid metal has to be closely controlled to provide the necessary upward displacement of liquid in a non-turbulent manner so as to avoid oxide entrapment, which could otherwise be detrimental to attainment of good mechanical properties. The dies are usually made from high quality die steels which are heated and lubricated by graphitic die coat.

In indirect squeeze casting process, metal is injected into the die cavity by a small diameter piston, by which mechanism the pressure is also applied during freezing. As can be seen Figure 7 and Figure 8 horizontal and vertical types of squeeze casting machines are used in composite productions. This technique is really a hybrid process between pressure die casting and squeeze casting. The cast product is not of the superior quality as is obtained from the direct squeeze casting process. A further disadvantage of this process is that material utilization is often inefficient in that runner and gating system is used and this material has to be removed and recycled. The only merit of the indirect squeeze casting process is that, because the casting forms inside a closed die, the dimensions of the casting are relatively easier to control and thus the use of a highly accurate metering system may not be necessary.

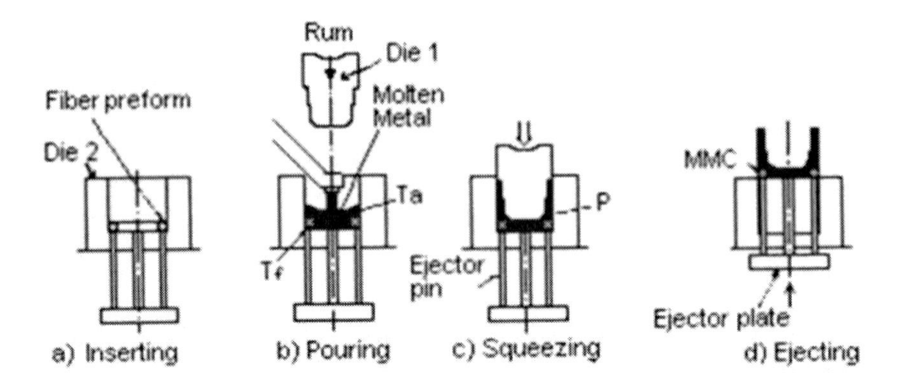

Figure.6. Schematic of direct squeeze casting method [24].

The direct process of the two squeeze casting methods is more common practice for the production of full integrity casting and MMC components. Melting, pouring and metering of liquid metal are inherently established within the individual direct and indirect processes. The other parameters that require attention are the casting variables such as pouring temperatures, die temperatures, the magnitude of pressure, and the dwell time of applied pressure. All of these parameters need optimizing for each different alloy system and for each casting shape. In general, a low pouring temperature for wide freezing range alloys is more effective in yielding good metallurgical quality, so as long as the applied pressures are sufficiently high to remove all remains of porosity.

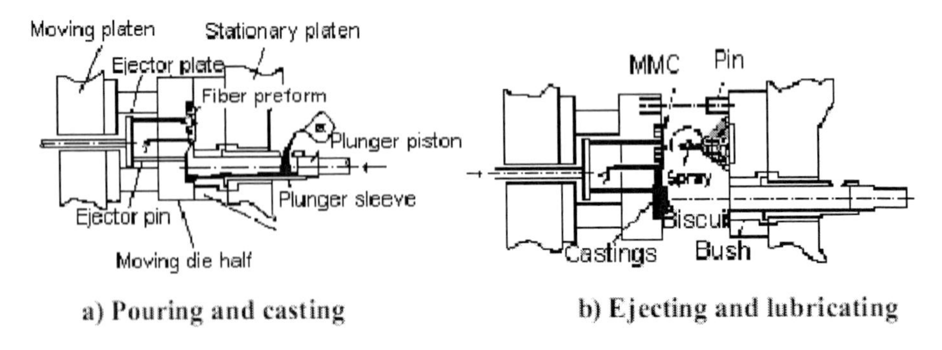

Figure 7. Indirect squeeze casting operations with a horizontal type of machine [24].

The cycle times for squeeze casting are generally much shorter than for either gravity or low pressure die casting because of the high heat transfer coefficients. The use of multi-cavity dies can further improve the efficiency of squeeze casting process if a machine of sufficient power rating is available.

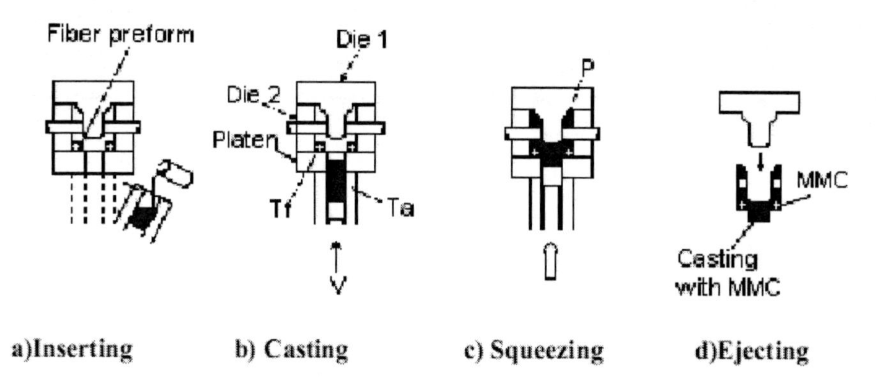

Figure 8. Indirect squeeze casting operations with a vertical type of machine [24].

Recently, the majority of squeeze casting tend to be in aluminum alloys, since these materials are very easy to handle in this process. There is a relatively low metal pour temperature and long die life possible when Aluminum is being cast. Aluminum has attractive properties for many applications due to its low density and corrosion resistance and aluminum squeeze castings have the additional advantage of forged-like properties. For certain applications, there would be a significant advantage in replacing steel forgings with Aluminum squeeze castings due to the weight savings and making design changes possible in other parts of mechanism.

4. Preparation and Properties of Cast Aluminum-Ceramic Particle Reinforced MMCs

Ceramic particle reinforced MMCs were generally made by conventional powder metallurgy methods. However, these production routes are very cumbersome and expensive. Therefore, recently, researches have been made to prepare these composites by liquid processing techniques where the particles are introduced into melt prior to solidification.

The advantages of casting fabrication methods are [1]:

i. simplicity, speediness and cheapness,
ii. production of composite parts of complex shapes or near-net shape,
iii. ease of fabrication of selectively reinforced components,
iv. suitably for various kinds of metal matrices and reinforcements.

The major difficulties in fabricating aluminum-ceramic particle reinforced composites by melt metallurgy techniques are wetting and bonding between matrix and reinforcements, and distribution of reinforcement in the matrix.

Absence of wetting between molten aluminum and ceramic particles (oxides or carbides) at temperatures used in conventional foundry practice is the most important factor. The lack of wetting leads to rejection of ceramic particles when they are added to the melt. Also density differences between Al and ceramic particles cause segregation of dispersed particles in the melt.

Dendrite segregation of particles during casting of discontinuously reinforced Al-matrix composites is another serious problem. In some situations, this segregation causes severe agglomeration and interparticle contact. The factors believed to influence dendrite segregation is the dendrite arm spacing (DAS) of matrix that is proportional to the dendrite growth rate, the size of the particles, the relative thermal conductivities, and the difference in contact angles between a particle/solid interface.

Several approaches have been employed to improve the wettability between the liquid Al and ceramic particles [18]:

1. Use of metal coatings such as Ni and Cu on ceramic particles
2. Heat treatment of the particles
3. Addition of surface active elements such as Mg, Ca, Ti, Zr, and P in Al alloy melt just prior to introduction of particles has been employed by several workers.
4. Ultrasonic vibrations of the melt

Metal coatings on ceramic particles increase the overall surface energy of the solid and improve wetting by changing the contacting interfaces to metal/metal, instead of metal/ceramic. However, coatings result in a change in the composition of matrix alloy that may not be desirable in certain cases. It also requires a source of metal coated ceramic particles.

Heat treatment of particles before their dispersion in the melt aids their transfer by causing desorption of absorbed gases from the ceramic surfaces (e.g. formation of wettable, oxygen deficient surfaces when Al_2O_3-type oxides are treated at high temperature in vacuum).

Use of reactive elements may promote wetting by reducing the surface tension of the melt, decreasing the solid/liquid interfacial energy of the melt, or inducing wettability by a chemical reaction. Very small quantities of reactive elements may be quite effective in improving wetting since they segregate either to the melt surface or at the melt/ceramic interfaces. However, use of reactive elements may lead to small changes in the composition of the matrix. Also, magnesium additions lead to the formation of magnesium oxide on periphery of the alumina particles that may be undesirable in certain applications.

Ultrasonic vibrations promote wetting of ceramics by metallic melts as a result of partial desorption of absorbed gases (mostly hydroxyls) from the surface of the particles; in addition, they supply the excess energy for melt cavitations which facilitates particle dispersion in the melt.

Also, several foundry techniques have been developed to promote wetting between Al melt and ceramic particles. In the compocasting technique requires specialized equipment to stir the alloys at temperatures between their liquidus and solidus temperatures. Mixing of ceramic particles in the liquid Al requires cleaning of ceramic particles in vacuum by ion bombardment of composite in vacuum.

4.1. Wettability of SiC with Al and Al Alloys

The wettability between reinforcing materials and base metal is an important factor for strength of composite. In literature, on quantitative basis, the wettability has mainly been determined from contact angle (θ) in sessile drop method and rarely from the adhesion strength or the rupture strength of produced composite.

The equilibrium configuration of a drop of liquid resting upon a flat solid surface is determined by Young's Equation, the balance between liquid-vapor interfacial energy (γ_{vl}), solid-vapor interfacial energy (γ_{sv}), and solid-liquid interfacial energy (γ_{sl}) as shown Figure 9.

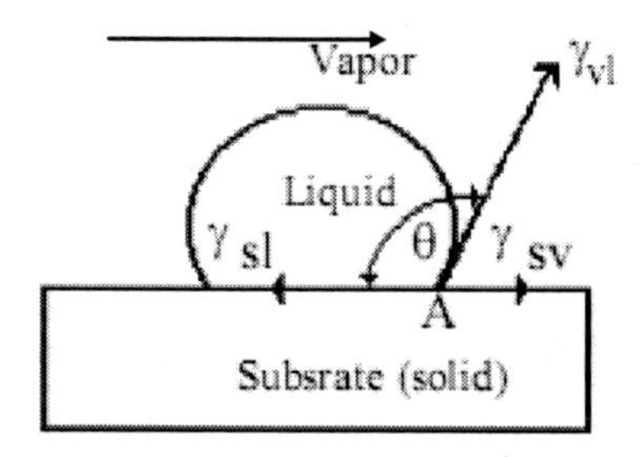

Figure 9. Schematic of sessile drop method [25].

$$\gamma_{sv} = \gamma_{vl} \cos\theta + \gamma_{sl} \quad (1)$$

Hence, wetting is achieved when $\theta < 90^{\circ}$ (i.e., when $\gamma_{sv} > \gamma_{sl}$)

However, for the wettability of SiC to Al, whose surface react with the atmosphere, it is difficult to explain the interfacial phenomena between the materials and type, or more importantly, the degree of bonding, only from the value ofθ. Moreover, the effects of atmosphere on the measuring vessel for the sessile drop method, or the presence of an oxide film on drop surface can not be ignored during the determination ofθ. On the other hand, the rupture or adhesion strength of the composite depends on the antecedents of production.

The contact angle between molten Al and SiC decreases with increasing temperature. For example, it is 154° at the melting point of Al and 40° at 1100°C. Although it has little temperature dependence below 900°C. The time dependence of θ, which decreases from the initial value of 110-120 ° to about 70 ° after waiting at 980 °C for 2100 s.

In some previous studies, it was possible to disperse heat-treated SiC particles in molten Al and prepare cast Al-SiC particle composites where as-received SiC particles that are not heat-treated were rejected by the melt. Other studies have been carried out on the formation of aluminum carbide Al_4C_3 at the SiC/Al interface, the prevention of this carbide formation by inserting a barrier of ZrC between the SiC and Al, effects of Mg and Si on the reaction between SiC and Al, and the reaction in the composite produced by exchange diffusion method.

The thermodynamics of chemical equilibrium between SiC and some common metals can be divided into reactive and stable types. In reactive systems, SiC react with the metal to form silicides and/or carbides and carbon. SiC and a metallic matrix alloy can coexist thermodynamically, that is, a two-phase field exists in stable systems. For example, SiC in alloys of Al, Au, Ag, Cu, Mg, Pb, Sn, Zn. This does not imply that SiC will not be attacked. In fact, a fraction of SiC in contact with molten Al can dissolve and react to give Al_4C_3:

$$Al + SiC \longrightarrow Al_4C_3 + (Si) \quad (2)$$

This reaction can happen because the section SiC-Al lies in a three-phase field. Such reactions can be avoided by prior alloying of Al with Si. The addition of Si to Al puts the overall composite composition in the two-phase Al-SiC field.

Also the reaction between SiC and Al initiating the middle kinetics has been studied detailed in literature. It was observed that the reaction proceeds in two steps. In a first stage, silicon carbide dissolves in the liquid metal:

$$SiC\ (s) \longrightarrow (Si)_L + (C)_L \qquad (3)$$

$$Al + (Si)_L + (C)_L \longrightarrow Al_4C_3 + (Si)_L \qquad (4)$$

Secondly, the Al_4C_3 compound precipitates growing in random islands on the SiC surface. The oxidation treatment of the SiC surface is found to have beneficial but limited effect on wettability in the Al/SiC system.

The effects of alloying elements on wettability were also investigated using elements having relatively small affinity with carbon such as Mn, Cu etc. and those having greater affinity such as VIA, VA and VIA group elements in the periodic table.

The effects of Si, Mn, Fe and Cu on the wettability are given in Figure 10. The incubation period is decreased by alloying 10-20% Si, Mg, or Fe with Al. This phenomenon results in some improvement in wettability. No compound could be detected at SiC/Al alloy interface by using EPMA analysis.

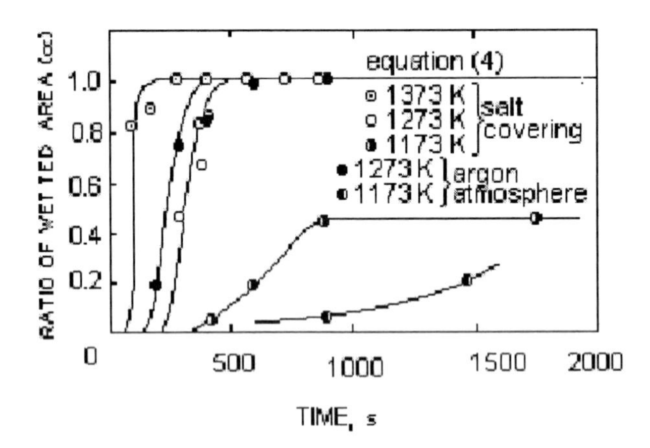

Figure 10. Effects of Silicon, Iron, Manganese, and Copper on wettability of SiC by liquid aluminum at 1000°C [1].

Minor alloying elements (e.g., Mn, Cr), commonly used in wrought alloys as grain refiners, are unnecessary in discontinuously reinforced MMCs. Furthermore, these additions should be avoided, since they might result in formation of coarse intermetallic compounds during consolidation and subsequent processing, thus impairing the tensile ductility of the composite.

On the other hand, the additions of reactive elements such as Li, Mg, Ca, Ti, Zr, and P to matrix material improve wetting characteristics of metal-ceramic systems through:

a) a reduction of the surface tension of the melt,
b) a reduction of the solid-liquid interfacial energy of the melt, or
c) by including a chemical reaction at the interface.

In Al alloys, wetting of certain ceramics can be enhanced through additions of elements which have a high affinity with oxygen, such as those in group I and II, (e.g., Li and Mg). For example, the presence of 3 wt% Mg decreased the surface tension of pure Al from 0.760 to 0.620 Nm^{-1}, the surface tension of Mg is 0.599 Nm^{-1} [1].

4.2. Casting Fluidity

Fluidity is simply the ability of molten metal (Aluminum) to fill very narrow spaces, whether in mould cavity (thin wall casting) or in the gap between the fibers or other discrete reinforcing components in cast composites. Casting constitutes the earliest known near-net shape components. They provide, as a result of a single manufacturing step, both thick sections that tax the capabilities of forging or powder metallurgy techniques, and thin sections otherwise obtainable only by either forming sheet metal or machining plates, extrusions, or forgings. However, excessive fluidity may sometimes be harmful. Since it contributes to defects such as flashing at mould joints, mould penetration in sand moulds, and surface roughness in die or permanents mould castings.

Among pure metals, Al has greater fluidity than Cd and Pb but less than Zn, Sn, and Bi for same degree of superheat. The pure metals, including Al, have considerable degree of fluidity when poured at just above their liquidus temperature (zero superheat). Also, the fluidity of alloyed Al decreases with decreasing purity.

At present, there are three commercially important groups of Al casting alloys:

- Al-Cu (Cu content nominally around 4.5%)
- Al-Si (Hypoeutectic Si level around 7%, eutectic Si level at 12% and hypereutectic Si level around 18%)
- Al-Mg (Mg content around 7%)

For both Al-Cu and Al-Mg binary alloys, fluidity at given pouring temperature initially decreases rapidly with increasing amount of alloying element. The lowest value corresponds roughly to a composition typical of most common commercial alloys in either case. Fluidity then increases again as the alloy composition approaches the eutectic (33% Al for both Al-Cu and Al-Mg binary systems). Incidentally the eutectic fluidity of Al-Cu alloys is higher than that of pure Al, but that of Al-Mg alloys is somewhat lower.

Probably the Al alloy group, Al-Si displays a similar trend. The lowest fluidity value corresponds to the 5-7% Si range, typical of most commercial Hypoeutectic Al-Si casting alloys. However, the highest fluidity value, about equal to that of pure Al, occurs not at 12% Si eutectic. But rather around 18% Si, typical of commercial hypereutectic Al-Si alloys.

Early work on fluidity has shown that oxide-coated Al has a considerably higher surface tension (840 dyne/cm) than oxide-free Al (520 dyne/cm). A lower metal head is thus required for oxide-free Al to overcome surface tension (i.e., to force the melt into a small-diameter hole). In this work, the surface tension of Al in a nitrogen atmosphere calculated to be about 590 dyne/cm. This is only slightly higher than that of oxide-free Al.

A commercial application of this finding involves low pressure inert gas (argon) die casting which greatly improves melt fluidity by a factor of four over gravity die casting (permanent mould) and a factor of two over atmosphere low pressure die casting. The castings themselves are very clean (e.g., oxide-free and low-hydrogen level) with improved mechanical properties, especially fatigue strength.

Increasing the metallurgical cleanliness of an Al melt has also increased fluidity. As shown in Figure 11, for Al-6Si-3.5Cu (AA319) alloy, the clean melt is about 20% more fluidity than dirty melt. Thus, clean melt can be poured at a 55°C (100°F) lower temperature than dirty melt. As a result, mechanical properties are increased beyond the point achievable through a reduced amount of inclusions alone.

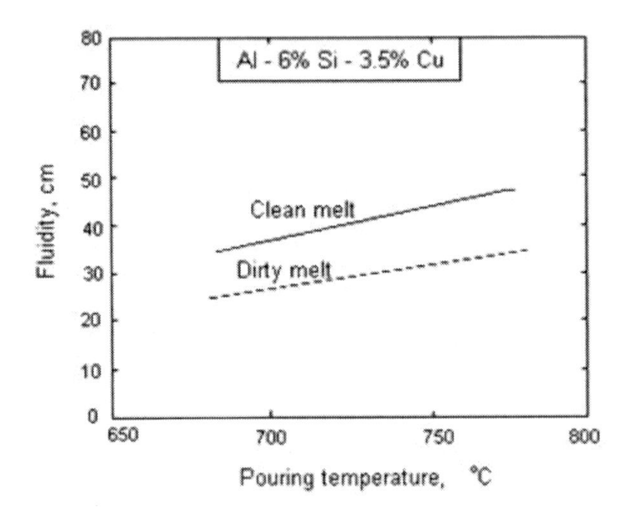

Figure 11. Influence of melt cleanliness on fluidity of Al-6Si-3.5Cu alloy [1].

The development of cast MMCs, on the other hand, is still in the early stages. However, some of the results indicate the key parameters controlling fluidity (how thoroughly fiber mats can be infiltrated) include: packing density of the fibers, preheating temperature of fibers, and any surface treatment intended to promote wetting. Melt superheat does not appear to be a major factor in most cases.

4.3. Effect of Pressure

The effect of pressure on solidification of metals and alloys has received attention in the past. From the point of view of the fabrication of cast MMCs, the ductility of matrix "as cast" is of great importance. The factors that can improve the ductility are a fine grained equiaxed structure, matrix homogeneity, and freedom from voids and discontinuities, e.g. shrinkage cavities.

One effect of external pressure on alloy phase equilibrium is to change the melting (liquidus) temperature. It should be theoretically possible therefore (for alloys that shrink upon solidification) to solidify a melt of the alloy without lowering the temperature of the melt, if a sufficient high pressure can be applied. Other effects may include shifts in phase boundaries and in the eutectic point.

Another effect of pressure is to increase the heat flow from the casting through the die walls by a considerable amount compared to an equivalent casting at a pressure of 1 atmosphere and can, therefore, increase potential for undercooling (supercooling) during solidification.

The application of pressure can affect the virtual elimination of shrinkage and other voids and discontinuities.

Under favorable conditions, therefore, pressure applied during solidification of an alloy can result in fine-grained equiaxed macro-structure with micro-structure being characterized by small dendrite arm spacing, small constituent particles and more homogeneous distribution of structural components [21].

5. Properties of Al/SiCp MMCs Produced by Direct Squeeze Casting Method

5.1. Experimental Details

From the available literature on MMCs, it is obvious that the morphology, distribution physical properties and volume fraction of the reinforcement phase are all factors that affect the overall properties of resultant composite [5]. In this paper the effect of direct squeeze pressure on Al-Cu-Mg based alloy and their composites is investigated. A cost effective perspective is taken in to account by analyzing microstructural and mechanical properties in Al-Cu-Mg -SiC_p composite.

In this study, direct squeeze casting system was used to production of specimens. A four-piece die which included main parts from cast iron and rams from tool steel. Al-4.5 Cu alloy was chosen as matrix alloy because of its widespread commercial applications. To increase casting fluidity and wetting between SiC particles and the matrix, 3 wt.% Mg was also added to the matrix melt. Matrix alloy was prepared from commercially pure aluminum (99.7% Al) and electrolytic copper (99.98 % Cu) and pure magnesium. Copper was added in the melt Al and finally Mg was dissolved into the Al-Cu alloy melt just before the poring of the alloy. Melting of alloy, degassing and additions of particulates were performed in an electrically resistance heated furnace at dry nitrogen atmosphere. Graphite crucible was used for melting of matrix alloy, and the addition and mixing of particulates were made into the melt in the crucible. The degassing was carried out by blowing dry nitrogen gas above and into the melt just before pouring. Then, matrix alloy was poured to the mould. For casting operations a constant 100 MPa pressure was used for unreinforced matrix alloy and the composites. Matrix alloy was additionally produced by a gravity casting method into a steel mould heated to 350 ^{o}C to compare the properties with squeeze cast unreinforced alloy. The mean value of spectrographic analysis of matrix alloy is given in Table 3.

Three composites were produced by introducing SiC particulates with 36 μm average size in 5, 10 and 15 vol.%. to the Al-Cu-Mg matrix alloy. Density of the SiC particulates was 3.2 g/cm^3 and chemical composition is given in Table 4. The particles, as received, are in clumps and could not be dispersed uniformly into the melt in this state. Furthermore the sheen on the received particles led to non-wetting and particle rejection from the melt. In order to avoid these problems the as received particles were heated at 700 °C temperature for two hours before addition into the melt.

Table 3. The chemical composition of the matrix alloy (wt. %)

Cu	Mg	Zn	Fe	Si	Mn	Sn	Pb	Cr	Ni	Ti
4.45	3.14	0.50	0.20	0.08	0.01	0.009	0.006	0.002	0.001	0.001

Table 4. The chemical composition of SiC particles (wt. %)

Element	Amount (%)
SiC	min. 98.80 - 98.40
SiO_2	min. 0.50 - 0.70
Free Si	max. 0.35 - 0.45
Free C	max. 0.20 - 0.30
Fe	max. 0.05

Unreinforced matrix alloys produced by both gravity and squeeze casting techniques and the composites produced by squeeze casting were cut to prepare samples for mechanical tests and metallographic examinations. Four samples of each cast were prepared further by using diamond blades in CNC machine for tensile testing. Tensile test samples, have 6.4 mm diameter with a gauge length of 26 mm, were prepared for testing in Hounsfield Tensometer.

The specimens were prepared for metallographic examinations using 220-320-500-1000 mesh emery papers, followed by polishing with diamond paste. Microscopic methods were used to study the composite structure, particle distribution and porosity. Microstructure and fracture surfaces (obtained by tensile test) were studied by SEM.

Hardness measurement was carried out using a Brinell hardness tester. Before testing, specimen surfaces were polished using emery papers down to 1000 mesh. At least 10 measurements were taken for each sample and the average was taken as the hardness value.

Density of composites and unreinforced matrix alloys were measured by water displacement technique. Specimens in 8 mm diameter and 20 mm heights were preheated to 120 °C for 1 h in a furnace. Subsequently, the weights of the materials were determined. This is followed by coating of the sample surfaces with a polymeric gel to close surface porosities. Specimens were suspended into double distillated water at 25 °C and their weights were determined by four digits balance. This weight was accepted as the volume of the specimen. Density was calculated according to formula given below:

$$\rho = \frac{W_{PH}}{W_{SW}} \tag{5}$$

where;

ρ is the density of specimen,
W_{PH} is the weight of preheated specimen and
W_{SW} is weight of specimen suspended into the water.

In order to determine the volume fraction of SiC particulates, Linear Analysis Method was applied. In this analysis, the total length of randomly placed lines within the phase of interest (ΣL_P) is divided by total line length (L_T) to obtain lineal fraction (L_L) or equivalently volume fraction V_V:

$$L_L = \frac{\Sigma L_P}{L_T} = V_V \tag{6}$$

In this study, 30 random lines' length of 70 mm drawn for each sample at magnification of X 250 and their average was taken as volume fraction of SiC.

Wear tests were carried out under dry sliding condition on a pin-on-disc apparatus. A steel disc (AISI D2) was used as counter surface and composites were made into pins having 6 mm diameter and 30 mm in height. The steel disc was heat treated to achieve 61 HRC, then polished with 600-grid emery paper. Average roughness (Ra %) of the disc was 1.2 μm. The pins were made to slide on the steel disc at 0.5, 1.0 and 2.0 m/s sliding speeds under the loads of 5, 10, and 15 N for a 1000 m sliding distance. The wear was measured in terms of weight loss of the materials measured to 0.0001 g resolution. The wear damage on the specimens was evaluated via wear rate (mm^3/m) calculated by using ASTM formulas.

Steady state friction coefficient values were also measured by a load cell-equipped with the pin-on disc apparatus. A complete wear microstructural characterization was carried out via scanning electron microscopy (JEOL JSM 5410 and JEOL JSM 5600). The wear tracks on the disc were also investigated with optical microscopy.

5.2. Results and Discussion

5.2.1. Microstructure

The most important aspect of microstructure is the distribution of reinforcing particles, and this depends on processing and fabrication routes involved. However, particles can modify other aspects of the matrix microstructure.

In powder processed material, the reinforcement distributions depend on the blending and consolidation procedures, as well as the relative size of the matrix and reinforcing particles. If the matrix powder is large relative to the reinforcement, the reinforcing particles will agglomerate in interstices or the coarse particles, and be very inhomogeneously distributed in the final product.

In composites processed by molten metal mixing methods, the situation is somewhat more complicated because the reinforcement distribution is influenced by several factors:

1. Distribution in the liquid as a result of the mixing.
2. Distribution in the liquid after mixing but before solidification.
3. Redistribution as a result of solidification.

The distribution during mixing will obviously depend on the mixing process used, and it is essential to produce as uniform a distribution as possible without any gas entrapment. Since any gas bubbles will be lined with reinforcing particles. After mixing before solidification, particles will segregate due to gravity. With relatively high volume fraction of particles and range of particle size, the settling will be hindered. Settling rate will be function of particle density and size, and there is also the possibility that particle shape will play a role. Particles of different size and shape will settle at different rates producing agglomeration.

The third factor that influences reinforcement distribution is the solidification process itself. Reinforcing particles do not generally nucleate the primary solidifying phase, through solidification nucleation may occur in some hypereutectic systems. If solidification nucleation does not occur the reinforcing particles are rejected at solid/liquid interface, and segregate to the interdendritic regions that solidify last.

The application of pressure can affect the virtual elimination of shrinkage and other voids and discontinuities. Therefore, pressure applied during solidification of an alloy can result in fine-grained equiaxed macro-structure with micro-structure being characterized by small dendrite arm spacing, small constituent particles and more homogeneous distribution of structural components. Effects of pressure on microstructure and grain size can be seen by comparing with Figure 12 a and b.

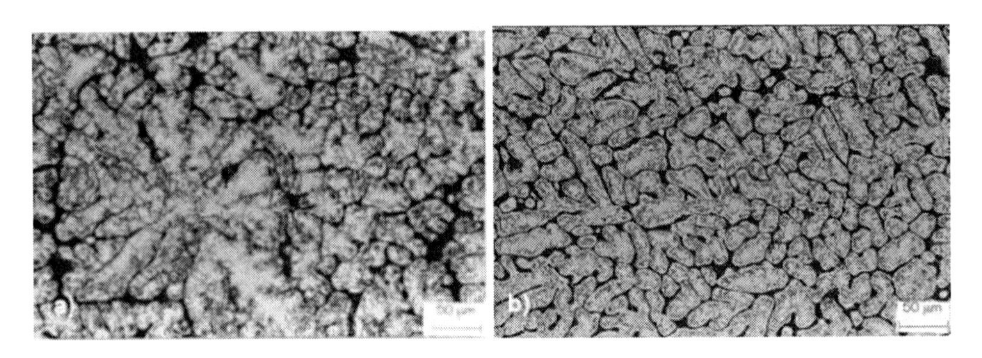

Figure 12. Optical micrographs of matrix alloy: a) Gravity Cast, b) Squeeze Cast.

The comparison of the microstructure between gravity cast and the squeeze cast specimen shows that the most important factor is pressure employed at squeeze casting. Squeeze casting sample has improved the microstructure. Dendrite arms were broken down and fine-grained equiaxed microstructure was obtained by squeeze casting. Matrix homogeneity and freedom from voids and discontinuities, i.e., shrinkage cavities compared to gravity cast sample are other advantages. Main constituents in gravity cast and squeeze cast matrix alloys are primary phase (A), eutectic region (B) and grain boundary lamellae (C) (Figure 13).

From the EDS results, it is clear that the chemical compositions of primary phase and eutectic regions are different for gravity cast and squeeze cast matrix alloy. The Al content of the primary α- Al phase in gravity cast is higher than that of squeeze casting. Inversely, solubility of Cu and Mg in α- Al primary phase at squeeze casting was increased because of higher heat transfer during solidification. This is an expected result. It is well known that increasing the heat transfer resulted in shifting of equilibrium curves during phase transformation. This could hopefully result in strengthening of primary α- Al dendrites in squeeze cast alloys.

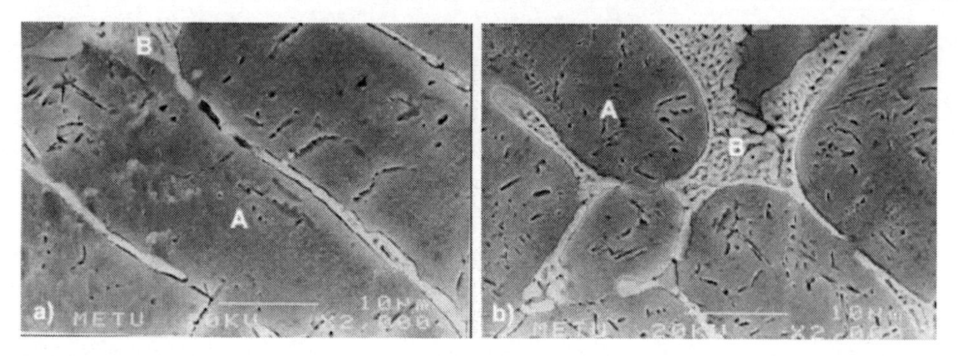

Figure 13. SEM micrograph of matrix alloy: a) Gravity Cast, b) Squeeze Cast. A denotes Primary Phase and B denotes Eutectic Phase.

The atomic percent values of the elements both in primary phase and the eutectic region in gravity cast and squeeze cast matrix alloy are given Table 5. High cooling of solid solutions is well known that resulted in obtaining supersaturated structure. The structures produce fine intermetallics by natural aging process. Since squeeze cast samples exhibited higher amount of solid solubility of Cu and Mg than that of gravity cast alloy, finer and higher amount of intermetallics can be expected to precipitate in squeeze cast matrix. Here an Al-Cu-Mg type precipitate phase can be normally formed in the Al matrix.

Table 5. The qualitative element identification of gravity cast and squeeze cast microstructures (wt. %)

Element	Primary Phase		Eutectic Region	
	Gravity Cast	Squeeze Cast	Gravity Cast	Squeeze Cast
Al	97,14	94,80	48.19	53.25
Mg	1,77	2.03	15.96	16.60
Cu	1.09	3,17	35.84	30.15

The requirements of low density, with reasonably high thermal conductivity, have made Al and Mg alloys most commonly used matrices. Regarding alloying additions, the results of several studies have shown that low matrix alloying additions result in MMCs with attractive combinations of strength, ductility and toughness. Minor alloying elements (e.g., Mn, Cr), commonly used in wrought alloys as grain refiners, are unnecessary in discontinuously reinforced MMCs. Furthermore, these additions should be avoided, since they might result in formation of coarse intermetallic compounds during consolidation and subsequent processing, thus impairing the tensile ductility of the composite.

On the other hand, the additions of reactive elements such as Li, Mg, Ca, Ti, Zr, and P to matrix material improve wetting characteristics of metal-ceramic systems through:

a. Reduction of the surface tension of the melt,
b. Reduction of the solid-liquid interfacial energy of the melt, or
c. By including a chemical reaction at the interface.

In Al alloys, wetting of certain ceramics can be enhanced through additions of elements which have a high affinity with oxygen, such as those in group I and II, (e.g., Li and Mg). For example, the presence of 3 wt% Mg decreased the surface tension of pure Al from 0.760 to 0.620 Nm^{-1}, the surface tension of Mg is 0.599 Nm^{-1} [26].

Interdendritic segregation of particles during casting of discontinuously reinforced Al-matrix composites is a serious problem. In some situations, this segregation causes severe agglomeration and interparticle contact. The factors believed to influence dendrite segregation is the dendrite arm spacing (DAS) of matrix that is proportional to the dendrite growth rate, the size of the particles, the relative thermal conductivities, and the difference in contact angles between a particle/solid interface. In this study, since solidification rate was very high in squeeze cast composites, so serious agglomerations have not been observed. Typical microstructures observed in composites containing 5, and 15 % vol. SiC particles are given in Figure 14.

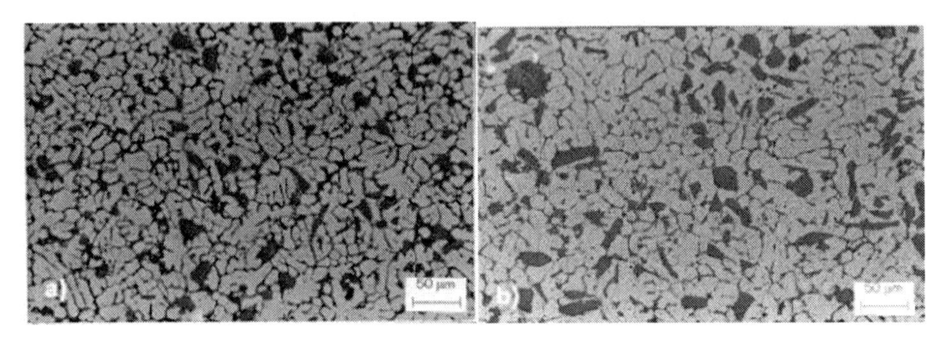

Figure 14. Optical micrographs of composite specimens a) 5 vol.% SiC b) 10 vol.% SiC.

As a general observation it is seen that distributions of SiC particles were fairly uniform with no agglomeration of SiC particles. Microstructure of composite consists of primary phase and eutectic regions. SiC particles were surrounded by eutectic region. SiC particles are behaving as a nucleation agent for Al and eutectic phase during solidification. In general, eutectic phase was formed on the SiC particles and/or SiC particles were segregated into last frozen eutectic regions.

Microstructure of Al-Cu-Mg- 5 vol. % SiC contained primary phase and eutectic regions. SiC particles were surrounded by eutectic region (Figure 14a). Local chemical analyses that performed at eutectic close and away from SiC particles are given in Table 6.

Table 6. The qualitative element identification of SC 5 composite microstructure (wt.%)

Element	Eutectic close to SiC	Eutectic away from SiC
Al	61.32	46.26
Mg	8.41	11.82
Cu	30.27	41.91

The eutectic region close to SiC and away from the particles had somewhat different composition. From this table it can be concluded that SiC particles are behaving as a nucleation agent for Al and eutectic phase during solidification. In general eutectic phase was formed on the SiC particles and/or SiC particles were segregated into last frozen eutectic regions. Additionally, there were relatively large islands which consisted solely of eutectic.

Figure 15 shows the optical microstructures of 15 % vol. SiC particle reinforced materials. Figure 5a represents relatively low magnified microstructure. It can be clearly seen that eutectic is accumulated into two regions (eutectic close to SiC and eutectic away from SiC). These two eutectic regions are shown in Figure 5b and 5c respectively. The local chemical analysis measured in Al-Cu-Mg-10 % vol. SiC and Al-Cu-Mg-15 % vol. SiC are given in Table 7.

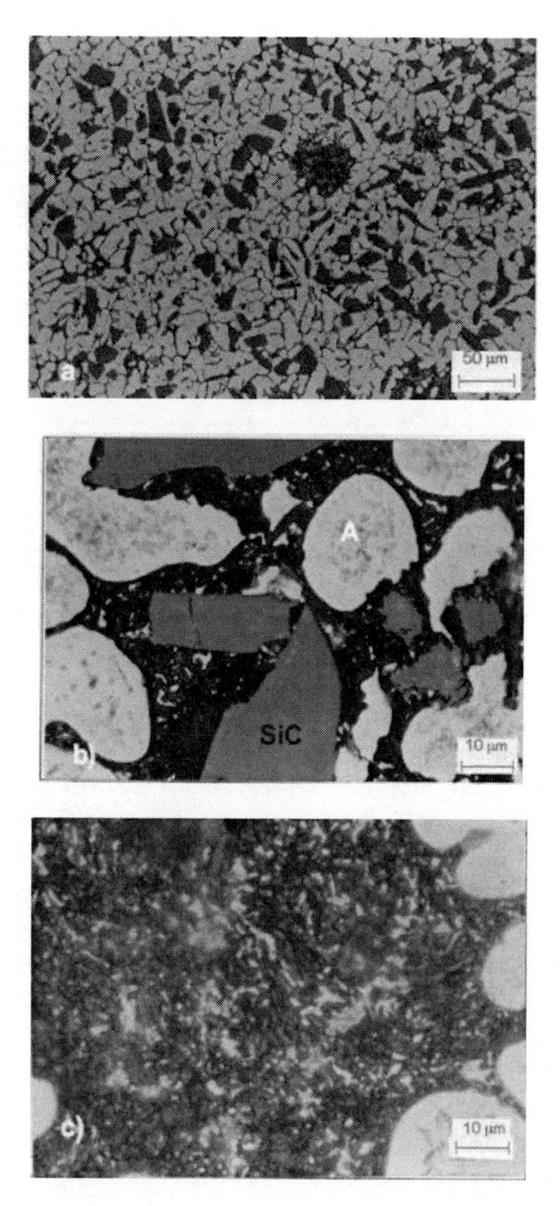

Figure 15. Optical micrograph of SC 15 (15 % vol. SiC): a) Relatively low magnification of microstructure, b) Eutectic region close to SiC particles, c) High magnification of eutectic region away from SiC.

Table 7. The qualitative element identification of SC 10 and SC 15 composite microstructures (wt. %)

Element	Eutectic Region		Island Region	
	SC 10	SC 15	SC 10	SC 15
Al	41.21	87.11	45.07	34.49
Mg	18.46	3.50	18.26	11.47
Cu	40.32	9.40	36.67	54.04

Although excellent microstructure and SiC distribution, some film inclusions have been detected in composite specimens. One of the film inclusions can be seen in Figure 16.

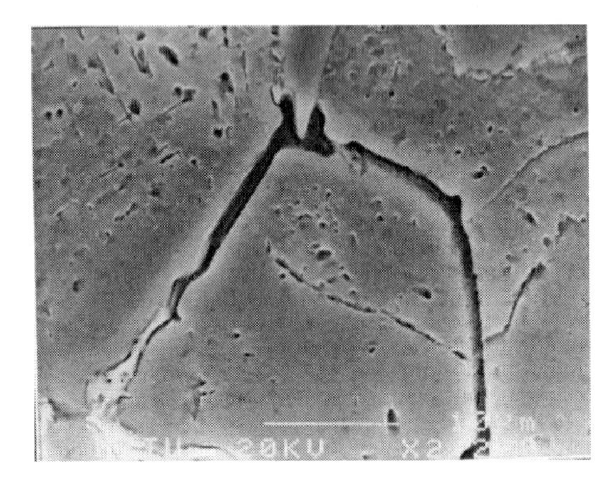

Figure 16. SEM micrograph of film inclusion observed in 10 % vol. SiC reinforced composite.

As can be seen from the micrographs, the solidification started at in the interparticle channels, and growing dendrites rejected solute-rich liquid to the particle surfaces, which were the last locations to solidify. Mg also segregated towards the particles that are generally believed to promote good bond strength through limited chemical reaction. Mg is expected to improve composite properties by both providing in limited chemical interaction at the particle surface, and by increasing matrix strength, while Cu has the opposite effect. This was also pointed out by than Zhang et. al. [27] alloying of Al with Cu resulted in formation of CuO between Al matrix and SiC particles and prevented wettability. Undissolved copper-bearing second phases are located mainly at the particle surfaces, were seen to have a negative influence on composite.

As it was demonstrated in Figure 6 that, film inclusions are segregated into grain boundaries. In some composites this segregation becomes dominant for mechanical properties and resulted in brittle fracture starts from grain boundary regions and caused to deteriorate strength. Ibrahim and co-workers [26] reported several results that indicating the diminishing effect of segregation.

5.2.2. Density

Density and porosity of the samples are given in Table 8. Experimental density was measured by water displacement and theoretical values calculated from Rule of Mixture. The porosity is calculated as

$$\text{Porosity } \% = \frac{\rho_T - \rho_E}{\rho_T} x100 \quad (7)$$

where, ρ_T is the theoretical density measured from Rule of Mixture (ROM), and ρ_E is the density measured by water displacement method.

Table 8. The results of density measurements

SAMPLE	THEORETICAL DENSITY (g/cm^3)	EXPERIMENTAL DENSITY (g/cm^3)	POROSITY %
GC 0	2,7361	2,5548	6,63
SC 0	2,7361	2,7063	1,09
SC 5	2,7574	2,7068	1,84
SC 10	2,7792	2,7092	2,52
SC 15	2,8006	2,7253	2,69

It is seen that gravity cast specimen has the highest porosity, 6.63%, because of microstructural defects raised from gravity casing. The squeeze cast matrix alloy has the lowest porosity, 1.09%. In the case of composites porosity increased with increasing volume fraction of SiC. The lowest porosity obtained in 5% SiC as 1.84 % and the highest porosity in 15% SiC as 2.69%.

5.2.3. Hardness

Hardness of gravity and squeeze cast matrix alloys as well as those of composites are tabulated in Table 9. It is seen that squeeze cast matrix alloy produced a higher hardness value than gravity cast. This is the main effect of the squeeze casting on microstructure and mechanical properties. Also hardness values gradually increase with increasing volume fraction of SiC particles in composite specimens. The maximum value was measured as 139 BHN in SC 15 specimen.

Table 9. The results of hardness measurements

HARDNESS VALUES (BHN)				
GC 0	SC 0	SC 5	SC 10	SC 15
94 ± 2.79	100 ± 1.94	112 ± 1.18	132 ± 2.19	139 ± 2.41

5.2.4. Tensile Test Properties

In a continuously reinforced composite, the matrix acts to transfer loads to the fibers and influence of matrix properties as determined from the rule of mixture (ROM) is usually small in the direction of reinforcement. However, the matrix plays a more important role in composite reinforced with particulates. As stated by Shorowordi and co-workers [2] in high strength matrix alloys the efficiency of particulates is decreasing. The effect of SiC volume fraction on composite yield strength, UTS, elongation and reduction in cross-section area values are given Table 10. UTS and elongation were obtained from load-elongation diagram.

The stress-strain curves of the specimens studied are presented in Figure 17. It gives a good comparison between the gravity cast, squeeze cast and the composites in terms of strain and stress behavior.

Table 10. The results of tensile tests

SAMPLE	Yield Strength (MPa)	UTS (MPa)	Elongation %	Reduction in area %
GC 0	168	195	1.78	1.68
SC 0	183	247	3.30	2.51
SC 5	187	216	1.53	1.11
SC 10	224	232	0.70	0.84
SC 15	-	239	0.45	0.44

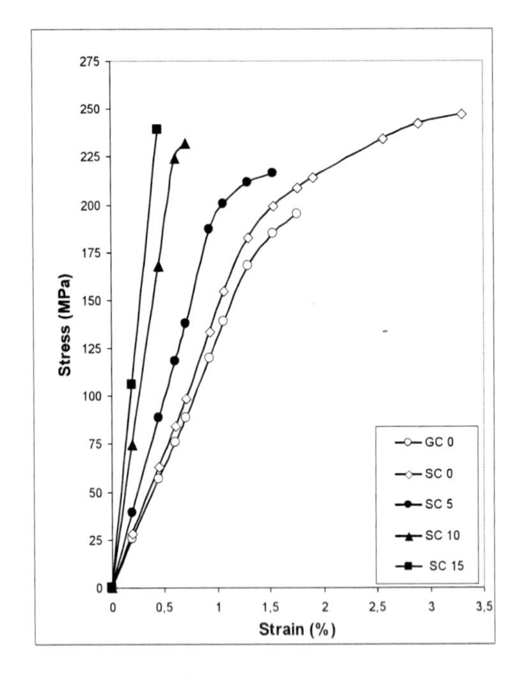

Figure 17. The stress-strain diagram of specimens.

Composite strength and ductility are influenced by both matrix alloy and the volume fraction and properties of reinforcement. Generally, the addition of reinforcement to a low strength matrix causes a large improvement in strength. Higher strength matrices show a smaller strengthening effect, although the resulting composite strength may be high. At a certain level of reinforcement the composite fails by brittle fracture at a low strain, and the strength achieved is limited. Therefore, an optimum level exists for a given matrix that, if exceeded, strength and ductility suffers [21]. As can be seen Table 10, SC 15 specimen (15 %vol. SiCp reinforced composite) shows no yielding and failure in elastic region. In composites, UTS values higher than gravity cast specimen but lower than squeeze cast sample. % 0.2 Yield Strength values of composites increase with increasing volume fraction of SiC particles and higher than matrix alloy. The elongation and strain percent decrease with increasing volume fraction of SiC.

5.2.4.1. Elastic Modulus

The effect of most discontinuous ceramic reinforcements on Young's Modulus (E) is quite significant. The predominant factor that influences this elastic characteristic is volume fraction of reinforcement (v) [13].

The quantitative value of the elastic modulus is somewhat dependent on the method of measurement, with dynamic measuring methods tending to give larger values than static measurements obtained from the elastic portion of the tensile stress-strain curve. Static values may also depend on whether the measurements are made in tension or compression. Most of these difficulties result from the presence of thermal residual stresses caused by differences in the coefficient of thermal expansion between the matrix and the ceramic particles. In the case of SiC and Al_2O_3 particle reinforced Al, matrix is in tension. This means that when composite is loaded, plastic flow occurs earlier in tension than in compression, and total strain will consist of both elastic and plastic components. The situation is further complicated by in homogeneity in reinforcement distribution that can also result in local plasticity. Comparison with theoretical expectation is also somewhat difficult due to uncertainty in the appropriate value for the modulus of the particle reinforcement [13].

The first extensive study on mechanical properties of discontinuously reinforced Al alloys was carried out by D.L. Mc Danels [28] who investigated mechanical properties and stress-strain behavior for several types of commercially fabricated Al matrix composites, containing up to 40 vol.% discontinuous SiC whisker, nodule, or particulate reinforcement. The elastic modulus of composites was found to be isotropic, to be independent of type of reinforcement, and to be controlled solely by the vol.% of SiC reinforcement present. The modulus of elasticity of 6061 Al-matrix composites increased with increasing reinforcement content (Figure 18).

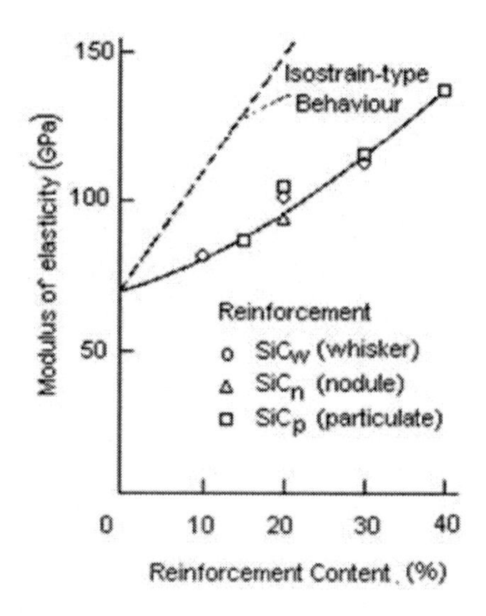

Figure 18. Effect of reinforcement content on modulus of elasticity of discontinuous SiC /6061 Al composites [28].

The dominant factor in controlling the elastic modulus is the volume fraction of reinforcement content in SiC/Al composites [12, 28]. It is relatively insensitive to the particle distribution, while variations in the type and shape of reinforcement can be accounted for by different expressions. For a given reinforcement content, the modulus tended to be isotropic, with nearly equal values obtained from tests in both longitudinal and transverse directions. The modulus of the composites was also independent of matrix alloy. Heat treatment of the composites may have had a slight effect on modulus.

5.2.4.2. Strength

D. L. Mc Danels [28] reported the factors influencing the yield and tensile strengths of SiC/Al composites are complex and interrelated. He reported up to 60% increase in yield and ultimate strengths, depending on the volume fraction of reinforcement, type of alloy, and the matrix alloy temper. Also the best way to evaluate this behavior is through isolation of variables and analysis of stress-strain curves and fracture behavior.

Effect of Matrix Alloy: One of the most important factors influencing the mechanical behavior of MMCs is the alloy matrix . The Al matrix used for SiC/Al was the most important factor affecting yield strength and UTS of SiC/Al composites stress-strain behavior is summarized in Figure 19. with the other parameters held constant.

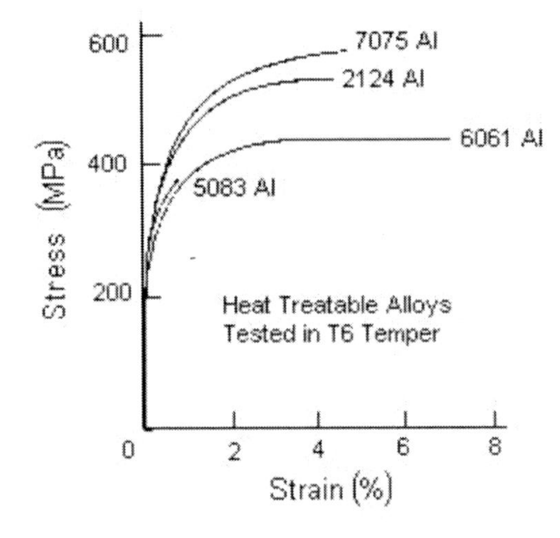

Figure 19. Effect of Al matrix alloy on stress-strain behavior of composites with 20 vol.% SiCw reinforcement (tested in direction parallel to final rolling direction) [28].

In this study, the composite contained 20 vol.% SiC_w as the reinforcement and the tests were done along longitudinal direction after T6-temper was applied. As can be seen in Figure 13. that SiC/Al composite with higher strength Al matrix alloys such as 2024, 2124 or 7075 Al, had higher strengths but lower ductilities. Composites with a 6061 Al matrix showed good strength and higher ductility while composites with a 5083 Al matrix failed in a brittle manner. The results showed that the yield and UTS strength of SiC/Al composites, with other parameters being constant, were primarily controlled by intrinsic yield/tensile strengths of the matrix alloys. These results also showed that, in general, the yield and UTS strengths of the composites with 20 vol.% SiC reinforcement, were higher than those of same heat treated matrix alloys without reinforcement. The same trends of increased yield/tensile strengths were also observed for composites with these matrices using other types of SiC reinforcement and at other reinforcement contents

Effect of reinforcement content: Reinforcement content is another important factor controlling the strength of SiC/Al composites. Tensile strengths, generally, increased with increasing reinforcement content only as long as the composite was able to exhibit enough ductility to attain full strength. As the content reached to 30-40 vol.% SiC the strength increase tended to taper off because the composites failed while still in the steeply ascending portion of the stress-strain curve. In this region, the matrix probably did not have sufficient internal ductility to redistribute the very high localized internal stresses, and the composites failed before being able to reach stable plastic flow and normal UTS.

Effect of reinforcement type and directionality: Composites containing acicular SiC whisker, irregular equiaxed SiC nodule or irregular jagged SiC particle reinforcements were studied to understand effect of reinforcement type and directionality on strength. Stress-strain curves for 6061 Al-matrix composites with 20 vol.% of various SiC reinforcements indicated that the yield strength and UTS of the SiC_w and SiC_p were similar while composites with SiC_n reinforcements were about 10% lower in yield and UTS. No significant effect on directionality on strength properties was observed with 6061 Al-matrix composites for any of the SiC reinforcements at contents from 10-40 volume percentage.

Some orientational differences were observed in SiCw/Al composites. Normally, any differences in yield strength orientation were minor, although significant differences were observed in UTS.

5.2.4.3. Ductility

The major limitation in the mechanical properties of composites is the rather limited ductility [13]. Ductility of SiC/Al composites, as measured by strain to failure, is again a complex interaction of parameters. However, the prime factors affecting these properties are reinforcement content, matrix alloy and orientation.

With increasing reinforcement content the failure strain of composites reduced rapidly, and the stress-strain curves also reflect a change in the fracture mode. Composites with ductile matrix alloys and lower reinforcement contents exhibited a ductile shear fracture with a 5-12% failure strain. As reinforcement content increased, the fracture progressed through a transition and becomes brittle, reaching 1-2% failure strain, at higher reinforcement contents [28].

Recent studies demonstrated that composite failure is associated with particle cracking and void formation in the matrix within clusters of particles. However particle fracture occurs from early strains, as demonstrated by a detectable decrease in modulus with increasing strain. It also apparent, in the melt processed composites that very different particle fracture behavior is not reflected in differences in the tensile elongination [13].

5.2.5. Fracture

Is it was demonstrated in Figure 16 that, film inclusions are segregated into grain boundaries. In some composites this segregation becomes dominant for mechanical properties and resulted in brittle fracture starts from grain boundary regions and caused to deteriorate strength. Ibrahim and co-workers [26] reported several results that indicating the diminishing effect of segregation.

The fracture surface micrographs of unreinforced matrix alloys are given Figure 20.

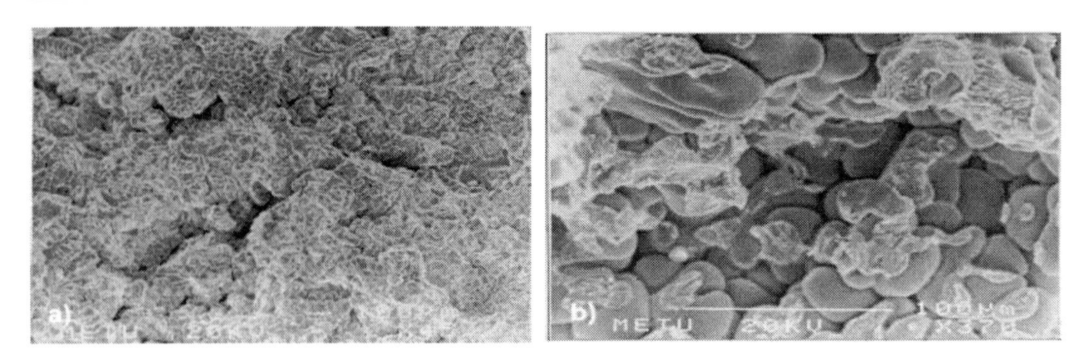

Figure 20. SEM micrographs of matrix alloy fracture surfaces: a) Gravity cast of matrix alloy, b) Squeeze cast of matrix alloy.

As can be seen Figure 20a, voids were detected at gravity cast sample fracture surface. However any void was not detected at squeeze cast specimen fracture surface at higher magnification (Figure 20b).

Figure 21a shows fracture surface of the 10 vol.% SiC particle reinforced composite and clearly shows that fracture started and initiated along the grain boundaries. Figure 21b represents the fracture surface of the 15 vol.% SiC particle reinforced composite and shows relatively brittle fracture and a limited amount of ductile fracture of the surrounding matrix. Figure 21c is a high magnification micrograph of the 15 vol.% SiC particle reinforced composite and shows fracture of SiC particle. The appearances of fracture surface indicate good bonding between the matrix and the SiC particles as there is no evidence for fracture initiation at the matrix particle interface.

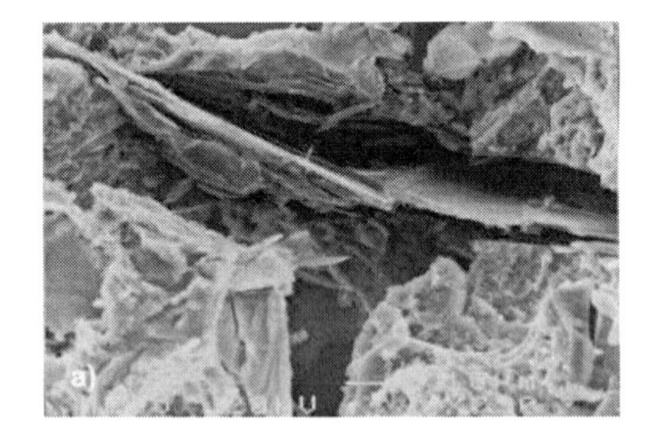

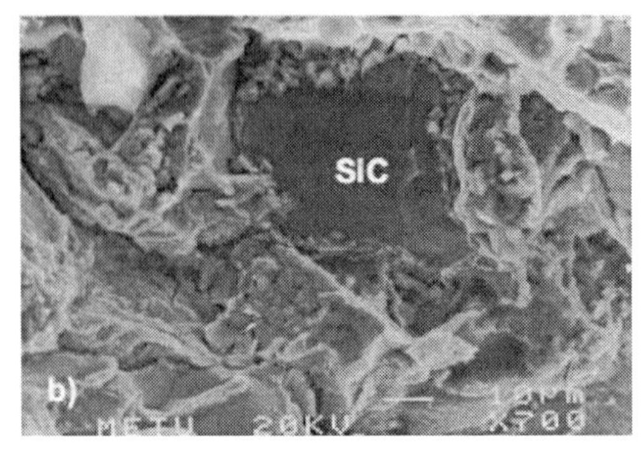

Figure 21. (Continued)

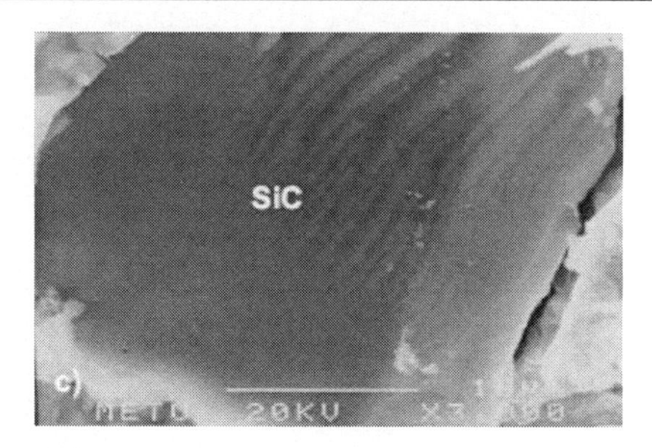

Figure 21. SEM micrograph of composites fracture surface: a) Fracture surface of SC 10 (10 % vol. SiC particle reinforced composite) b) Fracture surface of SC 10 (10 % vol. SiC particle reinforced composite) c) SEM micrograph of SiC particle failured in two pieces.

The extremely limited amount of damage below the fracture path noted on sectioned fractured specimens confirms that the nucleation of defects is the stage governing the fracture process in composites. Apparently, the growth and coalescence of fracture do not require significant remains of strain energy. Consequently, factors influencing the fracture of the MMCs must be sought among the aspects of nucleation. In the present investigation, initiation of voids was assumed to depend mainly on a particle cracking, particle /matrix debonding and grain boundary fracture. Both of these are complex phenomena influenced by a number of factors such as the local stress field, the presence of the interface inhomogeneities, structural alterations in the matrix close to the interface, or notch effects around and inside the brittle ceramic particles. From the tensile and fractographic results it may be inferred that completely ductile behavior was not totally achieved. The hardness results, also, supply this brittle fracture state.

3.3. Friction and Wear Behavior

Wear is one of the most commonly encountered industrial problems, leading to frequent replacement of components [29]. During the last two decades, superior wear performance of MMCs reinforced with ceramic particles has been reported [3, 4].

Extensive review papers on the dry sliding wear characteristics of composites based on aluminum alloys have been presented by Sannino and Rack [30], and abrasive wear behavior by Deuis et al. [31]. In their studies and discussions, the effect of reinforcement volume fraction, reinforcement size, sliding distance, applied load, sliding speed, hardness of the counter face and properties of the reinforcement phase which influence the dry sliding wear behavior of this group of composites are discussed in greater detail. Sliding wear rate and wear behavior were reported to be influenced by various wear parameters [32 35]. Lim et al. [36] studied the tribological properties of Al–Cu/SiC metal matrix composites. They reported that wear resistance increased drastically with increasing mechanical properties [37].

MMCs contain hard ceramic particles have high wear resistance and limited ductility. The wear resistance of composites increases linearly with hardness and above 5-10 vol.% of reinforcement the hardness is also a quasi-linear function of ceramic element content. C.S. Lee, Y.H. Kim and K.S. Han [38] investigated wear behavior of Al-matrix composites by dry spindle wear test under various conditions (volume fractions of reinforcements, sliding distances and speeds). They found that wear resistance of discontinuous MMCs is remarkably improved by addition of reinforcement. Weight loss strongly dependent upon hardness of counter materials. Dominant mechanism is the adhesive-abrasive wear at low and intermediate sliding speeds, and melt wears at high sliding speeds.

The machining characteristics of SiC reinforced Al alloy MMCs are different from unreinforced Al alloys. Very hard SiC ceramic particles in a relatively soft matrix are abrasive, creating higher temperatures between the tool and work piece; this causes faster tool wear. Modified machining parameters or special tool materials, such as compacted diamond or carbide, must be utilized. However, they produce close-tolerance, intricate parts with very good finish and surface integrity [39].

The composition of the MMC, in terms of type, shape, size, and volume fraction of reinforcement will affect machine ability. A matrix with a soft reinforcement, such as graphite, will be easier to machine than a matrix reinforced with Al_2O_3. The shape of the reinforcement may influence the ease with which the ceramic particles are sheared, but there is no information available on this. Increasing the reinforcement particle size and volume fraction will increase the abrasiveness of the composite, but the uniformity of the reinforcement distribution will also be important. Large clusters of ceramic particles will act as large inclusions with regard to tool damage [13].

Figure 22 presents the variation of friction coefficients of Al-Cu/SiC MMCs with applied loads at 0.5, 1and 2 m/s sliding speeds.

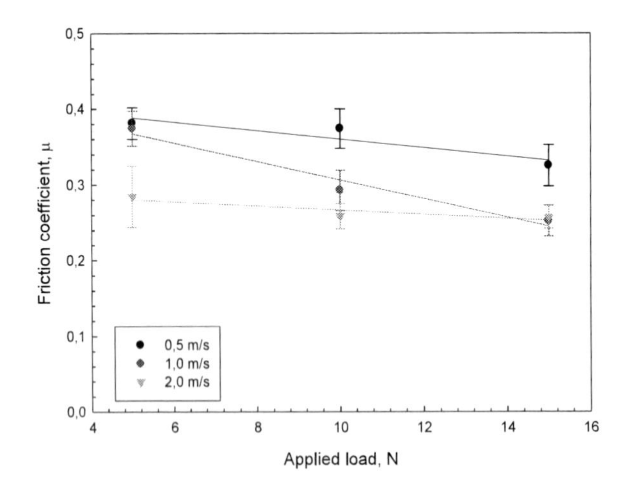

Figure 22. Steady state friction coefficient variations with applied load and sliding speed.

It can be seen that the friction coefficient of the composites decrease with increasing applied load. This is an agreement with the results obtained by Zhang [40] and Yalcin [41]. There is an average 16.9 %, 29.7 % and 11.9 % decrease in friction coefficient value at 200 % increase in applied load with sliding speeds respectively. The variations of friction coefficient with increasing normal applied load is more distinctly at sliding speed 1 ms^{-1} than other conditions. Moreover, there is an average 31.6 % decrease with the 300% increase in sliding speed.

Figure 23 represents the variation of wear rates of composites with applied load at 0.5, 1, and 2 m/s sliding speeds.

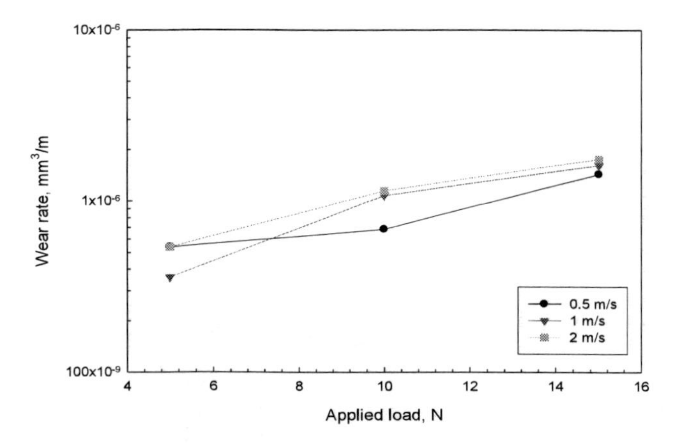

Figure 23. Variations in wear rates a function of applied load and sliding speed.

An increase in the normal applied load will lead to an increase in the wear rate. This is an agreement with studies of some researchers [29, 31, 37, 42, and 43]. Increasing sliding velocity also increases wear rate [31, 37]. Mild wear was observed for a small-applied load, but as the load was further increased up to 15 N, the wear rate of composite increased. There is an average 167 %, 350 % and 227 % increase in wear rate for a 200 % increase in applied load with sliding speeds respectively. Moreover, there is an average 30 % increase in wear rate for 300 % increase in sliding speeds. Heavy noise and vibration were observed during the process and transfer of the pin material to the disc was observed.

Specific wear rate of the MMC materials calculated from Figure 23 increases with the increase in sliding speed shown in Figure 24. There is an average 30.5 % decrease in specific wear rate with 200% increase in sliding speed.

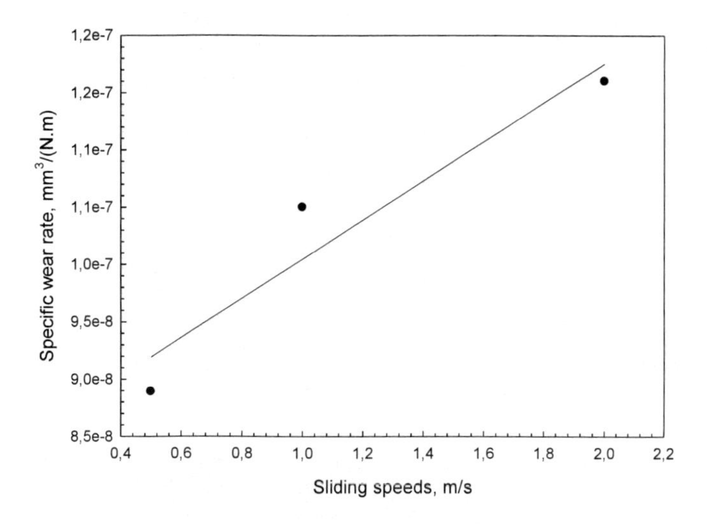

Figure 24. Variations in specific wear rate as a function of sliding speed.

The worn surfaces were examined with Scanning Electron Microscopy (SEM). As can be seen in Figure 25, the presence of grooves of varying sizes was observed frequently on the worn surfaces. The formation of such grooves during the sliding of composites had been, on numerous occasions, linked to the process of delamination [44-47] which according to Suh [48], is the preferential propagation of sub-surface cracks along the sliding direction, giving rise to the detachment of wear particles in the form of sheets or flakes. These grooves are also the evidence of abrasive actions during sliding. It is therefore proposed that the dominant wear mechanism operating during sliding over this range of loads at 1 m/s is a combination of abrasion and delamination [9].

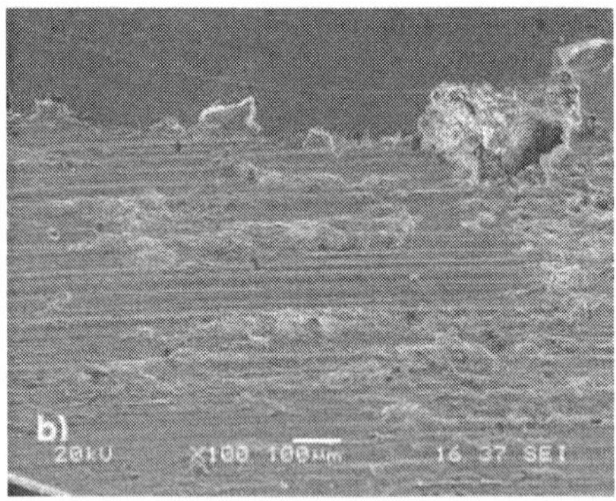

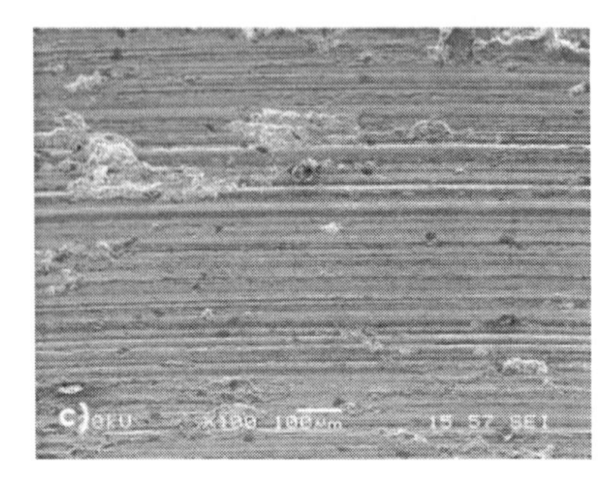

Figure 25. Micrographs of worn surface of composite materials: a) 0.5 m/s sliding speed and 10 N applied load b) 1m/s sliding speed and 10 N applied load. c) 2 m/s sliding speed and 10 N applied load.

At lower loads, the projected SiC particles in the composites will be in contact with the counterface during the course of wear. The asperities of the sliding composite material surface come into contact with the steel disc surface, and are work hardened under the applied load and speed due to cold working on the surface of the composite material [37].

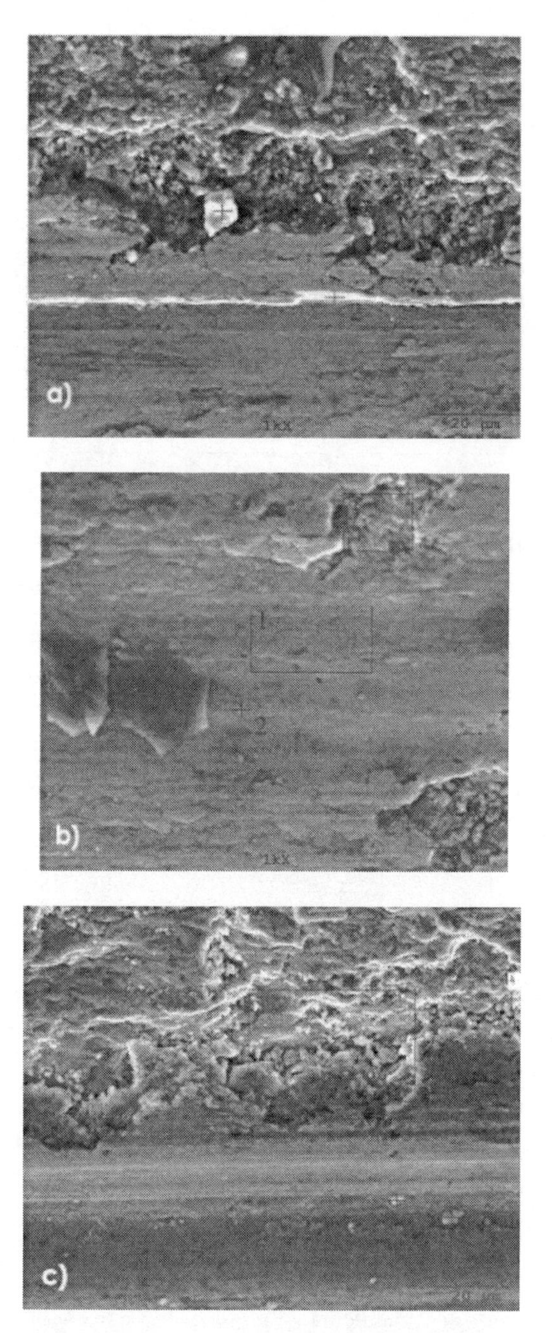

Figure 26. EDS analysis of worn surface of composite materials. a) 0.5 m/s sliding speed and 10 N applied load b) 1m/s sliding speed and 10 N applied load. c) 2 m/s sliding speed and 10 N applied load.

As the load increases, the morphology of the worn surfaces gradually changes from fine scratches to distinct grooves, and damaged spots in the form of craters can be seen. The asperities of both the composite material and counterface are in contact with each other and are subject to relative motion under the influence of applied load. Initially, both the surfaces are associated with a large number of sharp asperities, and contact between the two surfaces takes place primarily at these points. Under the influence of applied load and speed, when the asperities on each surface come in contact, they are either plastically deformed or remain in elastic contact.

At higher applied loads, high wear rates are observed. The wearing surface is characterized by a significant transfer of material between the sliding surfaces. A delamination wear mechanism has been inferred for this wear regime, where the tribolayer is removed by sub-surface plastic deformation and fragmentation of the SiC particles [50].

Similar transitions in wear behavior with increase in applied load were reported by Lim et al. [36] for an A356 alloy reinforced with SiCp. At a slow sliding velocity (0.5 ms^{-1}), the wearing surface was covered with a transferred material layer. At higher speeds, an oxide-like transferred layer formed at the sliding interface and reduced direct metallic contacts. At very high speeds, thermal softening of the matrix was reported. The oxide layer was observed to break down and allow greater direct metallic contact during sliding and SiC particles became dislodged and three-body abrasive wear was predominant.

The worn surface produced by adhesive wear, is defined as the transfer of material from one surface to another during relative motion by a process of solid-phase welding or because of localized bonding between contacting surfaces. Particles, which are removed from one surface, are either permanently or temporarily attached to the other surface resulting in a tribolayer (Figure 26).

This layer generally consists of plastically deformed and oxidized particles that reduce the wear rate of the material [50, 51]. The formation of wear debris was directly related to delamination wear active within this region. Studies of the tribolayer reveal that it is composed of a mechanical mixture derived from the wearing alloy and the counterface (Table 11).

Table 11. EDS analysis of tribolayer as a function of applied normal load and sliding speed (wt%)

	0.5 m/s			1.0 m/s			2.0 m/s		
	Fe	Cr	O	Fe	Cr	O	Fe	Cr	O
5 N	30,44	2,77	43,89	48,67	3,66	11,66	91,63	7,88	12,88
10 N	18,46	1,86	25,00	29,56	2,13	22,59	84,75	5,60	20,98
15 N	56,53	3,41	15,00	26,17	2,49	24,66	83,48	5,57	22,76

The transition from one wear mechanism to another as a function of applied pressure and sliding speed has also been documented in literature [52-56]. Moreover, some researchers developed wear mechanism maps for Al alloys and composites wear encountered in dry sliding of steel [57-61]. It was concluded that increases in sliding speed and/or normal load resulted in a transition in wear behavior from mild wear to severe wear and ultimately to seizure. At low loads, where a mild wear regime was reported, the wearing surfaces was described as relatively smooth and wear debris as small and brittle. Oxidation of aluminum was considered an important part of this wear mechanism even though no quantitative data regarding oxidation was given. In the severe wear regime with increased load and sliding velocity, the delamination wear mechanism was the dominant wear process. Formation of wear debris and associated transfer of material onto the steel counterface was related to the poor whisker/matrix interfacial bond strength. The wearing surface at low speeds exhibits a low wear rate and a stable tribolayer. In a defined sliding speed range, the extent of the tribolayer formation becomes significant and a reduction in direct sliding surface contact results in a lower wear rate. At very high speeds, thermal softening of the matrix indicates a breakdown in the tribolayer, corresponding to a high wear rate and oxidation (see Table 11).

Figure 27. Micrographs of worn surface of counterface (AISI D2 steel has 61 HRC): a) 0.5 m/s sliding speed and 10 N applied load b) 1m/s sliding speed and 10 N applied load. c) 2 m/s sliding speed and 10 N applied load

Worn surfaces of the counterface are given in Figure 27. Findings indicate that a proportion of the counterface material removed by the micro-machining effects of the reinforcement phase is subject to oxidation. This oxidized material is present as Fe, O, in the tribolayer of both wearing surfaces. Furthermore, this oxidized phase helps stabilize the tribolayer on the MMC surfaces.

5.5. Toughness and Impact Resistance

A general feature of resistance to crack propagation in discontinuously reinforced Al-matrix composite is a simultaneous reduction in toughness and impact resistance as the volume fraction of ceramic reinforcements is increased. Toughness depends mainly on the size, shape and distribution of ceramic elements. Since predominant mechanism of crack propagation is related to the coalescence of voids associated with these elements. As a result, the direction giving rise to easier crack propagation is the direction along which the distances between ceramic elements are the smallest .

5.6. Fatigue Resistance

Many of the potential applications for composites require a resistance to cycling loading, and several fatigue studies in MMCs have been performed. Consciously, one would expect the low cycle fatigue behavior of MMCs to be somewhat worse than unreinforced alloys because of the lower ductility in composites, whereas the high cycle performance should be improved because of the higher modulus [9].

The addition of ceramic elements, particularly whiskers, gives rise to a significant increase in fatigue life [27]. According to S.J. Harris [32], it is not entirely clear the influence of particles and their interfaces. In case of SiC particles fatigue cracks often propagate through the matrix regions without apparently involving the particles. When particles become involved, then it is usually the interface that breaks down although some evidence of particle cracking has been observed in short life specimens. There is some optimum particle size or volume fraction that promotes maximum crack brinding influences on crack growth behavior.

The fatigue performance of the composite is different to that of the unreinforced matrix, but whether the fatigue behavior is improved or not depends on the mode of testing. Most of the studies indicate that the fatigue performance can be comparable with or better than the unreinforced matrix, except at high stress or strain amplitudes where the reduced ductility of the composite influences the behavior. There are insufficient data to assess the influence of particle size, shape, etc., Fatigue growth models are not sufficiently well developed to make any predictions.

5.7. Elevated Temperature Properties

The use of particulate reinforced MMCs for high temperature applications requires a fundamental knowledge of the mechanisms affecting their creep behavior. Unfortunately, creep studies of particulate reinforced MMCs have been limited [5].

MMCs containing particulates, whiskers and short fibers show improvements in modulus and strength at elevated temperature over unreinforced alloys. Also, discontinuous SiC/Al composites show an advantage over conventional Al alloys at elevated temperatures. In case of Al alloy-matrix, composite strengths in excess of 200 MPa can be achieved at 300°C during short periods of exposure at this temperature. Humphreys [29] has suggested a transition in mechanical behavior when tested at temperatures in the range 200-300 °C. Below the transition, high work hardening rates, high flow stresses and low ductilities are found. Above transition, the work hardening rate reduces and increased ductilities apparent. Thermally controlled diffusional relaxation of stresses around particles has been put forward as the reason for this transition. Dislocations associated with the particles begin to climb at the elevated temperature, so that they do not accumulate around the particles as deformation takes place.

At temperatures in excess of 500 °C particulate composites begin to show evidence of grain boundary sliding. Under certain condition small particles and fine matrix grain size, super plastic deformation becomes possible. In the presence of large particles and coarse grains and higher strain rates, lower ductilities can results.

CONCLUSIONS

In this study a squeeze casting system was designed and constructed. System parameters were optimized and squeeze casting apparatus was operated successfully. For casting operations 100 MPa pressure and 350 °C die temperature were chosen. Investigation on gravity cast and squeeze cast Al-4.5Cu-3Mg alloy and 5, 10, 15 vol. %SiC particulate reinforced matrix alloys have shown followings:

1. Squeeze casting technique resulted in refining microstructure in squeeze cast matrix alloy the identically of Cu and Mg were increased in α-Al primary phase.
2. Microstructures of composites contain primary phase and eutectic regions. SiC particles were surrounded by eutectic regions. Also quantitative element identifications showed that eutectic region close to SiC particles had somewhat different composition than close to eutectic region away from the particles.
3. The microstructure of the SiCp-reinforced composite showed a reasonably uniform distribution of particles and good interfacial bonding of dispersed particles with the matrix alloy.
4. In squeeze cast matrix alloy hardness increased from 94 BHN to 100 BHN and density from 2.7528 g/cm^3 to 2.9165 g/cm^3 by applied pressure. In the composites, hardness increased with increasing SiC volume fraction and recorded a value of 139 BHN at 15 %vol. SiC.
5. Failures in composites, produced by squeeze casting, occur simultaneously both matrix and SiC particles implying that here is a good bonding between the matrix and the particles.
6. Composite material that possesses superior adhesive wear resistance is associated with a stable tribolayer. The creation of this layer depends on the magnitude of the applied load and sliding speed.

7. Mild wear was observed for a small-applied load, but as the load was further increased up to 15 N, the wear rate of composite increased mild to severe.
8. The amount of wear generally increases with increasing sliding speed and the extent of wear generally becomes greater with an increase in applied load.
9. The friction coefficient decreased with increasing applied load and sliding velocity. Variations of friction coefficient with increasing normal applied load are more distinctly at sliding speed 1 ms^{-1} than other conditions.
10. An increase in the normal applied load and sliding velocity increased wear rate. The specific wear rate also increased sharply from $8.9x10^{-8}$ to $1.21x10^{-8}$ mm^3/Nm with increasing sliding speed.
11. SEM analyses indicate that a proportion of the counterface material removed by the micro-machining effects of the reinforcement phase is subject to oxidation. This oxidized material is present as Fe, O, in the tribolayer of both wearing surfaces in EDS analysis. Furthermore, this oxidized phase helps stabilize the tribolayer on the MMC surface by pinning dislocations.

REFERENCES

[1] A. Onat, "Production and Microstructure Characterisation of Silicon Carbide Reinforced Aluminum-Copper Alloy Matrix Composites by Squeeze Casting Method", M.Sc. Thesis, The Graduate School of Natural and Applied Sciences of Middle East Technical University (METU), Metallurgy Engineering and Materials Science Department, 1995.

[2] K.M. Shorowordi, T. Laoui, A.S.M.A. Haseeb, J.P. Celis, L. Froyen, "Microstructure and interface characteristics of B4C, SiC and Al2O3 reinforced Al matrix composites: a comparative study", *Journal of Materials Processing Technology* ,142 (2003), 738–743.

[3] M. Kök, "Abrasive wear of Al2O3 particle reinforced 2024 aluminum alloy composites fabricated by vortex method", *Composites: Part* A, 37, (2006) 457–464.

[4] K.R. Suresh, H.B. Niranjan, P. Martin Jebaraj, M.P. Chowdiah, "Tensile and wear properties of aluminum composites", *Wear*, 255, (2003), 638–642.

[5] M.R. Ghomashchi, A. Vikhrov, Squeeze casting: an overview. *Journal of Materials Processing Technology* 2000; 101: 1-9.

[6] Z. Zhao, S. Zhijian and X. Yingkun, "Effect of Microstructure on Mechanical Properties of an Al Alloy 6061 -SiC Particle Composite", *Materials Science and Engineering* A, 132 (1991), 83-88.

[7] M. Mondali, A. Abedian, S. Adibnazari, "FEM study of the second stage creep behavior of Al6061/SiC metal matrix composite", *Computational Materials Science*, 34 (2005), 140–150.

[8] Y. Sahin, "The effect of sliding speed and microstructure on the dry wear properties of metal-matrix composites" *Wear,* 214 (1998), 98 –106.

[9] Y. Sahin, Preparation and some properties of sic particle reinforced aluminum alloy composites, *Materials and Design*, 24 (2003), 671–679.

[10] R. Debdas, B. Bikramjit, B.M. Amitava, "Tribological properties of Ti-aluminide reinforced Al-based in situ metal matrix composite", *Intermetallics*, 13 (2005), 733–740.

[11] T.S. Srivatsan, I. A Ibrahim, F.A. Mohammed, E.J. Lavernia, "Processing Techniques For Particulate-Reinforced Metal Aluminum Matrix Composites", *Journal of Materials Science,* 26, (1991), 5965-5978.

[12] M.M. Schwartz, Composite Materials Handbook, McGraw-Hill Book Company, 1984

[13] D.J. Lloyd, "Particle Reinforced Aluminum and Magnesium Matrix Composites, *International Materials Reviews*, 39 (1994), 1-23.

[14] Kelly, "Metal Matrix Composites-an Overview", Cast-Reinforced Metal Composites, Edited by S.G. Fishman and A.K. Dhingra, ASM/TMS Committee, World Materials Congress, 24-30 Sept., Chicago-Illinois, USA, pp. 1-5, 1988.

[15] P. Rohatgi, "Advance in Cast MMCs", *Advanced Materials and Processes*, 2 (1990), 39-44.

[16] http://www.accuratus.com/silicar.html.

[17] http://www.accuratus.com/alumox.html.

[18] S.M. Lee, "Metal Matrix Composites, Casting Processes", Handbook of Composite Reinforcements, VCH Publishers Inc., 1993, 358-367.

[19] P.K. Rohatgi, R. Asthana, S. Das, "Solidification, Structures, and Properties of Cast Metal-Ceramic Particle Composites", *International Materials Reviews*, 31 (1986), 115-139.

[20] K. Ichikawa, M. Achikita, "Production and Properties of Carbide Dispersion-strengthened Coppers by Compocasting, *ISIJ International*, 31 (1991) No:9, 985-991.

[21] A.A. Das, A.J. Clegg, B. Zantout, M.M. Yakoub, "Solidification Under Pressure: Zinc Alloys Containing Discontinuous SiC Fiber", Cast-Reinforced Metal Composites, Edited by S.G. Fishman and A.K. Dhingra, ASM/TMS Committee, World Materials Congress, 24-30 Sept., Chicago-Illinois, USA, 1988, 139-147.

[22] G.A. Chadwick, T.M. Yue, "Principles and Applications of Squeeze Casting", *Metals and Materials.*, Jan. 1989, 6-12.

[23] S.K. Verma, J.L. Dorcic, "Manufacturing of Composites by Squeeze Casting", Cast-Reinforced Metal Composites, Edited by S.G. Fishman and A.K. Dhingra, ASM/TMS Committee, World Materials Congress, 24-30 Sept., Chicago-Illinois, USA, 1988, 115-126.

[24] H. Fukunaga "Squeeze Casting Process for Fiber Reinforced Metals and Their Mechanical Properties", Cast-Reinforced Metal Composites, Edited by S.G. Fishman and A.K. Dhingra, ASM/TMS Committee, World Materials Congress, 24-30 Sept., Chicago-Illinois, USA, 1988, 101-108.

[25] J.M. Howe, "Bonding, Structure, and Properties of Metal/Ceramic Interfaces: Part 1 Chemical Bonding, Chemical Reaction, and Interfacial Structure", *International Materials Reviews*, 38 (1993), 5, 233-256.

[26] I.A. Ibrahim, F.A. Mohammed, E. Lavernia, "Particulate reinforced metal matrix composites – a review", *Journal of Material Science*, 27 (1990), 1137-1156.

[27] R. Zhang, L. Gao, J. Guo, "Effect of Cu2O on the fabrication of SiCp/Cu nanocomposites using coated particles and conventional sintering", *Composites: Part A* 35 (2004), 1301–1305.

[28] D.L. Mc Danels, “Analysis of Stress-Strain, Fracture and Ductility Behaviour of Aluminum Matrix Composites Containing Discontinuous Silicon Carbide Reinforcement“, *Metallurgical Transactions* A, 16A (1985), 1105.

[29] Y. Sahin, K. Özdin, “A model for the abrasive wear behavior of aluminum based composites”, *Materials and Design*, 29, (2008), 728–733.

[30] P. Sannino, H. J. Rack, Wear, “Dry sliding wear of discontinuously reinforced aluminum composites: review and discussion”, *Wear* , 189, 1-2, 1995 1-19.

[31] R.L. Deuis, C. Subramanian, J.M. Yellup, “Abrasive wear of aluminum composites- a review”, *Wear*, 201, 1-2 (1996), 132-144.

[32] M. Narayan, M. K. Surappa, B. N. Pramila Bai, “Dry sliding wear of Al alloy 2024-Al203 particle metal matrix composites”, *Wear*, 181-183, 2, 1995, 563-570.

[33] P.H. Shipway, A.R. Kennedy, A.J. Wilkes, “Sliding wear behavior of aluminum-based metal matrix composites produced by a novel liquid route”, *Wear*, 216, 2, (1998), 160-171.

[34] M.H. Korkut, “Effect of particulate reinforcement on wear behavior of aluminum matrix composites”, *Materials Science and Technology*, 20, 1, (2004), 73-81.

[35] B. Venkataraman, G. Sundararajan, “Correlation between the characteristics of the mechanically mixed layer and wear behavior of aluminum, Al-7075 alloy and Al-MMCs”, *Wear*, 245, 1-2, (2000), 22-38.

[36] S.C. Lim, M. Gupta, L. Ren, J.K.M. Kwok, “The tribological properties of Al–Cu/SiCp metal–matrix composites fabricated using the rheocasting technique”. *Journal of Materials Processing Technology*, 89–90, (1999), 591-596.

[37] S. Basavarajappa, G. Chandramohan, R. Subramanian, A. Chandrasekar, “Dry sliding wear behavior of Al 2219/SiC metal matrix composites”, *Materials Science*-Poland, 24, 2/1, (2006), 357-366.

[38] C.S. Lee, Y.h. Kim, K.S. Han, “Wear Behaviour of Aluminum Matrix Composite Materials”, *Journal of Materials Sciences*, 27 (1992), 793.

[39] P. Niskamen, W.R. Mohn, “Versatile Metal-Matrix Composites”, Advanced Materials and Processes inc. Metal Progress, 3 (1988), 39-41.

[40] S. Zhang, F. Wang, “Comparison of friction and wear performances of brake material dry sliding against two aluminum matrix composites reinforced with different SiC particles”, *Journal of Materials Processing Technology*, 182, (2007), 122–127.

[41] Y. Yalcin, H. Akbulut, “Dry wear properties of A356-SiC particle reinforced MMCs produced by two melting routes”, *Materials and Design*, 27, (2006), 872–881.

[42] S.C. Tjong, K.C. Lau, “Properties and abrasive wear of TiB2/Al-4%Cu composites produced by hot isostatic pressing” *Composites Science and Technology*, 59, (1999), 2005-2013.

[43] S. Das, D.P. Mondal, S. Sawla, N. Ramakrishnan, “Synergic effect of reinforcement and heat treatment on the two body abrasive wear of an Al–Si alloy under varying loads and abrasive sizes”, *Wear*, 264, (2008), 47–59.

[44] A.T. Alpas, J. Zhang, “Effect of SiC particulate reinforcement on the dry sliding wear of aluminum-silicon alloys (A356)”, *Wear*, 155, 1, (1992), 83-104.

[45] A.T. Alpas, J.D. Embury, “Sliding and abrasive wear behavior of an aluminum (2014)-SiC particle reinforced composite”, *Scripta Metallurgica et Materialia*, 24, 5, (1990), 931-935.

[46] H.L. Lee, W.H. Lu, S.L.I. Chan, "Abrasive wear of powder metallurgy Al alloy 6061-SiC particle composites", *Wear,* 159, 2, (1992), 223-231.
[47] O.P. Modi, B.K. Prasad, A.H. Yegneswaran, M.L. Vaidya, "Dry sliding wear behavior of squeeze cast aluminum alloy-silicon carbide composites", *Materials Science and Engineering: A*, 151, 2, (1992), 235-245.
[48] N.P. Suh, "The delamination theory of wear", *Wear*, 25, 1, (1973), 111-124.
[49] T. Ma, H. Yamaura, D. A. Koss, R. C. Voigt, "Dry sliding wear behavior of cast SiC-reinforced Al MMCs", *Materials Science and Engineering* A, 360, (2003), 116-125.
[50] S. Mohan, J.P. Pathak, R.C. Gupta, S. Srivastava, "Wear behavior of Ni-Cr-Mo-V steel under dry sliding", *Zeitschrift fur Metallkunde* (Germany). 93, 11, (2002), 1140-1145.
[51] R.A. Antoniou, L.R. Brown, J.D. Cashion, "The unlubricated sliding of Al-Si alloys against steel: Mijssbauer spectroscopy and X-ray diffraction of wear debris", *Acta Metallurgica et Materialia,* 42, 10, (1994), 3545-3553.
[52] S.C. Lim, Y. Liu, M.F. Tong, "The effects of sliding condition and particle volume fraction on the unlubricated wear of aluminum alloy-SiC particle composites", Proc. Conf on Processing Properties and Applications of Metallic and Ceramic Materials, Birmingham. 7-10 Sept. 1992, 485-490.
[53] Wang, H. J. Rack, "Transition wear behavior of SiC-particulate and SiC-whisker-reinforced 7091 Al metal matrix composites", *Materials Science and Engineering* A, 147, (1991), 211-224.
[54] H.C. Park, "Wear behavior of hybrid metal matrix composite materials", *Scripta Metallurgica*, 27, (1992), 465-470.
[55] J.K.M. Kwok, H.S. Goh and S.C. Lim, "Effects of mechanical alloying on the friction and wear characteristics of Al-4.5Cu-15SiC particulate composites", Proceedings of 4th International Tribology Conference (Austrib 1994), 5-8 Dec. 1994, Perth, Australia, 241-247.
[56] Z.F. Zhang, L.C. Zhang, Y.W. Mai, "Wear of ceramic particle-reinforced metal-matrix composites: Part I. Wear mechanisms", *Journal of Materials Science*, 30, (1995), 1961-1966.
[57] A.R. Rosenfield, "A shear instability model of sliding wear", *Wear,* 116, (1987), 319-328.
[58] D.Z. Wang, H.X. Peng, J. Liu, C.K. Yao, "Wear behavior and microstructural changes of SiCw-Al composite under unlubricated sliding friction", *Wear*, 184, (1995), 187-192.
[59] S.C. Lim, M.F. Ashby, "Wear-mechanism maps", *Acta Metallurgica*, 35, 1, (1987) 1-24.
[60] R. Antoniou, C. Subramanian, "Wear mechanism map for aluminum alloys", *Scripta Metallurgica*, 22, 6, (1988), 809-814.
[61] Y. Liu, R. Asthana, P. Rohatgi, "A map for wear mechanisms in aluminum alloys", *Journal of Material Science*, 1991, 26, 99-102.

In: Silicon Carbide: New Materials, Production ... ISBN: 978-1-61122-312-5
Editor: Sofia H. Vanger

Chapter 3

MICROSTRUCTURE OF SILICON CARBIDE NANOWIRES

Ryan Rich, Monika Wieligor and T. W. Zerda
Department of Physics and Astronomy, TCU, Fort Worth, Texas, USA

ABSTRACT

Silicon carbide nanowires were produced from carbon nanotubes and nanosize silicon powder in a tube furnace at temperatures between 1100°C and 1350°C. Diameters of nanowires were controlled by selecting precursors, adjusting temperature, vapor pressure of silicon, and sintering time. The mechanism of SiC nanowires formation remains controversial, but when metals were found in carbon nanotube precursors, we were able to identify the growth mechanism as the vapor-liquid-solid, or VLS, process. In the TEM images 'flowers' were observed at the end of nanowires, and EDS analysis confirmed the presence of heavy metals at these ends. But formation of SiC nanowires obtained from highly oriented pyrolytic graphite and carbon nanotubes without metal catalysts was possible assuming the vapor-solid process, the VS mechanism. At the present time the possibility that the VS process contributes to the growth of SiC nanowires cannot be excluded from nanotube-based samples.

All SiC nanowires possess large quantities of planar defects, among which twins appear to be most abundant. Twins were identified in TEM images. By analyzing the geometry of crystallographic planes in the adjacent crystallites and from electron diffraction patterns recorded in the bright and dark images, it was possible to identify the angle between crystallographic planes as 141 degrees. Other planar defects frequently reported to be present in SiC nanowires are stacking faults. Stacking faults can affect x-ray diffraction patterns. A small shoulder that accompanies the [111] reflection is often considered the evidence of stacking faults. Distinguishing stacking faults from twins in TEM images is not a simple task. Besides, such an analysis does not allow for quantitative evaluation of the defects. In this paper we report a method of simultaneous analysis of profiles of several x-ray diffraction lines. This analysis allowed us to determine the crystallite sizes and the population of twins. It appears by this analysis that the population of stacking faults in silicon carbide nanowires is low. This result is confirmed by the analysis of TEM images.

All nanowires have crystalline cores and are coated with amorphous layers. The thickness of that layer varies from 1 to 10 nm, and its composition depends on properties of the precursors and the history of post production treatment. EDS was employed to analyze the chemical composition of that layer and FTIR to characterize its molecular structure. SiO and C=C groups were found in the amorphous layer of some specimens. Treatment of the specimens in acids and its affect on the composition of that layer is discussed.

INTRODUCTION

Silicon Carbide (SiC) is an excellent candidate for replacing Si or other III-V semiconductors in high power, high temperature, and high frequency device production due to its remarkable properties, which are the result of covalent bonding between C and Si atoms. Because of its high melting point (2830°C), good thermal shock resistance, high thermal conductivity (4.9 W cm^{-1} K^{-1}), electric field breakdown strength (4.0×10^6 V cm^{-1}), high chemical stability, and wear resistance SiC has become a major component in the high-power transistors, high-voltage diodes, and other electronic devices.[1,2]

One-dimensional semiconductor nanomaterials deserve special attention, because they play a crucial role as building blocks of future molecular electronic applications.[3-5] SiC in the one-dimensional form, such as nanowires, nanotubes, nanorods, or nanowhiskers, possesses even more attractive attributes than in the bulk structure. Attempts to use SiC nanowires for nanoelectronic and optoelectronic device applications have been reported.[4,6-8] Hardness and tensile strength of SiC nanowires are greater than in the case of the larger SiC whiskers.[9-11] Therefore they may be applied as reinforcement to monolithic ceramics in order to improve fracture tolerance and toughness of the composites.[12-14]

SiC nanowires have already been obtained by various methods, such as the sol-gel process,[15] catalytic chemical vapor deposition,[16] vapor-liquid-solid growth mechanism [17], reaction between carbon nanotubes (CNT) and SiO vapor at high temperatures [6], reaction between CNT and liquid or vapor Si [18], and others. The major goal in all these techniques is the control of the desired parameters, such as size of grain, diameter and length of nanowire, and concentrations of various defects. Certain defect structures strongly affect the mechanical and electronic properties and are critical for the functionality of semiconductor devices. We should not underestimate their significance in produced nanowires.

Since 2004, we have produced silicon carbide nanowires from carbon nanotubes and silicon powder in a tube furnace at temperatures between 1100°C and 1350°C. [13,18-20] We controlled diameters of nanowires by selecting precursors, adjusting temperature, vapor pressure of silicon, and sintering time. The mechanism of SiC nanowires formation remains controversial, but when metals were present in carbon nanotube precursors, the growth mechanism was identified as the vapor-liquid-solid (VLS) process. [21] In the TEM images 'flowers' were observed at the end of nanowires, and the EDS analysis confirmed the presence of heavy metals at these ends. Formation of SiC nanowires obtained from highly oriented pyrolytic graphite and carbon nanotubes without metal catalysts was also observed. In that case the formation mechanism was explained assuming the vapor-solid process (VS) mechanism. These mechanisms and associated chemical reactions were discussed in numerous papers, for reviews see Powell et al. [22] and Cao [23].

All SiC nanowires possess large quantities of planar defects, among which twins and stacking faults appear to be most abundant. Both planar defects were identified in TEM images, Raman spectra and x-ray diffraction images [18,20]. However, distinguishing stacking faults from twins in TEM images is not a simple task. Besides, such analyses would not allow for quantitative evaluation of the defects. Fortunately simultaneous analyses of multiple lines in x-ray diffraction allow filling of that gap. As discussed in the section on x-ray profile analysis, this methodology allows determination of crystallite sizes and population of dislocations and twins.

All nanowires have a crystalline core and are coated with an amorphous layer. The thickness of that layer varies from 1 to 10 nm. Its composition depends on properties of the precursors and the history of post production treatment. EDS was employed to analyze chemical composition of that layer and FTIR to characterize its molecular structure. SiO and C=C groups were found in the amorphous layer of some specimens. Treatment of the specimens in acids and its affect on the composition of that layer is discussed in the subsequent section. Description of the experimental setup and reaction models are presented in the last section.

PLANAR DEFECTS

Defects may be separated into zero- up to three-dimensional groups. The intrinsic point defects due to missing atoms, the displacement of atoms out of the crystal lattice, and impurity atoms may be included into the zero-dimensional defects. The one-dimensional defects are dislocations. The other imperfections, such as stacking faults, grain boundaries, twins, or domain walls belong to the group of two-dimensional defects. The three-dimensional defects include precipitates and voids. Since defects play an important role in determining mechanical properties of the materials, qualitative and quantitative analysis of defects is of primary importance for materials scientists. In this article we focus on planar defects and their identification by direct TEM imaging, TEM electron diffraction, and profile analysis of x-ray powder diffraction.

The planar defects are typical for crystalline materials with non-uniform, planar structure. The common planar defects include interfaces, stacking faults, and grain boundaries, and twins. For SiC, twins and stacking faults are especially important.

The polycrystalline material is composed of many small crystallites, identical in composition, but different in orientation. The regions between these crystals are called grain boundaries, Figure1. Grain boundaries are slightly disordered, and usually contain dislocations, impurities, and atoms relocated from their original sites. The grain boundary is less dense than the rest of the bulk grain, resulting in higher mobility, diffusivity, and chemical reactivity. They strongly affect mechanical properties, especially fracture toughness, by limiting the range and motions of dislocations. According to Han et al. [24] elongation of SiC nanowires of up to 200% is possible. This is a remarkable result and indicates that SiC nanowires have mechanical properties dramatically different from those of the bulk material. Bulk SiC is a brittle material and easily breaks even under small strains, less than 0.2%. Stacking faults appear to play an important role during elastic deformations while dislocations are important during plastic deformations.[25-27] The observed transformation from cubic to amorphous structure without crack formation may be associated with increased dislocation activities.

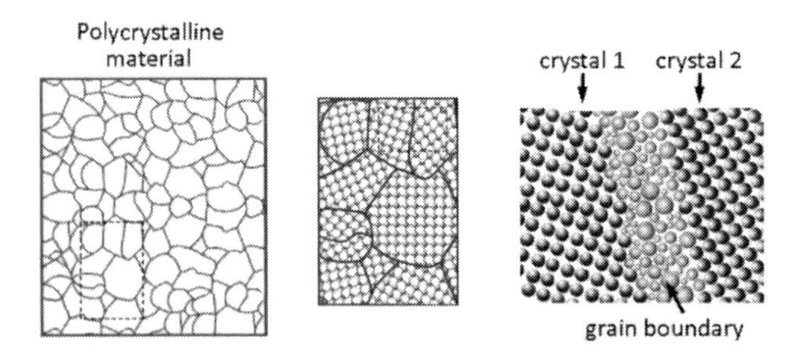

Figure 1. Polycrystalline material with grains showing different orientations and the grain boundaries.

We may describe a twin defect as a crystal structure reflection across a mirror plane, see Figure 2. The symmetry operations are necessary for twin construction, and formation of this defect occurs usually during crystal growth, thermal annealing, application of mechanical shear forces, or other deformations of the material. Twins are commonly found in spherical nanoparticles and in nanowires, and the fcc metallic nanocrystals usually have (111) twins.[28]

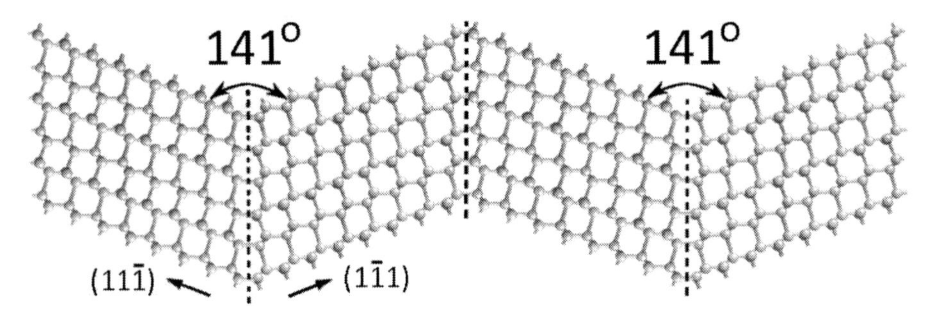

Figure 2. The example of twin boundaries in the [011] plane. Arrows show the (111) directions in adjacent crystals.

The additional reflections, observed in electron diffraction patterns, arise from plural scattering produced by the presence of twin domains, compare Figure 3b. The process of forming these extra spots is as follows: a diffracted electron beam that appears from one twin domain may act as a principal beam for a second twin domain that diffracts the beam again. Their intensities depend on two factors: electron path length through the twin domains and the arrangement of the domains. The angle between twins segments marked in the figure is 141°, which is twice the interplanar angle (70.5°) formed between two [111] planes characteristic for an fcc structure.[29] Indeed the most common twin structure in fcc crystals is the twin with rotation angle 70.53°, known as a first order, or the lowest order twin.[30] The higher order twins possess small interface angles and do not happen often. This is because they become unstable when their angle is below 20°, and they are not as energetically favorable as the first order twins.[30] This may explain why we observed only the twins with rotation angle 141° in almost all nanowires.

Another kind of planar defect is a stacking fault (SF), see Figure 4. It may be described as discontinuity in the stacking sequence. For example, the ABCABC arrangement in fcc crystal changes to ABCABABC. We may distinguish two forms of SF: intrinsic (Figure 4b), formed by an agglomeration of vacancies, and extrinsic (Figure 4c), created by inserting an additional plane of atoms. Simplifying, a stacking fault is a disturbance of the regular sequence of atomic plane positions in the crystal. In the defect-free crystal, the finite extension of a crystal plate results in a needle-shaped extension of the reciprocal-lattice points normal to the plate. The diffraction pattern will be conserved, and Bragg's reflections will appear over a large tilt range. But even small changes or tiny shifts in the position of diffraction spots can result from the intersection of the Ewald sphere with the needles. If the angle between the normal to the crystal plates and electron beam in near 90°, the needles at the reciprocal-lattice points create diffuse lines in the diffraction pattern. Therefore, the presence of extended streaks, instead of well-defined spots in the diffraction pattern is observed. This is considered the evidence for the presence of stacking faults in the specimens, or more specifically, in the section of the specimen probed by the electron beam. Sometimes, when stacking faults dominate in the specimen structure, streaks may extend from one spot to another.

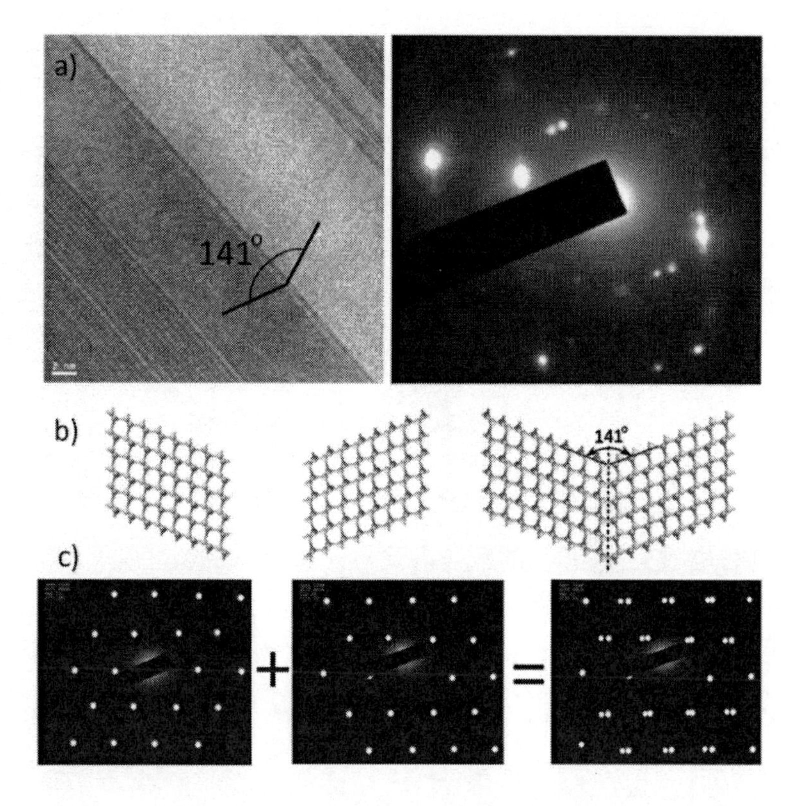

Figure 3. Stacking faults. a) TEM image of a SiC nanowire. The orientation of electron beam was the [011] direction. Next to the TEM image, an example of FFT pattern for twin structure is presented. b) The atomic models for the twin interface and periodic twin structure; open spheres represent the carbon atoms, filled spheres – silicon atoms. c) The scheme of a process formation of diffraction pattern from multiple scattering in twins domains. Images formed by each member of the twin pair is shown below the corresponding lattices. The resultant image in the right corner is an overlap of both images due to different orientations of the scattering lattices.

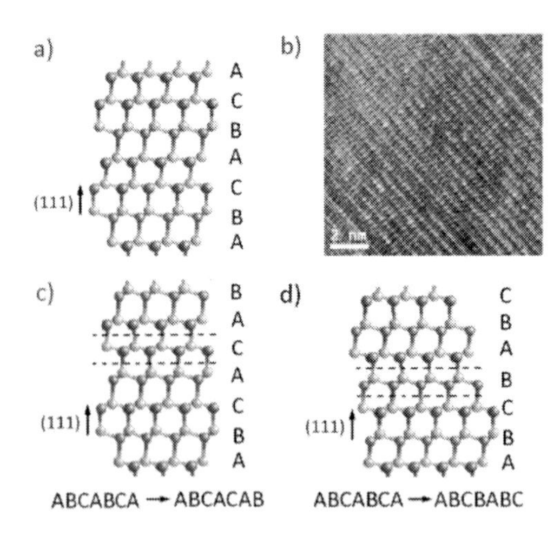

Figure 4. Stacking faults in the projection onto the (110) plane in the fcc crystal structure. a) a perfect structure; b) TEM image of SiC crystal with stacking faults; c) a model of intrinsic SF. Views from the (111) direction.

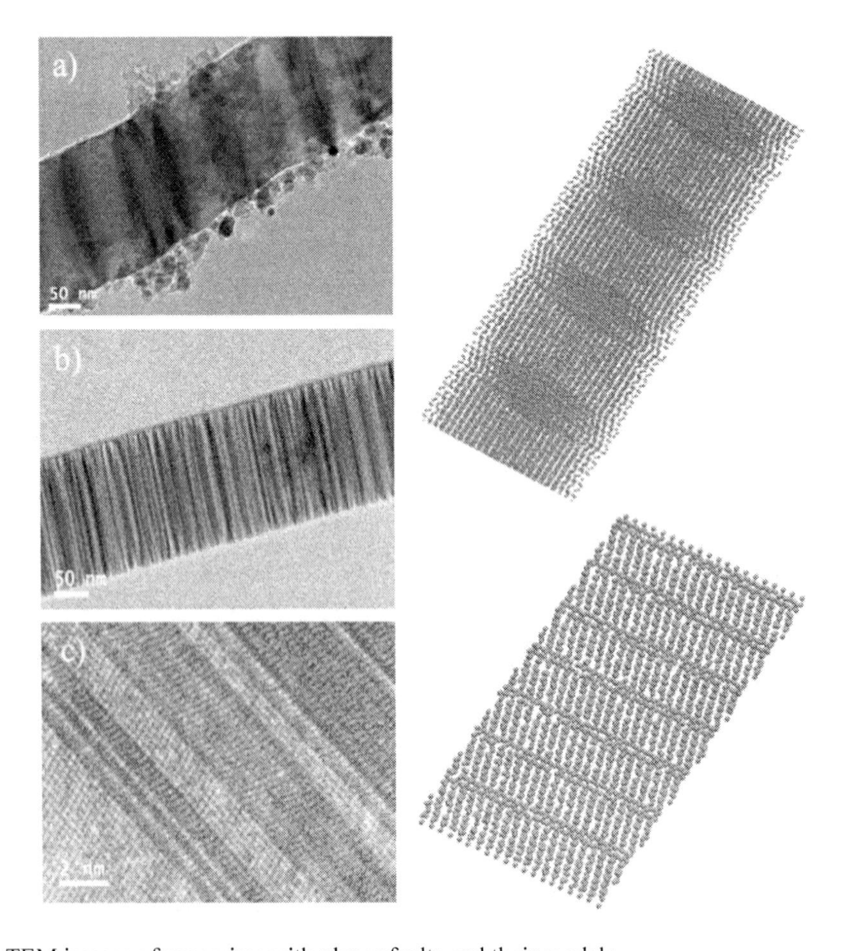

Figure 5. TEM images of nanowires with planar faults and their models.

Both twins and stacking faults are abundant in SiC nanowires, and it is difficult to identify them in conventional TEM images. The view from different directions creates the apparent effect of higher concentration of atoms within the shift area. Depending on the angle between the incident electron beam and planar faults plane, we may observe in TEM images different shapes of dark regions, disks, or steaks, compare Figure 5. Identification of planar defects from the TEM images is not straightforward and requires stage manipulation to allow for multiple observations of the same section of the specimen at different incident angles. However, recording a TEM image along with an electron diffraction pattern of the same region of interest allows one to simply identify the predominant planar defects in that particular region, compare Figure 6.

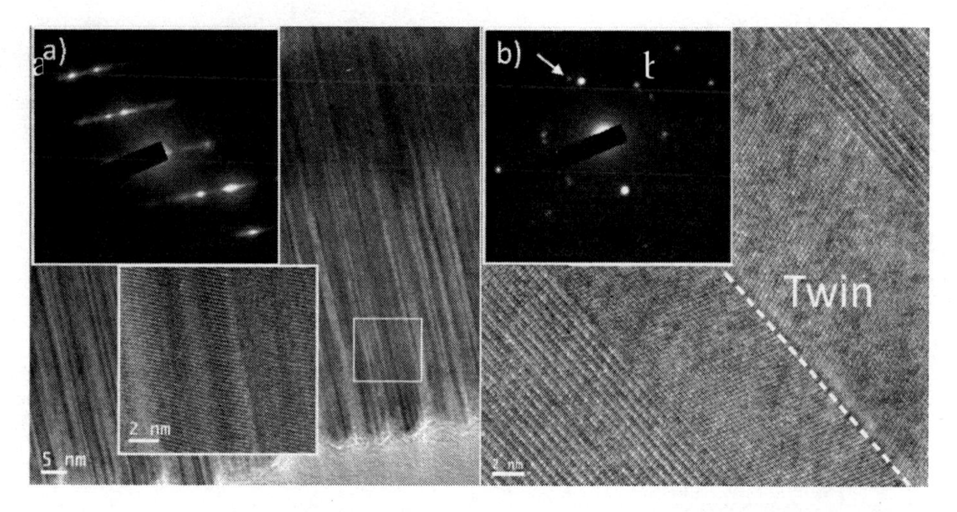

Figure 6. The HRTEM images and corresponding electron diffraction patterns for SiC nanowires. a) Stacking faults and their diffraction pattern with characteristic streaks. b) A twin and its electron diffraction pattern with the double diffraction spots. The twin plane is marked by a dashed line.

Figure 7 illustrates the presence of twins and stacking faults in the single SiC nanowire. In addition to sections with multiple faults there are large defect-free sections. The measured spacing between crystallographic planes is 2.52 Å, and it remains unchanged when measured in adjacent crystallites separated by a twin boundary. This value corresponds well to the known interplanar spacing of the [111] plane, and it is further seen that the angle between crystallographic planes in adjacent segments is 141°, as expected from geometrical considerations, see Figs. 2 and 7. Thus one may conclude that SiC nanowires grow in the (111) direction of the cubic structure. Since SiC exists in the form of many different polytypes, these planar imperfections may be incorrectly identified as the inclusion of an alpha phase inside the pure 3C structure. This observation plays a key role in Raman scattering experiments, because specific Raman bands can be associated with different polytypes of SiC.

As was previously mentioned, numerous sections of wires were free of planar defects. The causes for the presence of planar defects in some wires and its absence in the others or the formation of defect free zones within a nanowire are still unknown. Wang et. al. [31] interpreted the formation of stacking faults in nanowires in terms of surface energy. This approach is typical for theories attempting to explain the existence of stacking faults and twin boundaries as low energy structures that readily form to relieve strain.[31,32]

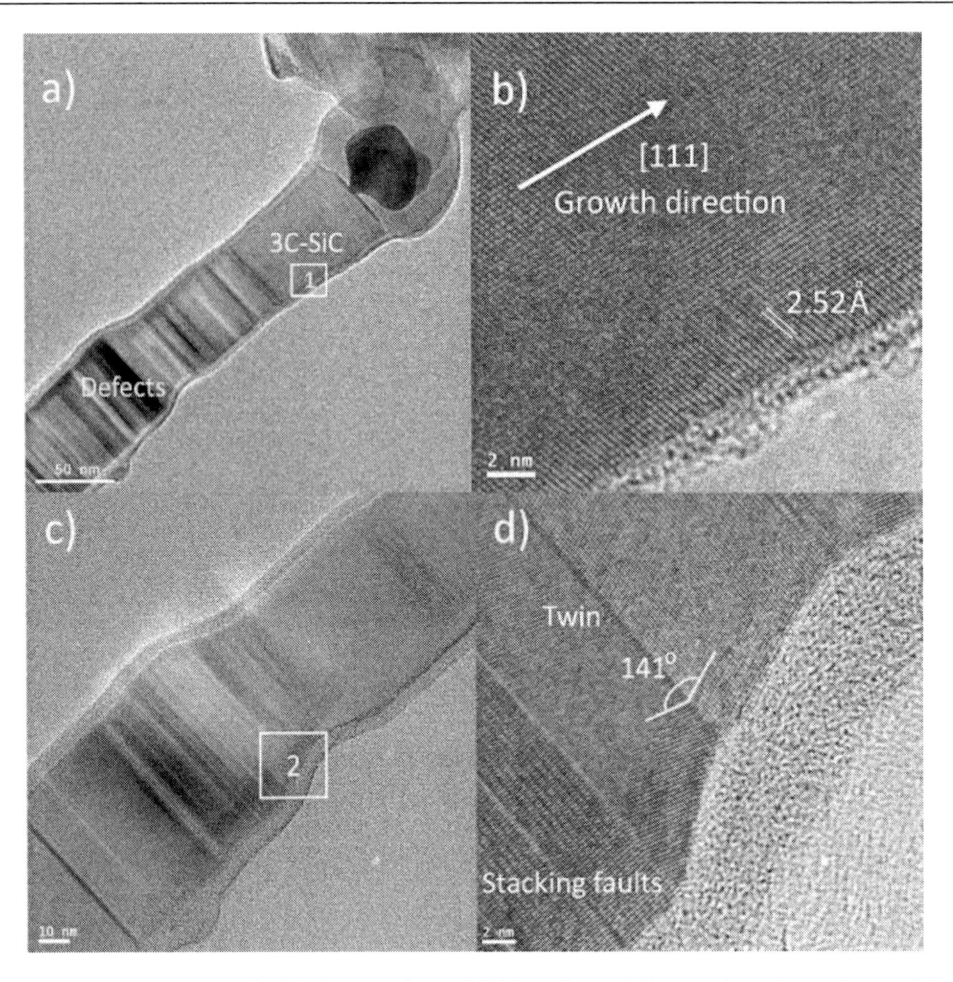

Figure 7. TEM image of a typical SiC nanowire, exhibiting planar defects and sections of pure, defect-free, cubic SiC. The rectangle 1 highlights the part of the wire shown in b. b) High-resolution image of the part of pure 3C-SiC. c) TEM image depicting multiple planar defects. The rectangle 2 highlights the part of wire shown in d. d) High resolution image with a random network of stacking faults and twin area.

In the cubic structure the (111) direction is energetically privileged, and the [111] planes of SiC are known to have lower surface energy than other [*hkl*] planes.[33].The cubic structure has an interesting feature: it possesses two interpenetrating face-centered sublattices. One lattice is formed by silicon atoms; the second one belongs entirely to carbon atoms. These lattices are separated by the displacement vector (a/4, a/4, a/4) (a is unit cell length and in 3C-SiC, a =4.36 Å). Thus, each lattice is occupied by different atoms, and opposite side surfaces [111] have different polarities. This creates the different surface energies, which means that the side with entirely silicon atoms has different energy than the side containing only carbon atoms. These differences at the nanosize scale often become a source of a bending and strain in the nanowires, due to the surface tension. In order to reduce the strain created during the nanowire growth, the polarity must be reversed, which is accomplished by the creation of twin structures. [31]

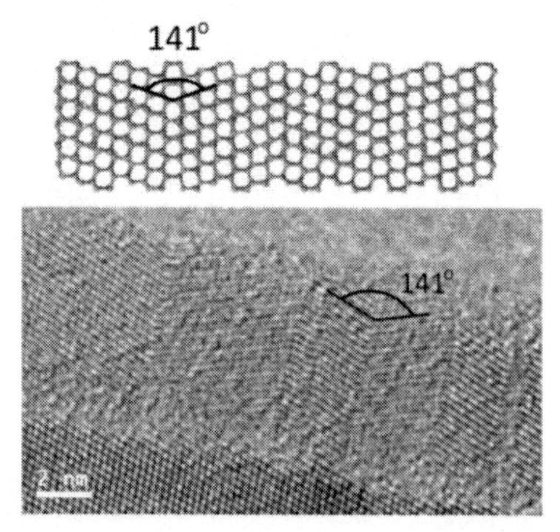

[Wieligor, M.; Rich, R.; Zerda, T.W. J. Mater. Sci. 2010, 45, 1725-1733].

Figure 8. TEM image of SiC nanowires with multiple twins. Nanowires were obtained from graphitized carbon black N990 with large graphitic crystallites. The angle subtended by crystallographic planes is 141°. Above is the atomic model of the corrugated structure of SiC nanowires caused by the twin defects, shown as broken lines.

It is interesting to note that usually twin boundaries in SiC nanowires are well separated-- The estimated distance between twins in TEM images in Figs. 6-8 is greater than 8 nm. The low energy of formation required for twin defects in bulk SiC would seem to contradict this observation. However, Zhang et al. [34] explained large separations of twin defects in nanowires by running classical molecular calculations and showing that the twin formation energy in nanowires depends on edge length and is 73 meV/nm. The twin formation energy is inversely proportional to the length of edges surrounding the twin boundary.

Stacking fault formation energy is higher than that of twins. For example, in case of III-V semiconductors the energy to form a stacking fault is approximately twice the twin formation energy. [35] Consequently, their populations appear to be smaller than that of twins. This formation energy is still very small and corresponds to the kinetic energy at room temperature. Thus, it is not surprising that nanowires manufactured at about 1200°C have abundant stacking faults and twins. These defects are always present in SiC nanowires, regardless of the manufacturing technique and precursors used.

TEM analysis indicates that densities of stacking faults and twins are dependent on the diameter of the wires and decrease with decreasing diameter. This conclusion is similar to that reached by Zhang et al. [36], Niu et al. [37], and Roy et al. [38], Yu et al. [39], and Kraft [40]. Unfortunately, analysis of TEM images does not allow for quantitative analysis of concentrations of either planar defects.

The most commonly employed method of microstructural characterization is transmission electron microscopy (TEM). TEM images quickly allow for direct measurement of crystallite sizes. High-resolution TEM makes it possible to view the lattice structure of the material, and thus imperfections in the crystal may also be directly observed. Stacking faults, for example, appear as a line of higher contrast, while twins are observed by the appearance of two sets of lattice planes joined together at an angle, compare Figs. 7 and 8. While the usefulness of TEM analysis is evident, great difficulties arise when attempting quantitative characterization.

Quantitative analysis of TEM images of nanosize materials is not a trivial task. The average size of the nanowire product, more specifically the diameter, can be determined by taking many TEM images of randomly selected portions of dispersed nanowires. By then directly measuring the diameters of hundreds of these wires, a size distribution may be obtained. Usually the experimental data are fitted to a lognormal distribution function, and a reasonable estimate of the average nanowire diameter can be obtained, see Figure 9. It is easy to see that obtaining a realistic average size via TEM becomes increasingly difficult as overlapping particles in the same orientation produce similar contrast features. From Figure 9 it is also easy to see that a strict control of sintering parameters is required to ensure manufacturing reproducibility and to obtain wires of desired sizes.

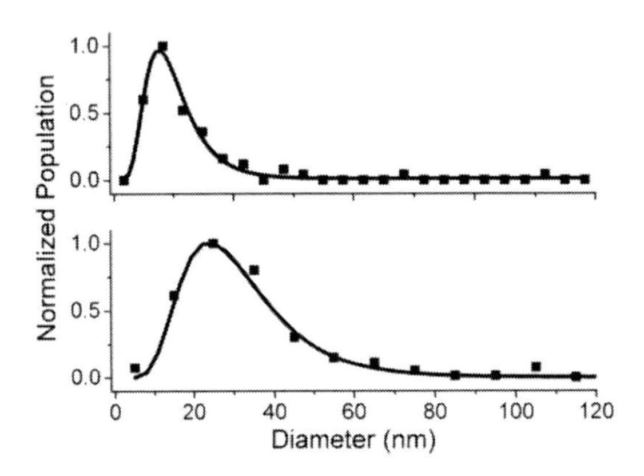

Figure 9. Fitting of the log-normal distribution function to the normalized distribution of SiC nanowire diameters. Both specimens were sintered from carbon nanotubes and silicon powder. Top – carbon nanotubes were kept at 1250 °C and silicon powder at 1200°C, reaction time 4 hours. Bottom – carbon nanotubes were kept at 1325°C and silicon powder at 1125°C for 16 hours. Unless specified, all data shown in this article were obtained for specimens of average size 30 nm, bottom curve.

Other characteristics of the nanowires, such as population of planar faults and dislocations, are even more difficult to obtain by TEM-based analysis. In the case of planar faults, for example, the hundreds of faults would have to be counted along each of hundreds of nanowires to get an adequate idea of their abundance. Planar faults do not even appear to be present when viewed edge–on, so the nanowire must be titled, thus creating the expected ring of higher contrast, compare Figure 5. Thus, quantitative analysis of population of planar faults would require multiple imaging of the same wire at various tilting angles followed by meticulous counting of hundreds of faults along hundreds of wires.

The type of planar fault can be determined if additional measurements are employed, such as visualization of bright and dark images, or by employing diffraction techniques. The latter technique has been described in the previous section and the former in publications by Huang [34,41]. In this study we employed a JEOL 2100 TEM microscope, which in nanobeam mode is capable of collecting electron diffraction data from 5 nm areas of the nanowire, making the distinction between stacking faults and twins potentially possible. However, this type of analysis is also making the characterization process much more tedious.

WHOLE PROFILE ANALYSIS OF X-RAY DIFFRACTION PATTERN

In recent decades, ever increasing computing capabilities combined with improved laboratory x-ray sources and dedicated synchrotron sources have led scientists to revisit x-ray line broadening as a solution to the aforementioned problems in microstructural analysis. It is not only reasonably fast to collect the diffraction data, but such data is collected from a macroscopic sample, thus providing a more accurate picture of the sample.

The use of x-ray diffraction to characterize microstructure began in earnest by Scherrer in the 1920s. When multi-crystalline samples contain domains less than 100 nm in size, the x-ray powder diffraction reflections begin to broaden. The average particle size, $\langle x \rangle$, is related to the width of the reflection as follows [42]:

$$\langle x \rangle = B\,\lambda / (\beta \cos(\theta)) \tag{1}$$

where λ is the wavelength, β is the full width at half maximum (FWHM) of a particular reflection, θ is the diffraction angle, and B is a constant called the shape factor, which is typically around 1. The validity of this equation holds as long as stress and strain are not present in the sample. Clearly another method was needed to address additional broadening introduced from stress and strain.

Williamson and Hall [43] introduced a method by which one may determine the dominant source of peak broadening. Peak broadening due solely to decreasing particle size will effect each reflection equally, i.e. there is no dependence of FWHM on the diffraction vector K

$$K = 2\sin(\theta)/\lambda \tag{2}$$

The anisotropic nature of stresses and defects, however, lead to a direct dependence of FWHM on K. If a plot of FWHM versus K is made, referred to as a Williamson-Hall plot, and a non-monotonic increase in FWHM is observed [44], this indicates that the source of peak broadening involves stresses due to crystal defects as well as crystallite size. Figure 10 depicts the Williamson-Hall plot of silicon carbide nanowires and for comparison purposes also the corresponding data for SiC compacts heated to 1800°C at two different pressures. The SiC phase in compacts has nanosize structure, and the population of stacking faults and dislocations depends on the applied pressure. [56,57] At high pressure, 5.5 GPa, the population of dislocations is orders of magnitude higher than in the compacts sintered at relatively low pressure of 2 GPa. Nauyoks et al. [57] also showed that the population of stacking faults at 2.0 GPa is much higher than in the sample obtained at 5.5 GPa. The Williamson-Hall plot for the nanowires follows the same trend as the 2.0 GPa sample, just with proportionally reduced FWHM due to different average crystallite size. It is an indication that the stacking faults play an important role in SiC nanowires and the effect of dislocations on the line width of the nanowires is small.

Although further developments in the Williamson-Hall technique were made [45], they will not be discussed here and instead we will focus on recently developed simultaneous analysis of available diffraction peaks. This technique is based on a multiple profiles fitting procedure.[46]

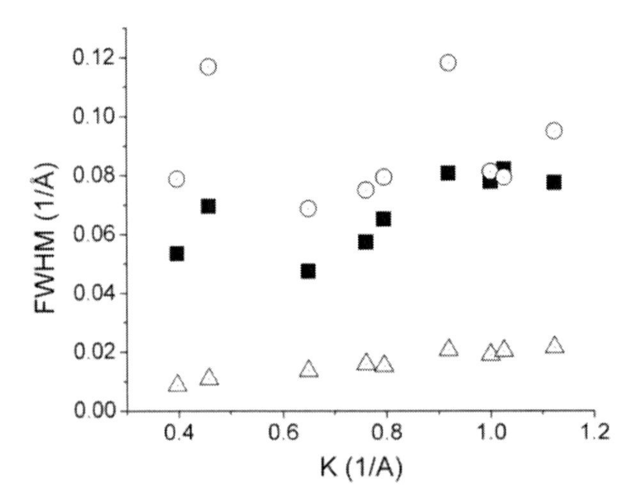

Figure 10. The Williamson-Hall plots for SiC nanowires of average diameter 30 nm, solid squares, and SiC compacts, open symbols. Circles represent data for compacts obtained at 2 GPa and triangles for compacts sintered at 5.5 GPa. Data on SiC compacts were taken from S. Nauyoks dissertation.[56].

The shapes of individual diffraction peaks are expressed as functions of crystallite sizes and their distribution, planar faults density, dislocation induced strain effects, and the dislocation type, screw or edge. The Convolutional Multiple Whole Profile (CMWP) procedure improves upon the traditional method by implementing the automatic deconvolution of overlapping peaks and instrumental effects.[47,48] The CMWP method assumes that the experimental profile is a convolution of functions I^d, I^f, I^s, and I^i representing broadenings due to dislocations, planar faults, limited crystalline sizes, and instrumental function, respectively, and also depends on the background, I^b

$$I^{exp}(\theta) = I^b(\theta) + I^s(\theta) * I^d(\theta) * I^f(\theta) * I^i \tag{3}$$

The stars in Eq. 3 represent convolution. In the first step of the analysis the background is removed from the experimental profile. Next, the experimental profiles are stripped of instrumental broadening. In our case we estimated instrumental broadening by recording diffraction pattern of LaB_6 powder. The LaB_6 powder free of crystal defects and composed of large crystal grains produces pattern that is broadened only by the instrumental function. Dividing the Fourier Transform of the experimental profile by the Fourier Transform of LaB_6, with the background subtracted from each, we get

$$A_L^{exp} = A_L^s A_L^d A_L^f \tag{4}$$

where L represents the Fourier variable. The difficulties in solving Eq. 4 is in defining the Fourier coefficients A_L^s, A_L^d, and A_L^f, which account for the size, distortions due to dislocations, and distortions due to planar faults, respectively.

The Fourier size coefficient, A^s, can be estimated assuming that the specimen is a mixture of spherical crystallites. The log-normal size function

$$f(x) = \frac{1}{(2\pi)^{1/2}\sigma x} \exp\left[\frac{-(\log(x/w))^2}{2\sigma^2}\right] \quad (5)$$

where σ is the variance and w is the median, describes the crystallite size distribution.

A^s can be calculated from [49,50]

$$A^S(t) = \int_{|t|}^{\infty}\left(1 - \frac{|t|}{M}\right) M^2 \left[\int_M^{\infty} f(x)dx\right] dM \quad (6)$$

$$A^s(L) \sim \frac{m^3 e^{4.5\sigma^2}}{3} erfc\left[\frac{ln(|L|/m)}{\sqrt{2\sigma}} - 1.5\sqrt{2\sigma}\right] - \frac{m^2 e^{2\sigma^2}|L|}{3} erfc\left[\frac{ln(|L|/m)}{\sqrt{2\sigma}} - \sqrt{2\sigma}\right] + \frac{|L|^3}{6} erfc\left[\frac{ln(|L|/m)}{\sqrt{2\sigma}}\right] \quad (7)$$

where erfc is the complementary error function. The area-weighted mean crystallite size is given by

$$\langle x \rangle_{area} = we^{2.5\sigma^2} \quad (8)$$

The size profile may also be scaled to account for non-spherical crystallites or subgrains.[51] One can find a similar expression for the volume-weighted mean but we will not use it in this paper and limit the discussion to the <x>area.

The Fourier coefficient due to dislocations was found by using the Warren-Averbach [46] method. The coefficient can be expressed as:

$$A_L{}^D = \exp\left(-2\pi^2 g^2 L^2 \langle \varepsilon_L^2 \rangle\right) \quad (9)$$

where L is the Fourier transform variable, g is the magnitude of the diffraction vector, and $\langle \varepsilon_L^2 \rangle$ is the mean square strain, which depends on how much the atoms are displaced due to strain compared to an ideal lattice arrangement.[46] Krivoglaz [52] developed a method to calculate the mean square strain caused by dislocations. Wilkens [53] improved Krivoglaz's method by assuming that dislocations have an outer cutoff radius, R_e^*. Using this parameter Wilkens derived the following equation for the mean strain:

$$\langle \varepsilon_L^2 \rangle = \left(\frac{b}{2\pi}\right)^2 \pi\rho C f\left(L / R_e^*\right) \quad (10)$$

where b is the magnitude of the Burgers vector, C is the contrast factor of the dislocations and f(η) is Wilkens function of strain as defined by:

$$f^*(\eta) = -\log\eta + \left(\frac{7}{4} - \log 2\right) + \frac{512}{90}\frac{1}{\pi}\frac{1}{\eta} + \frac{2}{\pi}\left(1 - \frac{1}{4\eta^2}\right)\int_0^\eta \frac{\arcsin V}{V} dV$$

$$-\frac{1}{\pi}\left(\frac{769}{180}\frac{1}{\eta} + \frac{41}{90}\eta + \frac{2}{90}\eta^3\right)\left(1-\eta^2\right)^{1/2}$$

$$-\frac{1}{\pi}\left(\frac{11}{12}\frac{1}{\eta^2} + \frac{7}{2} + \frac{1}{3}\eta^2\right)\arcsin\eta + \frac{1}{6}\eta^2, \ \eta < 1 \tag{11}$$

$$f^*(\eta) = \frac{512}{90}\frac{1}{\pi}\frac{1}{\eta} - \left(\frac{11}{24} + \frac{1}{4}\log 2\eta\right)\frac{1}{\eta^2}, \ \eta \geq 1 \tag{12}$$

where $f(L/R_e^*) = f^*(\eta)$ and $\eta = (1/2)\exp(-1/4)(L/R_e^*)$. The anisotropy of the dislocations is accounted for by the average contrast factor C. [44] In a cubic crystal, C is dependent on C_{h00}, the average contrast factor in the <h00> direction, which is dependent on the elasticity of the material and on the fourth-order polynomial of the hkl indices. The average contrast factor therefore is:

$$C = C_{h00}\left(1 - q\frac{h^2k^2 + h^2l^2 + k^2l^2}{\left(h^2 + k^2 + l^2\right)^2}\right) \tag{13}$$

The effect of faulting and twinning can be represented by a uniform profile function. This function was shown to be the Lorentzian, where the total profile of a faulted or twinned Bragg reflection is the linear combination of Lorentzians broadened and/or shifted according to well-defined hkl dependent rules. In order to determine the Fourier coefficient due to planar faults, A_L^f, the magnitude of peak broadening and peak location were first evaluated. Diffraction peaks were numerically calculated using the program: DIFFAX (diffraction of faulted crystals). DIFFAX [54] uses the recursion relationship of the wave function in crystals to simulate planar faults. Balogh and coworkers [48] calculated the first 15 Bragg reflections of the face-centered cubic crystals with extrinsic and intrinsic stacking faults and twins with probabilities up to 20%. They analyzed over 15,000 subreflections due to planar faults. It was found that they fit either a delta function or a Lorentzian function. The FWHM and the shift of the subreflections were dependent on the hkl indices and the planar fault probability which was approximated by a fifth order polynomial. These results have been summarized in the form of a table of the fifth order polynomials for each hkl set which is subsequently used by the extended CWMP program, eCMWP. [48,55] This program is an improvement of the CWMP program because it allows for quantitative characterization of planar faults.

The eCMWP fitting routine simultaneously analyzes the whole profile instead of individual peaks. For a cubic lattice it provides the following parameters: i) the mean size, ii) standard deviation, with which one can find $\langle x \rangle_{area}$, see equation (8), iii) β the fault or twin probability, iv) ρ the dislocation density, and v) q, a parameter that is dependent on the dislocation type, screw or edge.

The need for the eCMWP procedure is easily seen upon inspection of XRD data obtained from SiC nanowires. A typical diffraction pattern of SiC nanowires is shown in Figure 11. A shoulder is observed near the [111] peak at 33.6°, with copper beam at λ=1.54 , see the insert in Figure 11. This shoulder has been attributed to planar faults. Therefore the eCMWP method was applied to obtain meaningful information regarding microstructure.

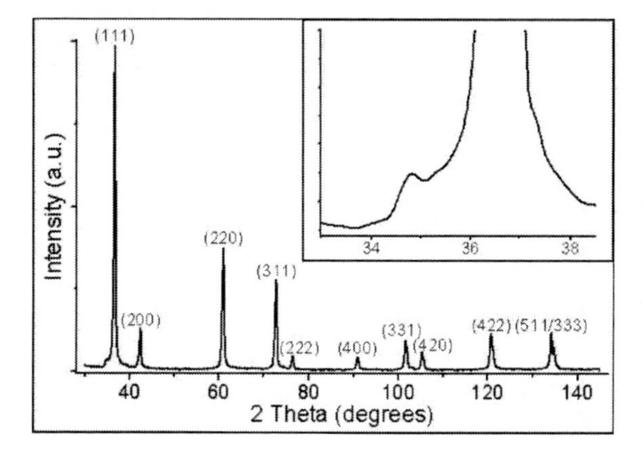

Figure 11. The X-Ray diffraction pattern for SiC nanowires of average diameter 30 nm. Reflections are assigned to different crystallographic planes. The insert shows details of the (111) peak with the shoulder due to stacking faults.

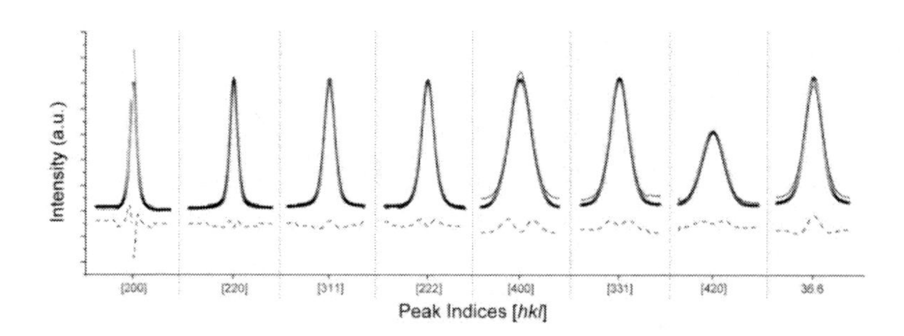

Figure 12. Example of the fitting procedure. Measured (open circles) and fitted (solid line) diffraction profiles for the 30 nm diameter SiC nanowires. The difference between the two sets of data is shown in the bottom part of the figure.

As explained in the previous section, upon inspection by TEM, twins appeared to be the most prevalent form of planar faults present in the SiC nanowires. Thus the eCMWP procedure was performed with this assumption in place. In Figure 12 we show the quality of the fitting procedure. When either extrinsic or intrinsic stacking faults were initially assumed in the eCMWP fitting process, the sum of the residuals was orders of magnitude larger, thus confirming the assumption regarding twins. The results of the x-ray profile analysis are summarized in Table 1. Probability of a twin defect was determined to be 2.2%, or in other words, there should be a twin about every 50 planes in the stacking sequence. SiC is known to grown along the (111) direction, and this is confirmed by TEM in our SiC nanowires. Assuming a one-dimensional wire, the distance between the [111] planes of SiC is 0.75510 nm, thus the distance between twins should be 37.8 nm. This back of the envelope calculation did not take into account that there are actually four equivalent (111) directions, and an average separation distance between two twins is expected to be less.

The eCMWP analysis shows the average area weighted crystallite size is 31 nm. This value is very similar to the mean value produced by the lognormal distribution of diameters directly measured from TEM. This result indicates that the nanowires must contain subgrains of lengths comparable to the diameter of the nanowires themselves. If this were not the case, nanowires oriented perpendicularly to the incident x-rays would have provided a larger coherence length, and thus a larger grain size in the eCMWP analysis.

Table 1. The calculated parameters of the microstructure of silicon carbide in different forms: nanowires, SiC compact sintered from nanosize powder at 5.5 GPa and 1800°C, and SiC phase in diamond/silicon carbide composites sintered at 8.0 GPa and 1820°C. $\langle x \rangle_{area}$ is the area-weighted mean crystallite size, β is the planar fault probability, and ρ is the dislocation density.

	SiC nanowire	SiC phase in diamond/SiC composite[a]	SiC compact[a]
$\langle x \rangle_{area}$	31 nm	20 nm	126 nm
β	2.2%	2.6%	0.17%
ρ	6.8×10^{15} m^{-2}	4.3×10^{15} m^{-2}	0.4×10^{15} m^{-2}

[a]From S. Nauyoks' dissertation, Ref. 56.

Typically, increased pressure and temperature cause growth of populations of planar faults and dislocations in ceramics materials, such as diamond or silicon carbide.[57-60] Often this process is accompanied by the increased crystallite sizes. In Table 1 we compare results obtained from the CMWP analysis for nanosize silicon carbide in three different materials. Diamond-silicon carbide composites were sintered from a mixture of nanosize diamond and silicon powders at high pressures and temperatures. Details of the sintering protocol can be found in Refs. 55-60. Silicon carbide compacts were obtained from nanocrystalline SiC powder of nominal grain size of 30 nm. They were sintered in the high pressure toroid system.[61] It is seen that the microstructure of SiC nanowires is similar to the microstructure of SiC in diamond composites. This is a surprising result. In diamond composites large population of planar faults and dislocation densities is caused by strain associated with different rates of thermal expansion and compressibility of diamond and silicon carbide phases. Comparable populations obtained for the nanowires indicate that its structural defects may be associated with stresses of similar magnitude as those in diamond-SiC composites.

CORE-SHELL MODEL OF SIC NANOWIRES

Surface atoms have different surroundings than atoms in the crystal interior. Contrary to the atoms in the bulk, surface atoms have asymmetrical interactions which of course affect their structure. In large crystals the fraction of atoms near the surface is low in comparison with the total number of atoms constituting the crystal. In nanosize crystals this situation is different. A considerable portion of atoms lie near the surface, for example in a 5 nm crystal about 50% of the atoms forms a surface layer about 1 nm thick. This effect explains different mechanical and chemical properties of nanomaterials. To account for different structures and different properties of nanosize materials, the surface layer must be treated as a separate phase, separate from the interior, the so called core.

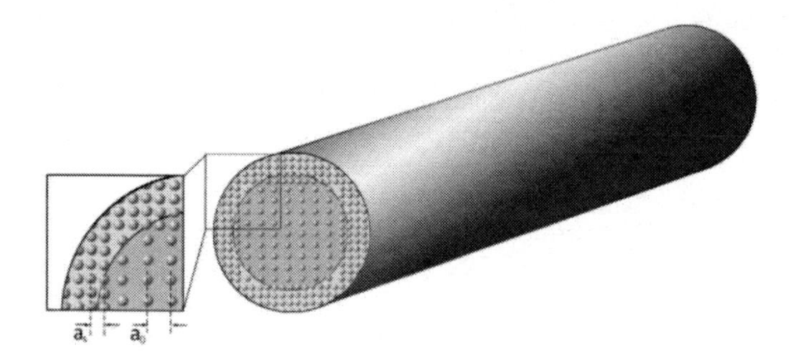

Figure 13. Schematic model of a cylindrical nanograin. The atoms in the surface shell are compressed relative to the relaxed lattice. a_o – lattice parameter of the core, a – lattice parameter of the shell.

The core-shell model of nanosize particles and wires has been recently introduced by Palosz and coworkers and has gained general acceptance.[62-66] One of the main successes of that model is explanation of modifications in x-ray diffraction patterns with decreasing particle sizes. Conventional powder diffraction techniques have been developed for materials with perfect 3-D periodic order in an infinite lattice, but a long range order in nanocrystals is limited by the size of the grain. Therefore, application of the methods derived for an infinite lattice is inadequate because the size of the crystallites may be smaller than the coherence length of the scatter beam. Figure 13 depicts a core-shell model of a cylindrical crystal. The core of radius R and the shell of thickness d_{shell} have the same crystallographic structure, but the lattice parameter in the shell is different than that in the core, which is indistinguishable from the infinite crystal lattice parameter, a_o. In comparison with the core, the shell may have expanded structure or compressed, and its thickness varies from 0.3 to 0.6 nm for different crystals. Reality is probably more complex and the lattice parameter changes continuously along the grain radius. It is a common phenomenon, and one example is shown in Figure 14.

In the alp method each Bragg reflection is uniquely defined by diffraction vector Q, and each reflection may provide a distinct lattice parameter, referred to as the *apparent lattice parameter*, alp. In contrast, each Bragg reflection from an infinite single crystal will produce an identical lattice parameter. It is important to note that x-ray beams incident with large diffraction angles will penetrate more deeply into the interior of the crystals. Thus a plot of reduced alp values, alp/a_0, versus Q will provide information about atomic spacings at increasing crystal depth. A depiction of such a plot obtained for NaCl grains of different sizes is shown in Figure 14. It is seen that with decreasing crystal size the shell structure becomes increasingly more relaxed. It means that the interatomic distances in the outer layers of nanosize NaCl grains are longer than the corresponding distances in the center of the same crystal.

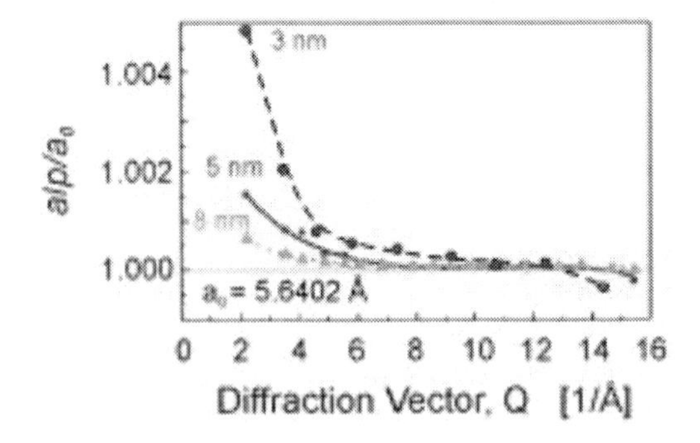

Figure 14. The reduce lattice parameter for 3 nm, 5 nm and 8 nm NaCl powders as a function of vector Q calculated by Palosz et al., Z. Kristallographie, 2007, 222, 580-594. The NaCl powders exhibit expanded shell. The alp values match the lattice parameter in bulk material at large Q. This graph illustrates the need of collecting data in the very wide range of Q values.

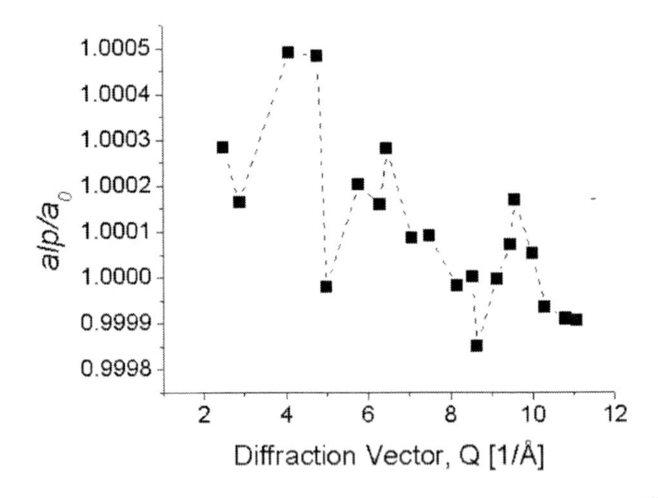

Figure 15. The alp-Q plot for 30 nm SiC nanowires. The a/a_0 values were calculated assuming a_0 of the perfect cubic structure, 4.356 Å. The broken line is shown to guide the eye.

In Figure 15 we plot lattice parameters calculated from different reflections for SiC nanowires of average diameter 30 nm. A standard x-ray diffractometer equipped with Cu or Mo lamps does not allow for observation of small changes in the structure of small crystals of diameters of a few nanometers. Typically the Q range for those sources is about 6 $Å^{-1}$ but the Q range required for a meaningful alp-Q plot is about 12 $Å^{-1}$. Data in Figure 15 were obtained on a custom made x-ray diffractometer equipped with a silver lamp with the incident wavelength of 0.55942 Å. Examination of these data is not straightforward and requires computer modeling of atomic structure of the nanowire. Such analysis was performed by Palosz et al. [64,68] for SiC 11 and 40 nm powders and reasonably good agreements between experimental and theoretical data were obtained when the structure of silicon carbide nanograins was approximated by the core-shell model with a shell about 0.5 nm thick. In that shell interatomic distances are by about 3% smaller than those in the core. This result is in good agreement with conclusions reached by Rich et al. [69] for 30 nm silicon carbide nanowires.

The compressed shell-core model was used to explain the pressure dependence of the lattice constant for nanosize SiC. High pressure in-situ experiments were conducted in a diamond anvil cell at pressures up to 60 GPa at Brookhaven National Lab with the incident x-ray beam of wavelength 0.404067 Å [69]. Figure 16 illustrates pressure induced shift of the [111] reflection. Lattice constant calculations were carried out on these data and later used to calculate the reduced constant, a/a_o, which is depicted in Figure 17. These results agree with the study by Liu et al. [70]. Broken lines represent the best fits between experimental data and the equation of state. Birch-Murnaghan approximation [71,72] was used for the equation of state:

$$P = \frac{3}{2}K\left[\left(\frac{a_0}{a}\right)^7 - \left(\frac{a_0}{a}\right)^5\right]\left\{1 + \frac{3}{4}(K'-4)\left[\left(\frac{a_0}{a}\right)^2 - 1\right]\right\} \tag{14}$$

where a/a_0 is the reduced lattice constant, P is the applied pressure, and K and K' are the fitting parameters, the bulk modulus and its derivative, respectively. The obtained bulk moduli of the 50 nm and 130 nm samples, 198 and 193 GPa, respectively, agree with literature data for bulk SiC – 206 GPa [69]. The 20 nm grains showed a significantly higher bulk modulus of about 260 GPa and for the nanowires we found 241 GPa.

The elevated bulk modulus of 20 nm grains is explained by the core-shell model. [62-66] In the core, the majority of the Si and C atoms are arranged in a manner consistent with bulk SiC and they exhibit all the properties associated with the bulk material, such as bulk modulus and lattice parameter. The shell atoms, on the other hand, possess unique qualities due to the fact that their interatomic distances are smaller compared to the bulk/core atoms. With nano-sized materials, a much larger percentage of the constituent atoms belong to the shell, and this may lead to unique properties dependent upon particulate size.

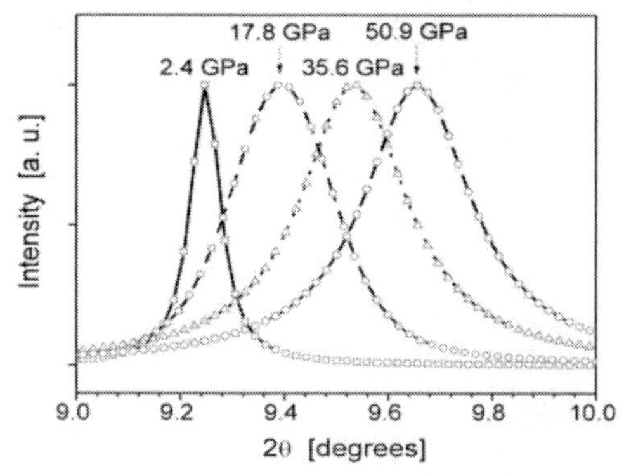

[Rich, R.; Stelmakh, S.; Patyk, J.; Wieligor, M.; Zerda, T.W.; Guo, Q. J. Mater. Sci. 2009, 44, 3010-3013].

Figure 16. Pressure induced shift of the (111) reflection for the nanowires. To guide the eye, experimental points recorded at different pressures are connected by lines.

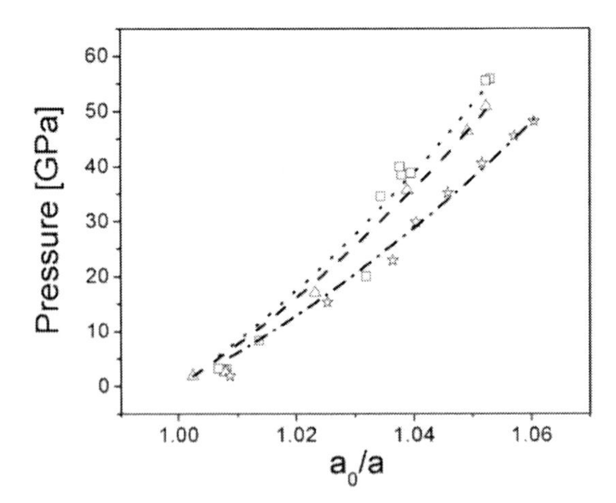

Figure 17. Experimental data points for the 130 nm SiC powder (stars), the 20 nm powder (squares) and the 30 nm nanowires (triangles). The dashed line represents the best fitting of Eq. 14 to the data obtained for the nanowires, the dotted line – the 20 nm SiC powder, and the dash-dotted line for the 130 nm SiC powder.

Amorphous Structure

Analysis of the alp-Q plot, Figure 15, indicated that the simple model composed of a single shell and a uniform core is just a first approximation and to better explain the microstructure of silicon carbide a more sophisticated model may be needed. Indeed, Palosz et al. [68] have pointed out that a more realistic model consisting of a modulated multi-shell structure is needed to explain details of the alp-Q plot for this material. In addition, in all TEM images, examples of which are depicted in Figs. 6-8, the amorphous layer is clearly observed. From our studies we learned that the thickness of the amorphous layer varies from 1 to 10 nm and depends on the manufacturing process and post production treatment. To account for the amorphous layer we introduce a model, see Fig 18, based on the core-shell model and consisting of the crystalline core of radius R_c, the shell of thickness d_{shell}, the amorphous layer of thickness d, and the overall radius of the nanowire R.

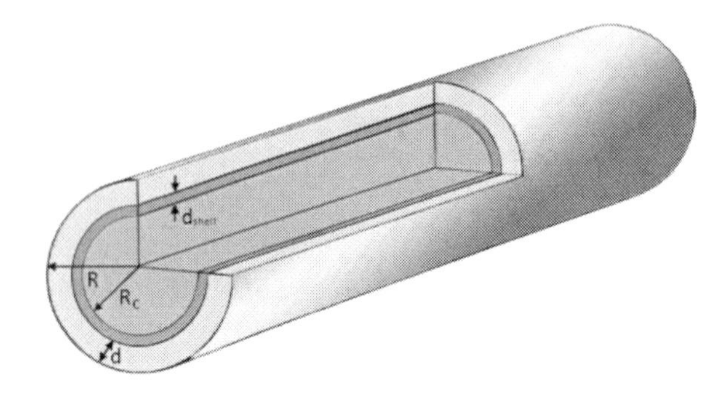

Figure 18. A model of a SiC nanowire. The core of radius Rc is surrounded by a shell of the same crystallographic structure but different lattice constant of thickness dshell. The outer layer of thickness d is the amorphous layer. The overall radius of the nanowire is R.

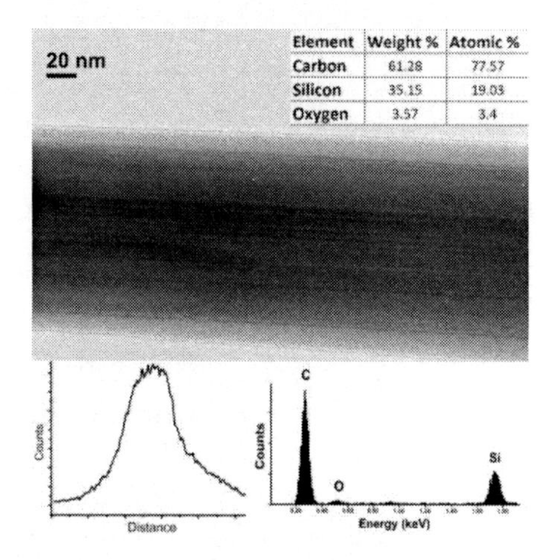

Element	Weight %	Atomic %
Carbon	61.28	77.57
Silicon	35.15	19.03
Oxygen	3.57	3.4

Figure 19. TEM image of a silicon carbide nanowire. The EDS spectrum is shown in the bottom right corner and in the insert in the left corner the solid line represents oxygen concentration measured across the nanowire diameter.

An important and still unresolved aspect of the structure of the nanowires is the atomic arrangement of the amorphous layer. Is it just an amorphous SiC, a carbon rich silicon layer, or maybe a mixture of silicon in a predominantly carbon layer? To answer this question we employed the energy dispersive x-ray spectroscopy, EDS and infrared spectroscopy. The EDS is an analytical technique that allows determination of chemical composition of a specimen mounted into a scanning microscope probe inside the TEM. Because the resolution of the microbeam used was limited to about 10 nm, for the analysis we selected only relatively wide wires, of diameters exceeding 100 nm. Elemental analysis of the amorphous structure clearly showed that beside silicon and carbon there was a significant concentration of oxygen, as indicated by the EDS results depicted in Figure 19. A line scan across the nanowire demonstrated that oxygen was almost uniformly distributed. We concluded that oxygen was present only in the outer layer of the amorphous structure. Although this conclusion is limited to the relatively large nanowires, we believe that it is also true for nanowires of smaller sizes, more representative for the samples.

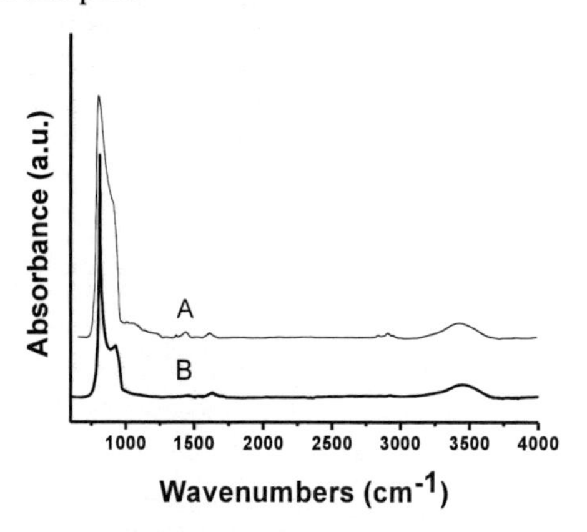

Figure 20. A - infrared absorption of SiC nanowires heat treated in oxygen atmosphere at 800oC, upper spectrum. B - the spectrum after etching in HF acid.

FTIR absorption studies have confirmed this conclusion. This technique is well suited for analysis of molecular composition of SiC specimens. In Figure 20 a typical spectrum of SiC nanowires is shown. This specimen was heat treated in air at 800°C for two hours to remove carbon impurities. In the high frequency region, the spectrum is dominated by a broad and asymmetric peak due to water at 3460 cm^{-1}. It intensity depends on the sample preparation, and decreases after heating the specimen at 200°C. Since isolated water molecules have O-H vibrations at higher frequencies, the position of the observed O-H stretching mode indicates that water molecules are hydrogen bonded and water exists in the form of clusters adsorbed on the surface of the nanowires. The three peaks at 2962, 2922 and 2851 cm^{-1} are assigned to CH_2 and CH_3 stretching vibrations. These three peaks and a number of small intensity bands in the frequency range from about 1300 cm^{-1} to 1700 cm^{-1} provide proof that carbon compounds are present in the specimen. The doublet at 1350 cm^{-1} and 1584 cm^{-1} are the so-called disorganized and organized vibrations of carbon in graphite-like arrangements. We have not found traces of Si-H bands, but Si-O and Si-OH bands were strong, indicating that glass-like structures terminated by hydroxyl groups were the main form of silicon containing species present in the specimens. These compounds were most likely assembled on the outer layers of the nanowires contributing to the amorphous structure. The most intense bands in the spectrum are due to two SiC phonons, the LO at 957 cm^{-1} and the TO at 792 cm^{-1}. [73] From this brief description of a typical infrared adsorption spectrum, one may conclude that various oxygen containing functional groups coexist with SiC crystalline structure. Since oxygen does not readily dilute in SiC we further conclude that these species are formed on the outer layers of SiC core-shell and contribute to the amorphous layer.

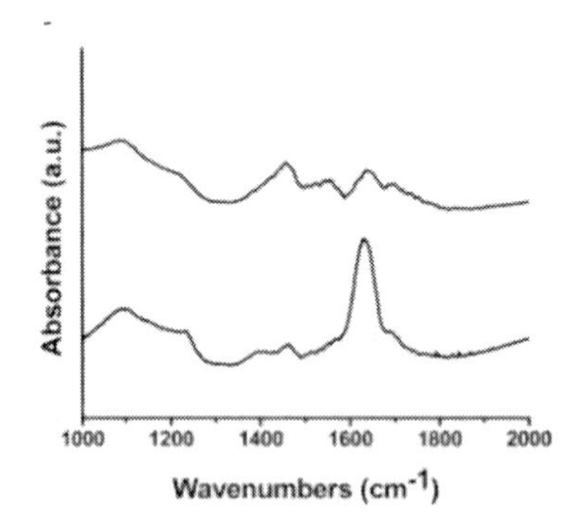

Figure 21 Infrared absorption spectra of silicon carbide nanowires as received (upper spectrum) and after heat treatment at 800°C for two hours (bottom spectrum). Both spectra were magnified to show details.

In the following paragraphs we will discuss changes in the absorption spectra induced by thermal treatment. As seen in Figure 21 the carbon doublet at 1350 and 1590 cm^{-1} disappeared after heating the sample to 800°C in air. At the same time two new peaks centered about 1620 cm^{-1} and 1650 cm^{-1} appeared. They overlap to form a contour with a maximum at about 1630 cm^{-1}. These peaks are assigned to the C=O stretch and the –COOH vibrations, respectively. To explain these effects we assumed that large agglomerations of carbon atoms, probably in the form of graphitic nanosize crystallites, 'burned out' leaving carbonyl and carboxyl groups attached to amorphous silicon carbide. It is important to note that the CH_2 and CH_3 peaks near 2900 cm^{-1} were not removed by the heat treatment. These are most likely other functional groups attached to amorphous silicon carbide.

Temperature treatment also affected the Si-O bands. The asymmetric stretching Si-O vibration was observed at about 1090 cm^{-1}. The bending mode O-Si-O at 810 cm^{-1} partially overlaps with the SiC phonon and is difficult to investigate. However, it is worthwhile to note that after a treatment in HF this peak disappeared, along with another peak due to silica at about 470 cm^{-1}, and as a result the two phonon bands due to TO and LO modes became better resolved and clearly seen as two separate peaks, compare Figure 20. The main focus of the analysis was on the 1090 cm^{-1} band. Assuming that this broad peak is caused by vibrations of structures similar to those present in glass, we ran a series of experiments to calculate the extinction coefficient for glass at this frequency range and next estimated concentration of SiO_2 from that peak intensity. The calculated amount accounted for about 1% of the total mass of a 2 nm thick amorphous layer wrapping around all structures. Therefore, we concluded that the thin film on the surface of SiC nanowires and particles is mainly amorphous silicon carbide with a small amount of SiO_x and amorphous carbon. The fact that the infrared absorption peak was blue shifted further indicated that it was not pure SiO_2 but an amorphous structure of random chemical stoichiometry. After heating the specimen at 800°C this peak intensity slightly increased indicating higher concentration of silica.

Etching in an acid or a strong base is another commonly used technique for removal of undesired functional groups and preparing specimens for subsequent surface functionalization procedures. In Figure 20 we show absorption spectra of the same sample before and after treatment in HF acid. Silica bands have decreased in intensity, or almost completely disappeared, indicating that their concentration dropped dramatically. However, after the treatment a peak centered at about 1630 cm^{-1} peak grew in intensity, compare Figure 21. It was accompanied by a small growth of a graphitic doublet at 1350 and 1590 cm^{-1}. This result indicates that HF etching promotes formation of graphitic crystallites and increases population of other C=C bonds. This result is in agreement with the conclusion of a study by Gogotsi et. al. [74] on hydrothermal treatment of SiC powders.

GROWTH MECHANISM OF SIC NANOWIRES

Chemical reactions between carbon and silicon containing precursors are involved in the growth of SiC nanowires. Spontaneous growth in the (111) direction is much faster than in other directions, and wires of uniform diameters can be obtained. Many techniques have been developed in the synthesis and formation of SiC nanowires. Some have attracted a lot of attention while others have not been extensively explored, but it is safe to assume that virtually every possible growth mechanism has been utilized to produce SiC nanowires. Among them the most commonly used are the chemical vapor deposition [16,75-80], the vapor-liquid-solid (VLS) growth [17,23,75,81,82], vapor to solid growth (VS) [23,83,84], the sol-gel process [15,85,86], electrospinning [87,88], template based growth [89-91], and laser ablation [92]. The list of potential precursors is also very long, but for practical applications only inexpensive chemicals are used.[93] Catalysts are often used, and they are especially important in the VLS method. [94] Nanowires obtained by the VLS method have one end capped with a sphere, which is an important characteristic of the VLS process.[35] Usually these spheres contain metal catalysts which reduce the melting temperature of the liquid on the tip of the nanowires. The size of the droplet controls the diameter of the nanowire.[23,35]

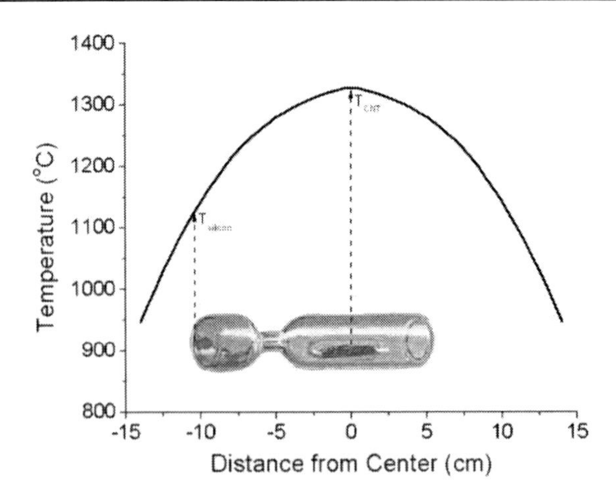

Figure 22. Temperature distribution along the tube furnace and the quartz ampoule used to sinter SiC. Zero indicates the center of the tube furnace. Carbon nanotubes were placed where temperature was highest and temperature of silicon powder was controlled by adjusting its position in the furnace. The precursors are sealed inside an evacuated quartz ampoule. By varying the separation of carbon nanotubes and silicon, i.e. the length of the quartz ampoule, one may vary the temperature of silicon. The carbon nanotubes are contained in a quartz vial within the ampoule to facilitate easy removal free of contamination by the silicon powder. The ampoule is crimped to keep the carbon nanotubes containing vial in place during setup.

Synthesis of SiC nanowires discussed in this article took place at high temperatures. We used carbon nanotubes and silicon powder as precursors. Although carbon nanotubes are not among the least expensive materials they have an important advantage - they allow confining the initial growth of silicon carbide to the volume of the nanotube. Both materials were placed at opposite ends of an ampoule which was evacuated, sealed, and heated at high temperatures for well defined time periods. Figure 22 depicts the experimental setup. We took advantage of the temperature distribution in a tube furnace, and by carefully adjusting position of the ampoule within the furnace, we ensured that both precursors, silicon powder and carbon nanotubes, were kept at 1125°C and 1325°C, respectively. Silicon sublimes at temperatures above 900°C and by controlling temperature of silicon powder we effectively controlled vapor pressure of silicon. At these conditions, the products were SiC nanowires of 30 nm diameter but varying lengths. The process is highly reproducible. The distribution of nanowire diameters was narrow and is shown in Figure 9. Details of the production protocol and precursors used can be found in Refs. [10,20,69,93]

The reaction between carbon nanotubes and silicon is not fully understood. In the presence of metal catalysts introduced to the system as leftovers from the carbon nanotube synthesis, it is clear that the growth proceeds through the VLS process. High concentration of metals in spherical structures found at the tips of SiC nanowires provides the necessary evidence. Intermediate steps of the reaction could involve formations of CO or SiO ions.[3,5,6] The possible reactions mechanisms have been discussed in many articles [22,23,95-97] and will not be repeated here. In the case when metal catalysts were present the reaction was relatively fast and the yield was high.

When metal catalysts were absent and concentration of oxygen was very low, the reaction could be explained in terms of the VS process. In this model, silicon vapor or SiO (more likely) attacks carbon atoms located on the outer surface of the nanotube, reacts with them to form a SiC layer. A perfect CNT is a graphene sheet seamlessly wrapped into a cylindrical tube. Experimental results however clearly indicate that CNTs possess many defects; for example, Mawhinney et al. [98] reported that several percent of carbon atoms on the carbon nanotube walls can be located at the defective sites. The most common defects are the so-called Stone-Wales (SW) defects, [25,99,100] which are pairs of adjacent 5 and 7 member rings, see Figure 23. The SW defects are formed unavoidably during nanotube synthesis process, since the additional paring of a heptagon with a pentagon is energetically favorable. High concentration of these defects may lead to tube bending and changed diameter. We have observed this behavior in TEM images of carbon nanotubes recorded prior to the reaction with silicon. It is generally accepted that SW defect sites trigger chemical reactions between carbon nanotubes and other materials. [101,102] At high temperatures the population of these defects is probably increased. Please note that the manufacturing of SiC nanowires takes place at temperatures between 1100°C and 1350°C. In addition, SW defects can glide, twist, and distort, and thereby some C-C bonds may break. It is then energetically favorable for these dangling bonds to react with Si atoms. Of course, other defects, such as vacancies, impurities, and kinks, may also be present and constitute nuclei for the reaction, and the combined effect of various defects could further lower the activation energy.

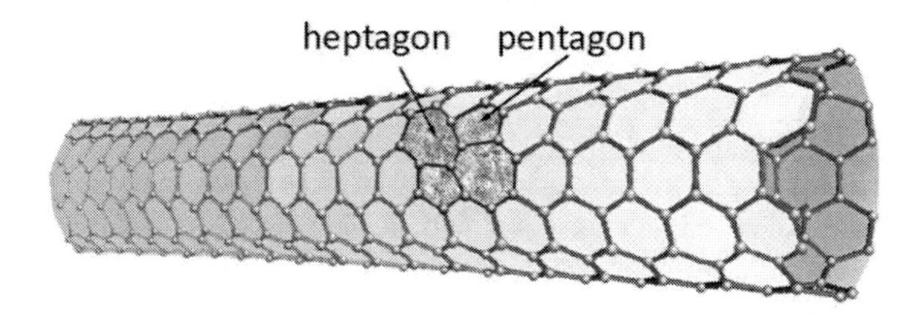

Figure 23. A model of a carbon nanotube with two sets of Stone-Wales defects.

We assume that the reaction with silicon starts at SW defects on outer walls of multiwall carbon nanotubes. Then SiC grows not only along the tube outer walls but also radially, and the high density of defects results in numerous nucleation sites. As the grains grow they coalesce. From the x-ray profile analysis, Table 1, we know that the average size of SiC domains is about 30 nm. Of course, continuing the reaction for an extended period of time at a high temperature facilitates growth of the crystallites. When specimens were allowed to reside at elevated temperatures for a long time, longer than 4 hours, a different process started, the Oswald ripening, and large diameter nanowires continued to grow at the expense of smaller wires. Smaller nanowires produce larger vapor pressure due to their larger surface area per unit mass. It logically follows that large nanowires produce lesser vapor pressures, and, in order to maintain equilibrium, smaller particles evaporate into surrounding medium. The specimens discussed in this article had average size of about 30 nm and were produced in times for which the Oswald ripening was not significant.

Once the SiC layer coats the exterior of carbon nanotube, to continue the reaction, carbon atoms must diffuse through the newly formed SiC layer to the outer surface to react with silicon, or silicon atoms must migrate toward the carbon rich area. As silicon diffuses into the interior of the coaxial structure it transforms carbon nanotubes into SiC and eventually the interior becomes solid SiC. The outer diameter of the nanowires also grows indicating that carbon atoms continue to diffuse through the SiC layers. As discussed in the previous section the outermost layers of SiC nanowires are composed of amorphous silicon carbide, SiO_x, and various carbon functional groups.

A layer of SiC formed on carbon substrate, initially thin, grows with reaction time. Diffusion coefficients of both carbon and silicon atoms in crystalline SiC are very small [103-107] and to explain the rapid growth of SiC we assume that it proceeds mainly by diffusion of those atoms along grain boundaries, stacking faults, and dislocations. Pantea et al. [108] observed this effect in diamond composites in which the diffusion of carbon and silicon atoms was enhanced by the presence of structural defects and extensive grain boundaries.

CONCLUSION

A systematic collection of experimental data on planar faults will aid in the search for a correlation between the population of twins, stacking faults and morphology of nanowires. Analysis of TEM images enables distinguishing between stacking faults and twins, but this is not a simple, routine task and thus not very often used. TEM images are excellent tools when only qualitative analysis is required, but when such information is not sufficient and quantitative analysis is needed, a different experimental technique should be used. In this article we demonstrated that simultaneous analysis of multiple x-ray diffraction peaks represents such alternative methodology. This analysis allowed us to conclude that twins are more frequent that stacking faults, and evaluate their concentration. The population of twins depends on nanowires diameter and decreases with decreasing size. In addition, population of dislocations and average crystallite sizes are also provided by that analysis. Importance of information on defects in nanowires may be illustrated by indicating how important these structures are in determining their electronic and mechanical properties. Recently, several groups have predicted theoretically multi quantum wells consisting of cubic SiC wells and hexagonal SiC barriers [109-113]. SiC wires with periodic twins have recently been obtained by Wang et al. [113]. Such periodic structures exhibit a bandgap modulation, rather than an average value of the potential energy [111]. Twin defects can be treated as antiphase boundaries. Superplasticity of SiC nanowires and their fracture toughness are also defined by the planar defects [114]. Generation and propagation of dislocations leading to amorphization is limited to the pure cubic phases of β-SiC. Basically nanocrystallites, free of planar defects, are responsible for these nonlinear elongations. However, segments of nanowires with high density of planar defects participate in deformations only through elastic processes. Thus, quantitative analysis of defects in nanowires may predict their mechanical behavior.

The mechanism of the reaction leading to SiC formation remains controversial. Because in the literature dominate descriptions of VLS reactions involving Si-O ions and C-O molecules these reactions have not been discussed in this article. Instead we focused on a less frequently used vapor to solid model of the reaction involving Stone-Welles defects in carbon nanotubes. Although we have not proof that this mechanism is actually responsible for SiC nanowires synthesis it provides an alternative route worth further exploration.

We developed a highly reproducible method of producing SiC nanowires on a large scale. Distribution of diameters was narrow and the average size of SiC nanowires was 30 nm. They were coved with an amorphous layer. Composition of that layer was established to be mainly amorphous silicon carbide with a significant concentration of carbon and small population of silica, less than 1%. Below that amorphous layer there is a slightly compressed shell and the core of interatomic distances identical to those found in bulk SiC.

ACKNOWLEDGMENTS

The authors thank Polite Stewart for his assistance in FTIR study, and Dr. B.Palosz, Unipress, Warsaw, Poland for recording x-ray diffraction pattern of SiC nanowires.

REFERENCES

[1] Choyke, W. J.; Matsunami, H.; Pensl, G. *Silicon Carbide: Recent Major Advances*; Advanced Texts in Physics; Springer-Verlag: Berlin Heidelberg, 2004; Vol. 79, pp 1-4.
[2] Harris, G. L. in *Properties of Silicon Carbide*; Harris, G. L. Ed.; Datareviews Series; The Institution of Electrical Engineers: London, UK, 1995; Vol. 13, pp.12-133.
[3] Cui, Y.; Lieber, C. M. *Science*. 2001, 291, 851-853.
[4] Rurali, R. *Phys. Rev. B*. 2005, 71, 205405.
[5] Lieber, C.M. *Nano Lett*. 2002, 2, 81-82.
[6] Dai, H.; Wong, E. W.; Lu, Y. Z.; Fan, S; Lieber, C. M. *Nature* 1995, 375, 769-772.
[7] Seong, H. K.; Choi, H. J.; Lee, S. K.; Lee, J. I.; Choi, D. J. *Appl. Phys. Lett*. 2004, 85, 1256 (3 pages).
[8] Yan, B.; Zhou, G.; Duan, W.; Wu, J.; Gu, B. *Appl. Phys. Lett*. 2006, 89, 023104.
[9] Wong, E. W.; Sheehan, P. E.; Lieber, C. M.; *Science* 1997, 277, 1971-1975.
[10] Rich, R.; Stelmakh, S.; Patyk, J.; Wieligor, M.; Zerda, T. W. *J. Mater. Sci*. 2009, 44, 3010-3013.
[11] Zhao, Y.; Qian, J.; Daemen, L. L.; Pantea, C.; Zhang, J.; Voronin, G. A.; Zerda, T. W.; *Appl. Phys. Lett*. 2004, 84, 1356-1358.
[12] Yang, W.; Araki, H.; Tang, C. C.; Thaveethavorn, S.; Kohyama, A.; Suzuki, H.; Noda. T. *Adv. Mater*. 2005, 17, 1519-1523.
[13] Wang, Y.; Voronin, G. A.; Zerda, T. W.; Winiarski, A. *J. Phys.: Condens. Matter*. 2006, 18, 275-282.
[14] Hao, Y. J.; Wagner, J. B.; Su, D. S.; Jin, G. Q.; Guo, X. Y. *Nanotechnology* 2006, 17, 367-375.
[15] Sun, J.; Li, J.; Sun, G.; Zhang, B.; Zhang, S.; Zhai, H. *Ceram. Int*. 2002, 28, 741-745.
[16] Lai, H. L.; Wong, N. B.; Zhou, X.T.; Peng, H. Y.; Au, F.C.K.; Wang, N.; Bello, I.; Lee, C. S.; Lee, S. T.; Duran, X. F. *Appl. Phys. Lett*. 2000, 76, 294-296.
[17] Bootsma, G. A.; Knippengerg, N. F.; Verspui, G. *J. Crystal Growth*. 1971, 11, 3, 297-309.
[18] Wieligor, M.; Wang, Y.; Zerda, T. W. *J. Phys.: Condens. Matter*. 2005, 17, 15, 2387-2395.
[19] Wang, Y. *Silicon Carbide Nanowires and Composites Obtained from Carbon Nanotubes*. Ph.D. Thesis, Texas Christian University, 2006.
[20] Wallis, K.L. *Production and Characterization of Nanostructured Silicon Carbide*. Ph.D. Thesis, Texas Christian University, 2008.

[21] Cimalla, V.; Foerster, C.; Cengher, D.; Tonisch, K.; Ambacher, O. *Phys. Stat. Sol.* 2006, 243, 7, 1476-1480.

[22] Powell, J. A.; Pirouz, P.; Choyke, W. J. in *Semiconductor Interfaces, Microstructures, and Devices: Properties and Applications,* Z. C. Feng, Ed. Bristol, United Kingdom: Institute of Physics Publishing, 1993, pp. 257-293.

[23] Cao, G. *Nanostructures and Nanomaterials*. Imperial College Press: London, 2008; pp 110-168.

[24] Han, W; Fan, S.; Li, Q.; Liang, W.; Gu, B.; Yu. D. *Chem. Phys. Lett.* 1997, 265, 374-378.

[25] Wang, Y.; Zerda, T. W. *J. Phys. Condens. Mater*. 2006, 18, 2995.

[26] Seeger, T.; Redlich, P. K.; Ruhle, M. *Adv. Mater*. 2000, 12, 279-282.

[27] Veprek S, in *Handbook of Ceramic Hard Materials*; Riedel, R. Ed.; Wiley-VCH: Weinheim, Germany, 2000, pp 104-139.

[28] *Handbook of Nanophase and Nanostructured Materials*; Wang, Z. L.; Liu, Y.; Zhang, Z.; Ed.; New York, NY, 2002; Vol. 2. pp 37-41.

[29] Shechtman, D.; Feldman, A.; Vaudin, M. D.; Hutchison, J. L. *Appl. Phys. Lett.* 1993, 62, 487-489.

[30] Daulton, T. L.; Bernatowicz, T. J.; Lewis, R. S.; Messenger, S.; Stadermann, F. J.; Amari, S. *Geochimica et Cosmochimica Acta* 2003, 67, 4743-4767.

[31] Wang, D. H.; Xu, D.; Wang, Q.; Hao, Y. J.; Jin, G. Q.; Guo, X. Y.; Tu, K. N. *Nanotechnology* 2008, 19, 215602 (1-7).

[32] Hyde, B.; Espinosa, H. D.; Farkas, D. *JOM-J. Min. Met. Mater. Soc*. 2005, 57.

[33] Wang, L.; Wada, H,; Allard, L. F. *J. Mater. Res*. 1992, 7, 148-163.

[34] Zhang, Y.; Shim, H. W.; Huang, H. *Appl. Phys. Lett*. 2008, 92, 261908 (1-2).

[35] Davidson, F. M.; Lee, D. C.; Fanfair, D. D.; Korgel. B. A. *J. Phys. Chem. C* 2007, 111, 2929-2935.

[36] Zhang, Y.; Wang, N.; Gao, S.; He, R.; Miao, S.; Liu, J.; Zhu, J.; Zhang, X. *Chem. Mater*. 2002, 14, 3564-3568.

[37] Niu, J. J.; Wang, J. N. *J. Phys. Chem. B* 2007. 111, 4368-4373.

[38] Roy, S.; Portail, M.; Chassagne, T.; Chauveau, J. M.; Vennegues, P.; Zielinski, M. *Appl. Phys. Lett.* 2009, 95, 081903 (1-3).

[39] Yu, Q.; Shan, Z. W.; Li, J.; Huang, H.; Xiao, L.; Sun, J.; Ma, E. *Nature*. 2010, 463, 335- 338.

[40] Kraft, O. *Nature Mater*. 2010, 9, 295-296.

[41] Shim, H. W.; Huang, H.; *Appl. Phys. Lett.* 2007, 90, 083106.

[42] Warren, B. E. *X-Ray Diffraction*. Dover: New York, NY, 1990; p 253.

[43] Williamson, G. K.; Hall, W. H. *Acta Metall.* 1953, 1, 22-31.

[44] Ungar, T.; Tichy, G. *Phys. Stat. Sol. A*. 1999, 171, 425-434.

[45] Langford, J. I.; Louer, D. *Rep. Prog. Phys*. 1996, 59, 131-234.

[46] Warren, B.E.; Averbach, B. L. *J. Appl. Phys*. 1950, 21, 595-610.

[47] Ribarik, G.; Gubicza, J.; Ungar, T. *Mat. Sci. Eng. A*. 2004, 343, 387-389.

[48] Balogh, L.; Ribarik, G.; Ungar, T. *J. Appl. Phys*. 2006, 100, 023512 (1-10).

[49] Guiner, A. *X-Ray Diffraction*; Freeman: San Fransisco, CA, 1963; pp 1-378.

[50] Bertaut, E. F. *Acta Cryst*. 1950, 3, 9-14.

[51] Ungar, T.; Gubicza, J.; Ribarik, G.; Pantea, C.; Zerda, T. W. *Carbon* 2002, 40, 929-937.

[52] Krivoglaz, M.A. *Theory of X-ray and Thermal Neutron Scattering by Real Crystals*; Plenum: New York, NY, 1969; pp 95-122.

[53] Wilkens, M. in *Fundamental Aspects of Dislocation Theory*; Simmons, J.A.; de Wit, R.; Bullough, R.; Ed.; NBS Spec. Publ.: Washington, DC, 1970; Vol. 2, 317.

[54] Treacy, M. M. J.; Newsam, J. M.; Deem, M.W.; *Proc. Roy. Soc. Lond. A.* 1991, **433,** 499-520.

[55] Balogh, L.; Nauyoks, S.; Zerda, T.W.; Pantea, C.; Stelmakh, S.; Palosz, B.; Ungar, T. *Mater. Sci. Eng. A.* 2008, 487, 180-188.

[56] Nauyoks, S. E. *Microstructure of Nano and Micron Size Diamond-SiC Composites Sintered Under High Pressure High Temperature Conditions*, Ph.D. Thesis, Texas Christian University, 2009.

[57] Nauyoks, S.; Wieligor, M.; Zerda, T.W.; Balogh, L.; Ungar, T.; Stephens, P. *Composites A Appl. Sci. Manufact.* 2009, 40, 566-572.

[58] Voronin, G.A.; Zerda, T.W.; Qian, J.; Zhao, Y.; He, D.; Dub, S.N. *Diamond Relat. Mater.* 2003, 12, 1477-1481.

[59] Gubicza, J.; Ungar, T.; Wang, Y.; Voronin, G.A.; Pantea, C.; Zerda, T.W. *Diamond Relat. Mater.* 2006, 15, 1452-1456.

[60] Pantea, C.; Gubicza, J.; Ungar, T.; Voronin, G.A.; Zerda, T.W. *Phys. Rev. B* 2002, 66, 094106.

[61] Gubicza, J.; Nauyoks, S.; Balogh, L.; Labar, J.; Zerda, T.W.; Ungar, T. *J. Mater. Res.* 2007, 22, 1314-1321.

[62] Palosz, B.; Stelmakh, S.; Grzanka, E.; Gierlotka, S.; Palosz, W. *Z. Kristallographie*, 2007, 222, 580-594.

[63] Stelmakh, S.; Grzanka, E.; Wojdyr, M.; Proffen, T.; Vogel, S.C.; Zerda, T.W.; Palosz, W.; Palosz, B. *Z. Kristallographie*, 2007, 222, 174-187.

[64] Palosz, B.; Stelmakh, S.; Grzanka, E.; Gierlotka, S.; Nauyoks, S.; Zerda, T.W.; Palosz, W. *J. Appl. Phys.*, 2007, 102, 074303.

[65] Stelmakh, S.; Grzanka, E.; Zhao, Y.; Palosz, W.; Palosz, B. *Z. Kristallographie (Supplement)*, 2006, 23, 331-336.

[66] Palosz, B.; Grzanka, E.; Gierlotka, S.; Stel'makh, S.; Pielaszek, R.; Lojkowski, W.; Bismayer, U.; Neuefeind,J.; Weber, H.P.; Palosz, W. *Phase Transitions* 2003, 76, 171-185.

[67] Palosz, B.; Grzanka, E.; Gierlotka, S.; Stel'makh, S.; Pielaszek, R.; Bismayer, U.; Neuefeind,J.; Weber, H.P.; Palosz, W. *Acta Phys. Pol. A* 2002, 102, 57-82.

[68] Palosz, B.; Grzanka, E.; Gierlotka, S.; Wojdar, M.; Palosz, W.; Proffen, T.; Stel'makh, S. in *Proceedings IUTAM Symposium on Modeling Nanomaterials and Nanosystems;* Pyrz, R; Rauche, J.C. Eds.; IUTAM Book series 13; Springer; Aalborg, Denmark, 2009, 75-88.

[69] Rich, R.; Stelmakh, S.; Patyk, J.; Wieligor, M.; Zerda, T.W.; Guo, Q. *J. Mater. Sci.* 2009, 44, 3010-3013.

[70] Liu, H.; Hu, J.; Shu, J.; Hausermann, D.; Mao, H. Appl. Phys. Lett. 2004, 85, 1973-1975.

[71] Birch, F. *Phys. Rev.* 1947, 71,809-82.

[72] Murnaghan, F. D. *Proc. Natl. Acad. Sci.* 1944, 30, 244-247.

[73] Merie-Mejean, T.; Abdelmounim, E.; Quintard, P. *Molec. Struct.* (1995) 349, 105-108.

[74] Gogotsi, Y.G.; Nickel, K.G.; Bahloul-Hourlier, D.; Merle-Mejean, T.; Khomenko, G.E.; Skjerlie, K.P. *J. Mater. Chem*. 1996, 6, 595-604.
[75] Wagner, R. S.; Ellis, W.C. *Appl. Phys. Lett.* 1964, 4, 89-90.
[76] Wagner, R. S.; Ellis, W.C.; Jackson, K.A.; Arnold, S.M. *J. Appl. Phys*, 1964, 35, 2993-3000.
[77] Givargizov, E.I. *J. Vac. Sci. Technol.* B, 1993, 11, 449-453.
[78] Zhang, Y.F.; Gamo, M.N.; Xiao, C.Y.; Ando, T. *J. Appl. Phys*. 2002, 91, 6070.
[79] Zhou, X.T.; Wang, N,; Lai, H.L.; Peng, Y.; Bello, I.; Lee S.T. *Appl. Phys. Lett*. 1999, 74, 26, 3942–3944.
[80] Wu, Z.S.; Deng, S.Z.; Xu, N.S.; Chen, J.; Zhou, J.; Chen, J. *Appl. Phys. Lett.* 2002, 80, 3829-3831.
[81] Zhou, X.T.; Wang, N,; Au, F.C.K.; Lai, H.L.; Peng, Y.; Bello, I.; Lee, C.S.; Lee S.T. *Mater. Sci. Eng. A*, 2000, 286, 119-124.
[82] Korgel, B.A.; Hanrath, T.; Davidson, F.M. in *Encyclopedia of Chemical Processing*, Lee, S. Ed.; Taylor and Francis; New York, NY; 2006, vol. 5, pp 31-3203.
[83] Zhang, Y.; Wang, N.; Gao, S.; He, R.; Mia, S.; Liu, J.; Zhu, J.; Zhang, X. *Chem. Mater*. 2002, 14, 3564-3508.
[84] Wang, Z.L.; Liu, Y.; Zhang, Z. *Handbook of Nanophase and Nanostructured Materials, Characterization*; Springer, New York, NY, 2002.
[85] Cerneaux, S.; Xiong, X.; Simon, G.P.; Cheng, Y,; Spiccia, L. *Nanotechnology* 2007, 18, 055708.
[86] Hasegawa, T.; Nakamura, S.; Motojima, S. *J. Sol-Gel Sci. Technol.* 1997, 8, 577-579.
[87] Li, J.; Zhang, Y.; Zong, X.; Yang, K.; Meng, J.; Cao, X. *Nanotechnology* 2007, 18, 245606
[88] Ye, H.; Titchenal, N,; Gogotsi, Y.; Ko, F. *Adv. Mater.,* 2005, 17, 1531-1535.
[89] Li, Z.; Zhang, J.; Meng. A.; Guo, J. *J. Phys. Chem. B*, 2006, 110, 22382.
[90] Pan, Z.; Lai, H.L.; Au, F.C.K.; Duan, Z.; Wang, N.; Lee. C.S.; Wong, N.B.; Lee, S.T.; Xie, S.; *Adv. Mater.* 2000, 12, 16, 1186-1190.
[91] Niu, J. J.; Wang, J.N. *J. Phys. Chem. B*, 2007, 111, 4368-4373.
[92] Morales, A.M.; Lieber, C.M. *Science*, 279 (1998) 208-211.
[93] Wieligor, M.; Rich, R.; Zerda, T.W. *J. Mater. Sci.* 2010, 45, 1725-1733.
[94] Ho, G.W.; Wong, A.S.; Kang, D.J.; Welland, M.E. *Nanotechnology*, 2004, 15, 996-999.
[95] Zhang, E.; Tang, Y.; Zhang, Y.; Guo, C. *Phys. E,* 2009, 41, 655-659.
[96] Li, F.; Wen, G. *J. Mater. Sci.* 2007, 42, 4125-4130.
[97] Li. X, Chen, X.; Song, H. *J. Mater. Sci.* 2009,44,4661-4667.
[98] Mawhinney, D.; Naumenko, V.; Kuznetsova, A.; Yates, J. T.; Liu, J.; Smalley, R. E. *Chem. Phys. Lett.* 2000, 324, 213-216.
[99] Miyamoto, Y.; Rubio, A.; Berber, S.; Yoon, M.; Tomanek, D. *Phys. Rev. B* 2004, 69, 121413 (R).
[100] Ma, J.; Alfe, D.; Michaelides, A.; Wang, E. *Phys. Rev. B* 2009, 80, 033407 (4 pages).
[101] Chakrapani, N.; Zhang, Y. M.; Nayak, S. K.; Moore, J. A.; Carroll, D. L.; Choi, Y. Y.; Ajayan, P. M. *J. Phys. Chem. B* 2003, 107, 659308.
[102] Picozzi, S.; Santucci, S.; Lozzi, L.; Valentini, L.; Delley, B. *J. Chem. Phys.* 2004, 120, 7147.
[103] Cimalla, V.; Wohner, T.; Pezoldt, J.; *Mater. Sci. Forum.* 2000, 321, 338-342.
[104] Ghoshtagore, R.N.; Coble, R.L.; *Phys. Rev.* 1966, 143, 623-626.

[105] Hon, M.N.; Davis, R.F.; *J. Mater. Sci.* 1979, 14, 2411-2421.
[106] Hon, M. N.; Davis, R. F.; *J. Mater. Sci.* 1980, 15, 2073-2080.
[107] Birnie, D. P.; *J. Am. Ceram. Soc.* 1986, 69, C-33-C35.
[108] Pantea, C.; Voronin, G.A.; Zerda, T. W.; Zhang, J.; Wang, L.; Wang, Y.; Uchida, T.; Zhao, Y. *Diamond Relat. Mater.* 2005, 14, 1611-1615.
[109] Liu, J.; Liu, Z.; Ren, P.; Xu, P.; Chen, X.; Xu, X. *Acta Phys.Chim. Sinica* 2008, 24, 571-575.
[110] Fissel, A.; Schroter, B.; Kaiser, U.; Richter, W. *Appl. Phys. Lett.* 2000, **77,** 2418-2420.
[111] Fissel, A.; Kaiser, U.; Schroter, B.; Richter, W.; Bechstedt, F. *Appl. Surf. Sci.* 2001, 184, 37-42.
[112] Bechstedt, F.; Fissel, A.; Furthmüller, J.; Kaiser, U.; Weissker, H. -Ch.; Wesc, W. *Appl. Surf. Sci.* 2003, 212, 820-825.
[113] Wang, D. H.; Xu, D.; Wang, Q.; Hao, Y. J.; Jin, G. Q.; Guo, X.; Tu, K. N. *Nanotechnology* 19 (2008) 215602.
[114] Zhang, Y.; Han, X.; Zheng, K.; Zhang, Z.; Fu, J.; Ji, Y.; Hao, Y.; Guo, X.; Wang, Z. *Adv. Funct. Mater.* 2007, 17, 3435-3440.

In: Silicon Carbide: New Materials, Production ... ISBN: 978-1-61122-312-5
Editor: Sofia H. Vanger

Chapter 4

DUCTILE REGIME MATERIAL REMOVAL OF SILICON CARBIDE (SIC)

***Deepak Ravindra*[1] *and John Patten*[2]**
Manufacturing Research Center, Western Michigan University
Kalamazoo, Michigan, USA

ABSTRACT

Advanced ceramics such as silicon carbide (SiC) are increasingly being used for industrial applications. These ceramics are hard, strong, inert, and light weight. This combination of properties makes them ideal candidates for tribological, semiconductor and MEMS, and optoelectronic applications respectively. Manufacturing SiC without causing surface or subsurface damage is extremely challenging due to its high hardness, brittle characteristics and poor machinability. Often times, severe fracture can result when trying to achieve high material removal rates during machining of SiC due to its low fracture toughness. One of the most critical steps before machining SIC is to establish the Ductile to Brittle Transition (DBT) depth, also known as the critical depth of cut (DOC). The DBT depth is determined via scratch testing either with a diamond stylus or a single crystal diamond cutting tool. In this chapter, a study on the DBT of a single crystal SiC wafer is discussed. This study also demonstrates that ductile regime Single Point Diamond Turning (SPDT) is possible on SiC to improve its surface quality without imparting any form of surface or sub surface damage. Machining parameters such as depth of cut and feed used to carry out ductile regime machining are discussed. Sub-surface damage analysis was carried out on the machined sample using non-destructive methods such as Optical Microscopy, Raman Spectroscopy and Scanning Acoustic Microscopy to provide evidence that this material removal method (i.e., SPDT) leaves a damage-free surface and subsurface. Optical microscopy was used to image the improvements in surface finish whereas Raman spectroscopy and scanning acoustic microscopy was used to observe the formation of amorphous layer and sub-surface imaging in the machined region respectively. All three techniques complement the initial hypothesis of being able to remove a nominally brittle material in the ductile regime.

[1] deepak.ravindra@wmich.edu.
[2] john.patten@wmich.edu.

INTRODUCTION

Although silicon carbide (SiC) has been around since 1891, it was not until the mid 1990's that this material was introduced into the precision manufacturing industry. SiC is well known for its excellent material properties, high durability, high wear resistance, light weight and extreme hardness. However, SiC is also well known for its low fracture toughness, extreme brittleness and poor machinability. SiC is one of the advanced engineered ceramics designed to operate in extreme environments. This material is pursued as both a coating and structural material due to its unique properties, such as:

- Larger energy bandgap and breakdown field allowing it to be used in high-temperature, high-power and radiation-hard environments
- Mechanical stiffness, expressed by its high Young's modulus [1]
- Desirable tribological properties, such as wear resistance and self-lubricating [2]

SiC is commercially available in various forms/phases (polytypes) such as single crystal, polycrystalline (sintered and CVD) and amorphous. The most common polytypes of SiC are 2H, 3C, 4H, 6H, and 15R. The numbers refer to the number of layers in the unit cell and the letter designates the crystal structure, where C=cubic, H=hexagonal, and R=rhombohedral. In this study, only two polytypes will be discussed: 4H and 3C. The 4H polytype is a single crystal whereas the 3C CVD has a polycrystalline beta structure.

DUCTILE REGIME MACHINING

Materials that are hard and brittle, such as semiconductors, ceramics and glasses, are amongst the most challenging to machine. When attempting to machine ceramics, such as silicon carbide (SiC), especially to improve the surface finish, it is important to carry out a 'damage free' machining operation. This often can be achieved by ductile mode machining (DMM) or in other words machining a particular hard and brittle material in the ductile regime. Material removal processes can be considered in terms of fracture dominated mechanisms or localized plastic deformation. A fracture dominant mechanism for ceramics, i.e., brittle fracture, can result in poor surface finish (surface damage) and also compromises on material properties and performance [3].

The insight into the origins of the ductile regime during single point diamond turning (SPDT) of semiconductors and ceramics was provided by the research done by Morris, et al [4]. A detailed study of machining chips (debris) and the resultant surface was studied (analyzed using a TEM) to evaluate evidence of plastic material deformation. This seminal research concluded that the machining chips were plastically formed and are amorphous (not due to oxidation) due to the back transformation of a pressure induced phase transformation, and the machining debris (chips) contain small amounts of micro-crystalline (brittle) fragments.

According to the grinding research carried out by Bifano et al. [5], there are two types of material removal mechanisms associated with the machining process: ductile; plastic flow of material in the form of severely sheared machining chips, and brittle; material removal through crack propagation. This research discusses several physical parameters that influence the ductile to brittle transition in grinding of brittle materials. The researchers were successful in performing ductile mode grinding on brittle materials. However, these researchers did not propose or confirm a model or suitable explanation for the origin of this ductile regime. Bifano et al. also proposed a model defining the ductile to brittle transition of a brittle material based on the material's brittle fracture properties and characteristics. A critical depth of cut model was introduced based on the Griffith fracture propagation criteria. The critical depth of cut (d_c) formula is as follows:

$$d_c = (E \cdot R) / H^2 \quad (1)$$

where E is the elastic modulus, H is the hardness and R is the fracture energy.

The value of the fracture energy (R) can be evaluated using the relation:

$$R \sim K_c^2 / H \quad (2)$$

where K_c is the fracture toughness of the material. The above two equations can be combined to represent the critical depth (d_c) as a measure of the brittle transition depth of cut:

$$d_c \sim (E / H) \cdot (K_c / H)^2 \quad (3)$$

The researchers were successful in determining a correlation between the calculated critical depth of cut and the measured depth (grinding infeed rate). The estimated constant of proportionality was estimated as to be 0.15 and this is now added into Equation (2.3) to generate a more accurate empirical equation:

$$d_c \sim 0.15 \cdot (E / H) \cdot (K_c / H)^2 \quad (4)$$

Chip Formation

A critical depth, d_c is determined before any ductile mode machining operation is carried out. Any depth beyond or exceeding the critical depth, which is also known as the Ductile to Brittle Transition (DBT) depth, will result in a brittle cut. Since the equipment used in the current study (Universal Micro-Tribometer by CETR) is a load controlled (and not a depth controlled) machine, thrust force calculations were carried out for corresponding required depths of cuts. The Blake and Scattergood [6] ductile regime machining model (as shown in *Figure 1*) was used to predict the required thrust force for a desired depth of cut. In this model it is assumed that the undesirable fracture damage (which extends below the final cut surface) will originate at the critical chip thickness (t_c), and will propagate to a depth, y_c. This assumption is consistent with the energy balance theory between the strain energy and surface energy [5].

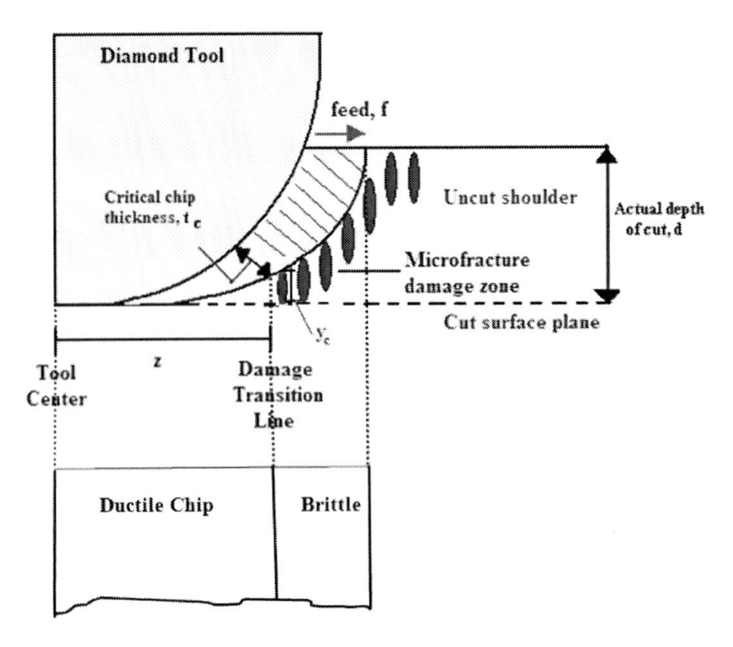

Figure 1. Model for ductile regime machining.

In general, the ductile-to-brittle transition (DBT) is a function of variables such as tool geometry (rake and clearance angle, nose and cutting edge radius), feed rate, and depth of cut.

HIGH PRESSURE PHASE TRANSFORMATION (HPPT)

Although SiC is naturally very brittle, micro/nanomachining this material is possible if sufficient compressive stress is generated to cause a ductile mode behavior, in which the material is removed by plastic deformation instead of brittle fracture. This micro-scale phenomenon is also related to the High Pressure Phase Transformation (HPPT) or direct amorphization of the material [7]. *Figure 2* shows a graphical representation of the highly stressed (hydrostatic and shear) zone that result in ductile regime machining.

Patten and Gao [7] state that ceramics in general undergo a phase transformation to an amorphous phase *after* a machining process. This transformation is a result of the High Pressure Phase Transformation (HPPT) that occurs when the high pressure and shear caused by the tool is suddenly released after a machining process. This phase transformation is usually characterized by the amorphous remnant that is present on the workpiece surface and within the chip. This amorphous remnant is a result of a back transformation from the high pressure phase to the atmospheric pressure phase due rapid release of pressure in the wake of the tool. There are two types of material removal mechanisms: ductile mechanism and the brittle mechanism [5]. In the ductile mechanism, plastic flow of material in the form of severely sheared machining chips occur, while material removal is achieved by the intersection and propagation of cracks in the brittle fracture mechanism. Due to the presence of these two competing mechanisms, it is important to know the DBT depths (or critical size) associated with these materials before attempting a machining operation.

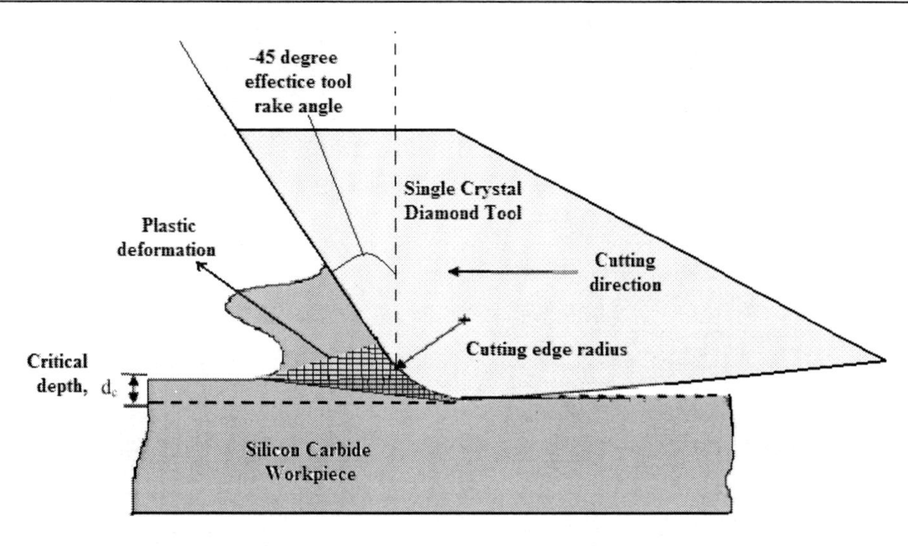

Figure 2. A ductile machining model of brittle materials.

Figure 2 shows a ductile cutting model showing the high compressive stress and plastically deformed material behavior in brittle materials. A -45° rake angle tool is demonstrated in the above schematic as a negative rake angle tool yields in higher compressive stresses at the tool-workpiece interface.

Ductile to Brittle Transition (DBT) of a Single-Crystal 4H-SiC Wafer

Introduction

Silicon carbide, like other brittle materials, is known for its poor machinability. However, ductile-regime machining is possible under certain conditions. This can be achieved if machining occurs at depths less than the critical depth of cut. Beyond this value, a Ductile-to-Brittle Transition (DBT) occurs and the material behaves in a brittle-fracture manner. One of the initial steps done before carrying out any form of ductile mode machining on brittle materials is to define a DBT depth. Once the DBT is defined for a particular material, all machining parameters such as depth of cut and feed can be altered in order not to exceed the DBT depth. Exceeding the DBT/critical depth during a machining process will result in material fracture which then generates poor surface finish and at times catastrophic failure of the material/part. The purpose of this study is to determine the DBT depth for a single-crystal 4H-SiC wafer by performing Nanometric cutting (Nanocut) experiments. The depth of cut is adjusted over a range of 100nm to1000nm in order to cover the entire ductile to brittle-regime and the corresponding material removal behavior. The Nanocuts were carried out using the Nanocut II, a second generation prototype experimental machining instrument. The Nanocuts were imaged and measured using an Atomic Force Microscope (AFM) and the height profile from the scanned images were used to determine the nature of the cut (i.e., ductile or brittle) and the DBT depth [3].

In this study, nanometric cutting experiments were performed to determine the DBT depth for a single crystal 4H-SiC (from Cree Inc.). A polished single crystal wafer was used in this study, which provides for an excellent reference surface for determining the DBT and to establish the corresponding critical depth. These wafers are also of high purity with few defects, which will help in predicting the actual brittleness of the material [8]. The 4H single crystal SiC wafer is typically used for:

- Optoelectronic Devices
- High-Power Devices
- High Temperature Devices
- High-Frequency Power Devices
- III-V Nitride Deposition

Experimental Details

The Nanocut II (a second-generation prototype) was used to carry out the nanometer level cuts. The Nanocut II was designed to perform nanometer depth cuts based on the commanded depth by the operator as executed by the control program. The main components of this equipment are the frame, PZT tube (provides x, y and z displacement), capacitance gage (displacement feedback), force sensors, sample holder, tool holder, and hysteretic positioners. The PZT tube is used to position the sample relative to the tool, to establish the depth of cut. The Z (depth of cut) position is determined via the capacitance gage, and the two orthogonally placed force sensors measure the cutting and thrust forces. There are four dual axis flexures used in the device to decouple the cutting and thrust forces and to support the tool stage positioning wedge actuators [9]. *Figure 3* shows a top view of the Nanocut II.

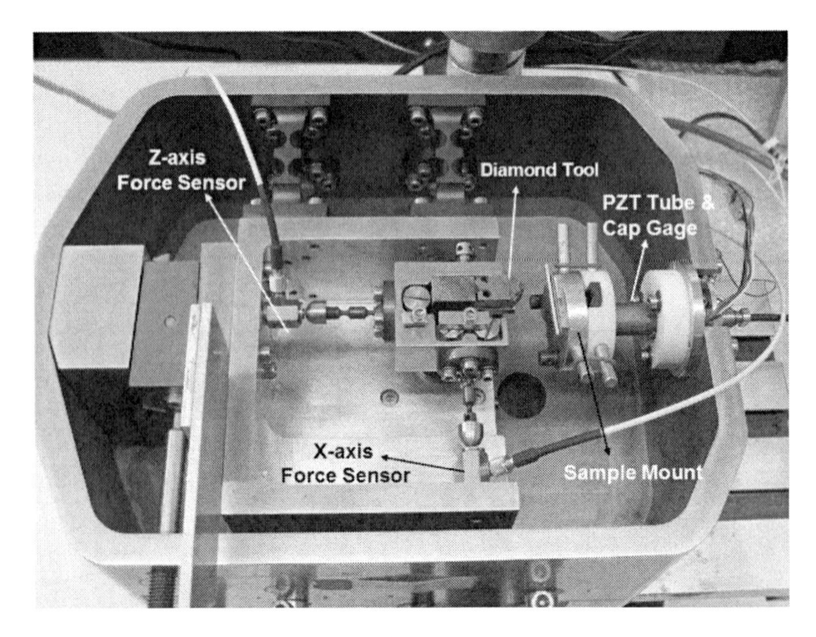

Figure 3. A top view of the Nanocut II used to perform the Nanocuts.

A series of cuts were performed on rectangular pieces, which were cleaved from a 3" (76.2mm) 4H-SiC wafer. The wafer was cleaved relative to its primary flat which is in $\{10\bar{1}0\}$ plane with the flat face parallel to the $<11\bar{2}0>$ direction. The rectangular sample size obtained from the wafer was about 12mm by 6mm. Identifying the crystal orientation is important to determine the preferred cutting direction. The SiC sample was mounted on the sample holder with an adhesive. To minimize any external vibration, noise and distortion, the Nanocut II was placed on an air vibration isolation table. A single-point diamond tool with a rake angle of -45 degrees and a clearance angle of 5 degrees, and 10 mm nose radius (cutting edge radius is estimated to be 50 nm) was used to perform the cuts. For the results reported here, the diamond tool was oriented such that the Nanocuts were performed parallel to the primary flat (in the $<11\bar{2}0>$ direction).

Two sets of experiments were carried out on different rectangular samples, taken from the same wafer and positioned at the same crystallographic orientation. The first set contained cuts with commanded depths of 100nm and 500nm. The second set contained cuts with commanded depth of 1000nm. The expected DBT was in the range of 100nm to 1000nm, and these experimental cut depths covered this entire region. Three cuts were done for each commanded depth to obtain comparable data. The second set of deeper cuts (1000nm and greater if necessary) would only be carried out if there were no brittle characteristics in the first set of cuts (100nm and 500nm). The cuts were done in a particular array pattern to help with imaging. Since the Nanocuts in this experiment are fairly small (approximately 20μm in length and 120μm in width), identifying them under the microscope can be challenging. *Figure 4* shows a schematic representing the pattern created with the Nanocut II on the rectangular SiC samples; the 100ñm and 500nm cuts were performed on one sample and the 1000nm cuts were performed on a second sample. *Figure 5* represents an actual image of three of the cuts (two at 100nm and one at 500nm) obtained from an optical microscope at 400X magnification. The other 100nm and 500nm cuts are outside the field of view at the magnification shown.

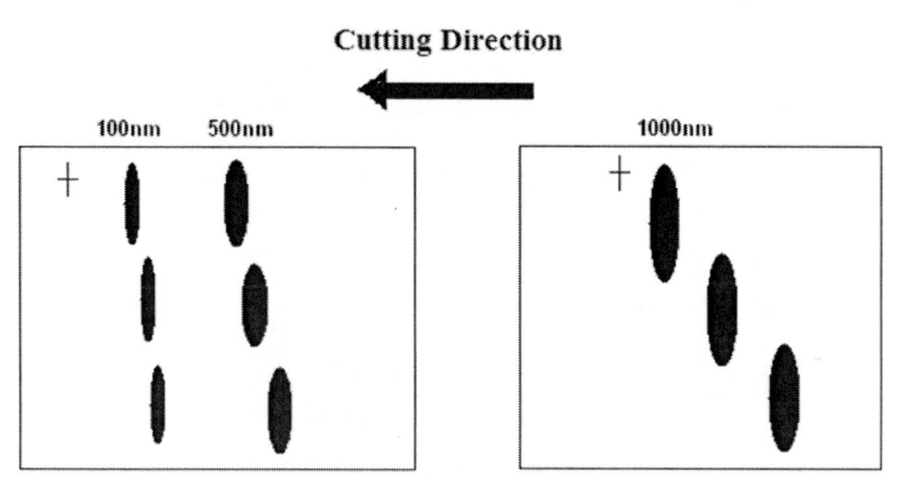

Figure 4. Nanocut matrix of cuts (100nm, 500nm and 1000nm).

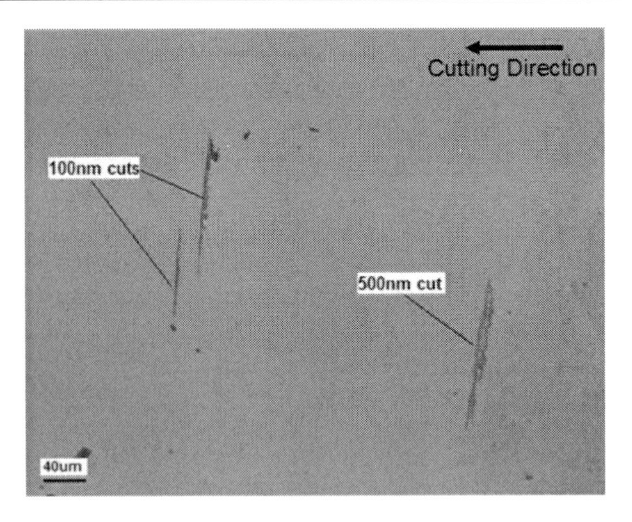

Figure 5.100nm and 500nm cuts made by the Nanocut II.

The cuts seen in *Figure 5* are made from right to left. The cuts are wider than they are long due to the geometry of the tool (10mm nose radius) and the maximum stroke of the PZT in the cutting direction (20-30 μm), which establishes the length of cut.

Results and Discussion

Figure 6 shows the thrust force and the cutting force for each corresponding depth of cut. The cutting forces are measured to be more than the thrust forces for all of the three reported depths of cuts. Both the cutting forces and the thrust forces increase as the depth of cut increases. The 50nm depth of cut is not reported in *Figure 6* as only one cut at that depth was successful and it could not be imaged. Each of the other reported depths had three successful/ repeatable cuts, for which the force data were averaged.

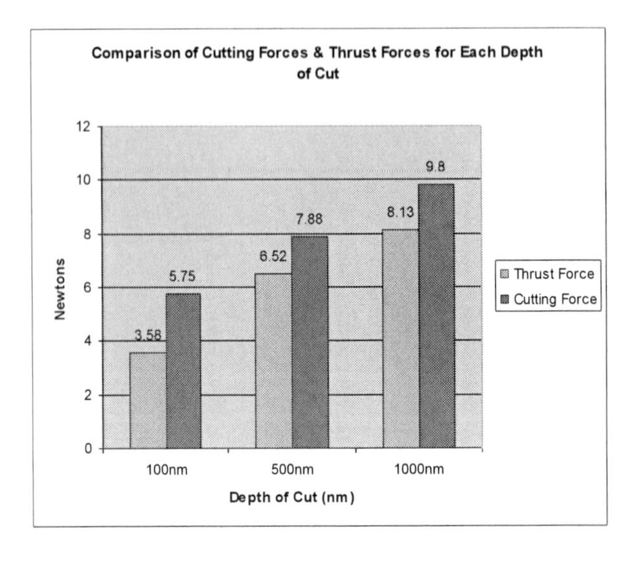

Figure 6. Cutting forces and thrust forces for three different depths of cuts.

A typical height profile obtained from the ductile regime is generally "V" or "U" shaped representing an imprint of the diamond tool cutting edge. *Figure 7* shows a typical height profile, obtained from an AFM, of a cut performed in the ductile region. The depth of this cut was measured to be 816nm and the programmed depth of cut was 1000nm. The actual depth of cut was measured to be less than the commanded or programmed depth of cut; this is due to the elasticity of the material and compliance of the instrument, and is consistent with previous results [9], [10]. The "V" shape seen in the cross-sectional profile of *Figure 7* is a typical characteristic of a fully ductile cut.

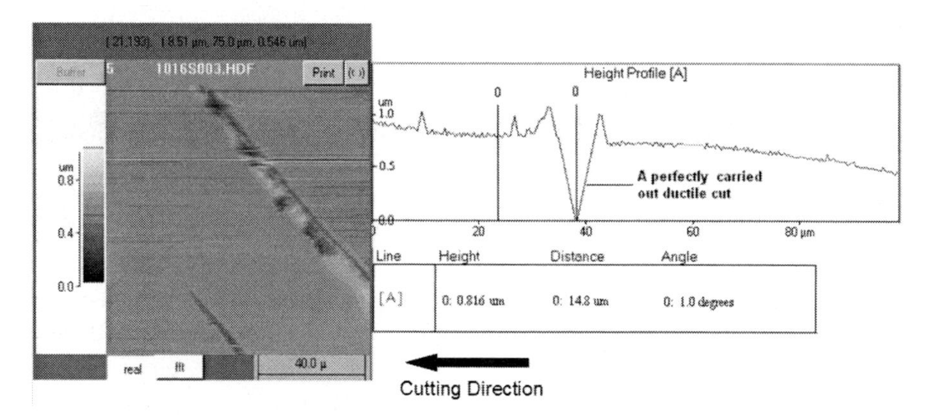

Figure 7. An AFM scanned section of a cut along with its height profile. The programmed or commanded depth of this cut was 1000nm, at the cross-section indicated the measured depth is 816nm.

Figure 8 shows a height profile of a cut where there is an indication of a DBT. This is evident by the poorly defined tool imprint seen in the cross-sectional profile. The depth of cut at this point was measured to be 836nm using an AFM and the programmed depth of cut is 1000nm.

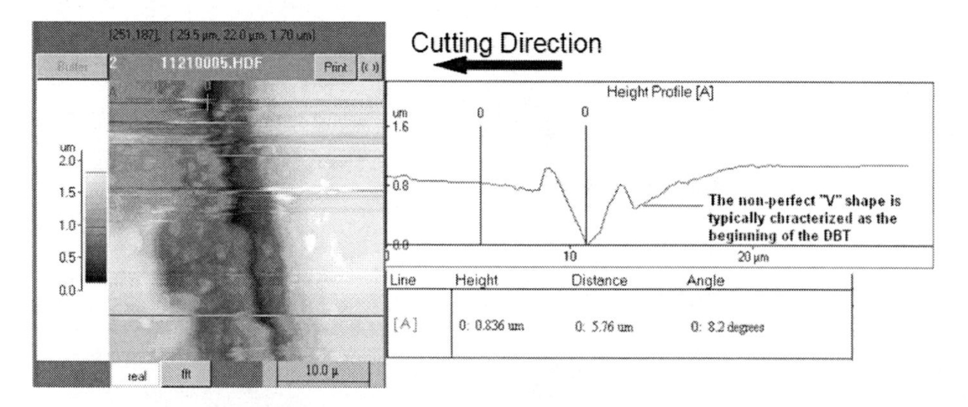

Figure 8. An AFM scanned section of a cut where the brittle characteristic of the material is visible.

Figure 9 confirms that the material removal is in the brittle-regime and there is no clear definition in the cross-sectional profile of this cut. The measured maximum depth of cut is 952nm and the programmed depth of cut of 1000nm. It can be seen in *Figure 9* that there could be more than one peak/valley in the brittle region as the fracture process results in uncontrolled material removal.

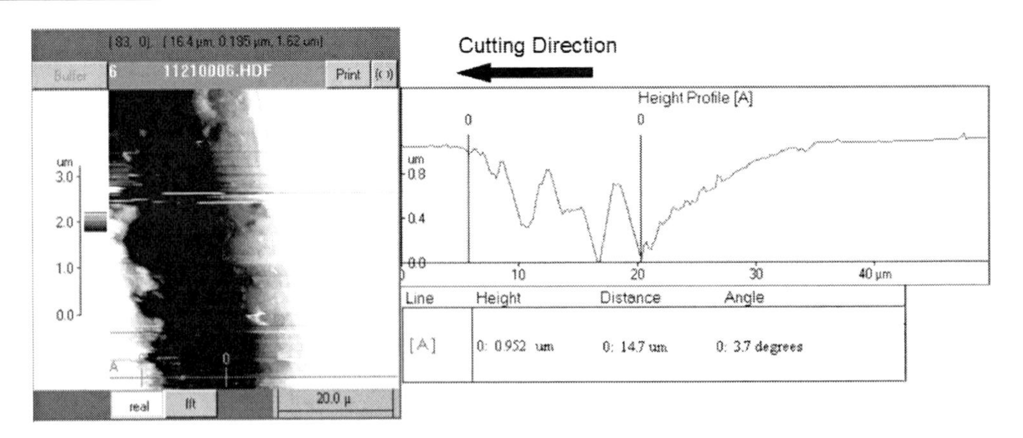

Figure 9. An AFM scanned section of a cut where the brittle characteristic of the material is observed.

The edges along the cuts can also be observed to determine the nature of the cut. A ductile cut has a much more defined and straight edge as seen in *Figure 7* compared to the jagged edges seen in a brittle cut shown in *Figures 8 and 9.* A clearer picture of the uneven edges along a brittle cut is shown in *Figure 10.* The jagged edges and chipped material along the left edge of the cut are caused due to crack propagation, and uncontrolled material removal in the brittle regime. This brittle material is clearly seen in *Figure 11,* where the cross section of a brittle cut is analyzed using the Wyko RST interferometric microscope. The actual depth varies from zero at the ends (top and bottom of the cuts, outside the field of view) to a maximum in the middle. The cutting direction for the cut below is from right to left as indicated.

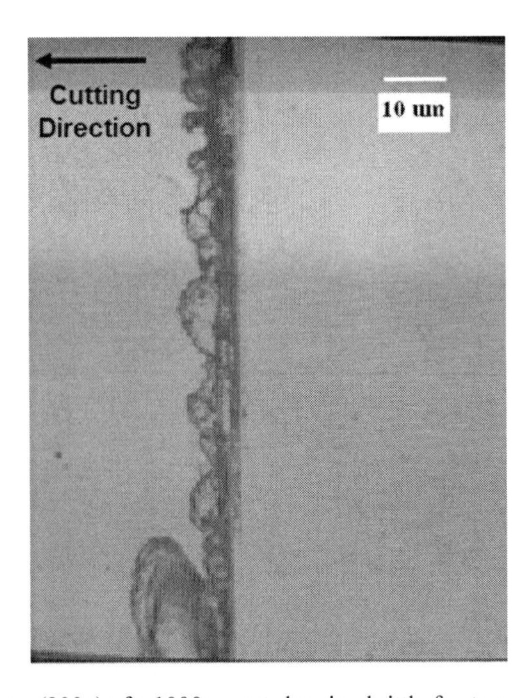

Figure 10. An optical image (200x) of a 1000nm cut showing brittle fracture.

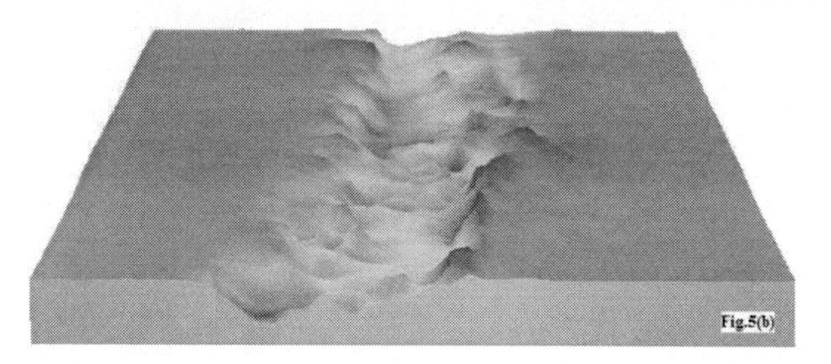

Figure 11. White light interferometric microscope image of 1000nm (programmed depth) brittle cut.

As seen in *Figure 12*, the brittle characteristics are clearly shown by the height profile graph from an AFM image. The programmed depth for this particular cut is 1000nm (1μm) but the maximum depth measured in this cut is 1.17μm. This characteristic (programmed depth < actual depth) was also observed by Hung and Fu in their experiment to study the ductile-regime machining of silicon, where the micro-cracks could extend deeper than the depth of cut below the machined surface [11].

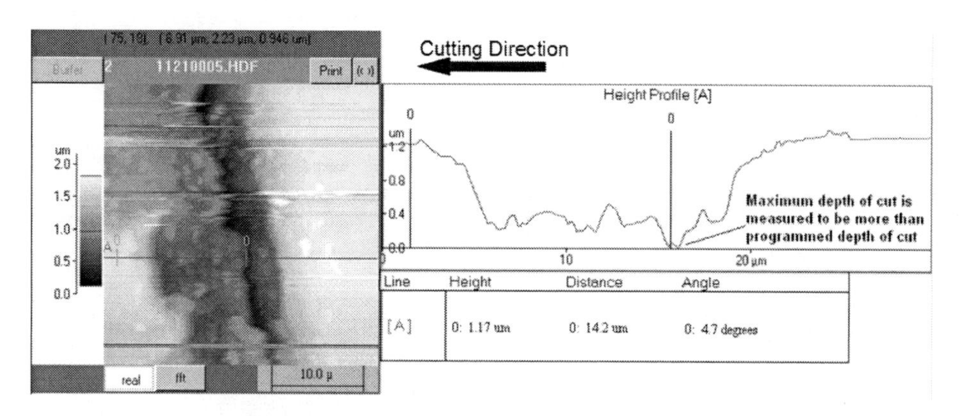

Figure 12. AFM scanned section of a purely brittle cut identified by the cross-sectional profile.

Purely brittle characteristics are observed where the maximum measured depth of cut (1170nm) is more than the commanded/programmed depth of cut (1000nm). In the case of a cut under extreme brittle conditions of SiC, there is no direct control of resultant material removal; i.e. the command depth does not provide a one-to-one correspondence to the actual depth of cut. It is this stage (beyond the DBT) where crack initiation, propagation and growth occur. These events can lead to catastrophic brittle failure of the material.

Summary

Although SiC is well known for its brittle characteristics, it is still possible to plastically deform this material at small scales (nominally less than a micrometer) to achieve ductile machining [7], [12], [13]. In order to machine a semiconductor or ceramic in the ductile-regime, it is crucial to know its DBT depth. The DBT depth was found to be between 820nm-830nm as measured on a single crystal 4H-SiC wafer, cut parallel to the wafer's primary flat orientation i.e. the $\{10\bar{1}0\}$ plane. Beyond this depth, or at greater depths, the cut produced became brittle. The DBT is an important parameter as it defines the border between where the material fails catastrophically by fracture from that at which it yields by plastic deformation [14]. The cutting forces and thrust forces increase as the depth of cut increases as expected. In this particular study, the force data did not specifically contribute to the determination of the DBT. However, the force data can often be used to indicate the nature of the cut (i.e., ductile or brittle) [15]. At the nano/micro scale, the cutting forces are extremely sensitive to the nature of the cut where instant instability in force patterns are observed at the DBT of the material. The fracture characteristics observed beyond the depth of 830nm are a result of brittle machining conditions [11]. This fracture then leads to pitting and micro-cracks resulting in significant and uncontrolled subsurface damage [16].

DUCTILE REGIME SINGLE POINT DIAMOND TURNING OF CVD-SIC

Introduction

The main goal of this study is to improve the surface quality of a chemically vapor deposited (CVD) polycrystalline SiC, to be used as optics devices (mirrors), via single point diamond turning (SPDT). In order to fulfill the requirements to be qualified as a good candidate for optical mirrors, the surface finish for this material has to be close to mirror finish (Ra < 20nm). This is not easy to achieve from an as received CVD coated SiC (coating approximately 250μm thick) work piece that has a fairly rough surface (in some cases the as received SiC disk (SuperSiC-2) [17] had a Ra value of approximately 3.6μm and a peak-to-valley of approximately 15 μm). Several machining techniques such as lapping, polishing, grinding, laser machining, and diamond turning are being used today in an attempt to micro/nano machine this brittle material to a high quality, low roughness and damage free surface.

CVD coated SiC is used in specialized industries due to its excellent mechanical properties such as high strength to weight radio, extreme hardness, high wear resistance, high thermal conductivity, optical characteristics, high electric field breakdown strength and high maximum current density [3]. The fully dense cubic (beta) polycrystalline silicon carbide (manufactured by POCO Graphite) CVD coating (≈250μm thick) is a potential candidate to be used as mirrors for surveillance, high energy lasers (such as airborne laser), laser radar systems, synchrotron x-ray, VUV telescopes, large astronomical telescopes and weather satellites [18] The primary reasons CVD coated silicon carbide is preferred for these applications is that the material possesses high purity (>99.9995%), homogeneity, density (99.9% dense), chemical and oxidation resistance, cleanability, polishability and thermal and dimensional stability. Machining silicon carbide is extremely challenging due to its extreme hardness (≈26 GPa) and brittle characteristics. Besides the low fracture toughness of the material, severe tool wear of the single crystal diamond tool also has to be considered.

The mechanics of material removal in SiC can be classified in two categories: brittle fracture and plastic deformation. Good optical quality surfaces can be achieved by removing the material in a ductile manner. The work of past researchers suggests that extremely brittle materials such as ceramics and glasses do not necessarily behave as a brittle material (even at room temperature) especially at the nanometric scale [19], [20], [21]. The strength, hardness and fracture toughness of the work piece material are the governing factors that control the extent of brittle fracture [22]. Some studies include observations of a small (nm to μm) amount of plastic deformation in brittle materials during a precision machining operation [23].

Previous researchers have successfully been able to precisely grind CVD-SiC (using high precision grinding) but this process is very expensive and the fine abrasive wheels often result in an unstable machine/process [24]. Although SiC is naturally brittle, micromachining this material is possible if sufficient compressive stress is generated to cause a ductile mode behavior, in which the material is removed by plastic deformation, instead of brittle fracture. This micro-scale phenomenon is also related to the High Pressure Phase Transformation (HPPT) or direct amorphization of the material [7] The plastic deformation or plastic flow of the material, at the nano to micro scale, occurs in the form of severely sheared` machining chips caused by highly localized contact pressure and shear stress.

Single point diamond turning (SPDT) was chosen as the material removal method as SPDT offers better accuracy, quicker fabrication time and lower cost when compared to grinding and polishing [25].Sub-surface damage analysis was carried out on the machined sample using non-destructive methods such as Raman spectroscopy and Scanning Acoustic Microscopy. Surface roughness (Ra) values of less than 100 nm (for CVD coated SiC), without sub surface damage were obtained. In addition to improving the surface roughness of the material, the research also emphasized increasing the material removal rate (MRR) and minimizing the diamond tool wear. Machining parameters that make this manufacturing process more time and cost efficient were also identified.

Experimental Details

Equipment Setup

The equipment used to carry out all of the machining experiments was the Micro-Tribometer (UMT) from the Center for Tribology Research Inc. (CETR). This equipment was developed to perform comprehensive micro-mechanical tests of coatings and materials at the micro scale. The tool fixture is mounted onto a dual axis load cell that is capable of measuring the thrust and cutting forces. When machining a fairly rough surface, like the as-received CVD SiC, a suspension is used (the tool fixture is mounted onto the suspension) for damping purposes enabling a much more uniform cut. *Figure 13* shows the equipment setup for the 6" CVD coated SiC disk. The MASTERPOLISH 2 Final Polishing Suspension (contains alumina and colloidal silica with a pH ~9) from Buehler, Inc. was used as the cutting fluid for all experiments involving diamond turning of SiC [26].

Diamond Tooling

A single crystal diamond tool with a 3mm nose radius, -45 degree rake angle and 5 degree clearance angle was used for the cutting tests. Diamond, the hardest material known to mankind, is an ideal candidate for tooling due to its ultra-hardness (≈100GPa), high resistance to wear and good thermal conductivity. This extreme hardness of diamond is necessary when attempting to cut an extremely hard and abrasive material like SiC (H≈ 26GPa). The general hardness ratio (tool/workpiece) preferred in industry is about 10 although in the case of diamond and SiC we only have a hardness ratio of about 4. This fairly low hardness ratio, between the tool and workpiece, results in tool wear which will be discussed in the later section of this chapter. The outstanding crystalline structure (tetragonal arrangement of the C-C covalent bonds) of diamond makes it possible to fabricate tool tips with very sharp cutting edges (20 to 50 nm). The ability to fabricate single crystal diamond tools with extremely smooth finish (Ra ≈ 3nm), makes it ideal for SPDT of ceramics to improve surface finish with a high degree of accuracy [27]. Synthetic single crystal diamond (approximately same properties as natural diamond) tools from Chardon Tools Inc. were used for all projects discussed in this chapter. These synthetic diamonds are usually grown at high temperature and pressure from carbon feedstock. The effect of the brittleness and propensity of diamond to chip (one mode of tool wear) is discussed in a later section. This limitation of the diamond necessitates larger included tool angle (large negative rake) for SPDT of hard materials, such as ceramics, which is also conducive for generating the requisite HPPT.

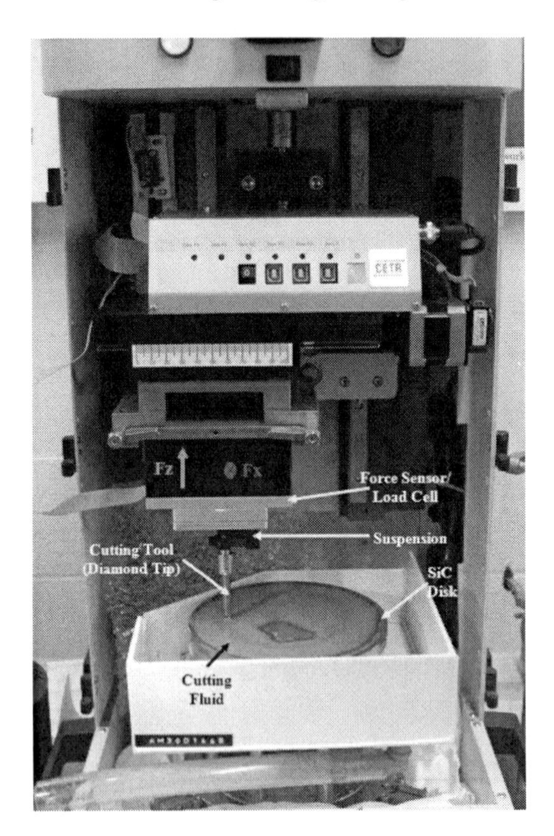

Figure 13. Single Point Diamond Turning Setup for the 6" CVD coated SiC disk.

Single Point Diamond Turning (SPDT)

A good manufacturing technique should provide both high product quality and cost efficiency. There are several manufacturing techniques such as turning, milling, polishing, lapping and honing. Traditionally, turning and milling are considered roughing/semi-finishing operations. Polishing is used to obtain the final surface finish; this operation is carried out to remove micro-cracks, scratches and voids (these damages are usually caused by previous manufacturing operations). Lapping and honing generally are carried out for obtaining form and shape accuracy such as flatness.

SPDT was a chosen as the material removal method as it offers better accuracy, quicker fabrication time and lower cost when compared to grinding and polishing [28]. The high demand in optical and electronic industry has been consistently pushing and breaking the barriers in various nanotechnology areas such as SPDT. Such advances are essential in order to economically produce high quality semiconductor, ceramic and glass parts. SPDT is well known for providing mechanical and optical advantages due to the level of precision of the equipment. SPDT was an ideal material removal process for these projects (discussed in this chapter) as it is capable of:

- machining in the ductile regime: able to control several parameters precisely such as depth of cut, cutting speed and feed, allowing ductile mode machining,
- very controlled material removal process: in the nanometer range),
- cost efficient (tools can be reused after relapping),
- time efficient: the limit of ductile regime machining can be pushed to minimize the number of passes unlike polishing where only very little material can be removed at once, and
- it provides good surface finish: best surface roughness, Ra, achieved during preliminary cuts was approximately 30nm Ra (without polishing).

Surface Finish

The surface characteristics and surface finish (roughness) are of major concern for all projects in this chapter. Surface characteristics/finish are extremely important in this study as the material requires a mirror finish in order to have good optical qualities for high power laser devices (optical mirrors and windows). It is essential to understand the surface behavior, properties, characteristics and topography before carrying out any machining. In general, a surface roughness measurement is done to understand the surface topography. For the CVD-SiC sample, the surface roughness was measured using a Mitutoyo surface profilometer before carrying out any machining operations. Four main surface roughness parameters were recorded for analysis purposes. These parameters include Ra (Roughness Average), Rq (Root Mean Square (RMS) Roughness), Rz (Average Peak-to-Valley of the Profile) and Rt (Maximum Height of the Profile). However, only the two most important surface roughness parameters are discussed in this chapter; Ra and Rz.

The initial goal when trying to improve the surface roughness of a ceramic is to reduce the peak-to-valley (Rz) values of the workpiece. The effect of Rz values can be easily visualized in the tool-workpiece model depicted in *Figures 14, 15 and 16* where the peak-to-valley of a small section of the workpiece is schematically modeled (for the ease of visualization, the model is not drawn to scale). If the average of the peaks is fairly high (in this case Rz $\approx$ 9μm for the 6" CVD coated SiC), then several machining passes have to be carried out in order to smooth the surface (where the maximum depth of cut is determined by the DBT depth $\approx$ 550nm for CVD coated SiC by scratch tests performed using a diamond stylus) [18].

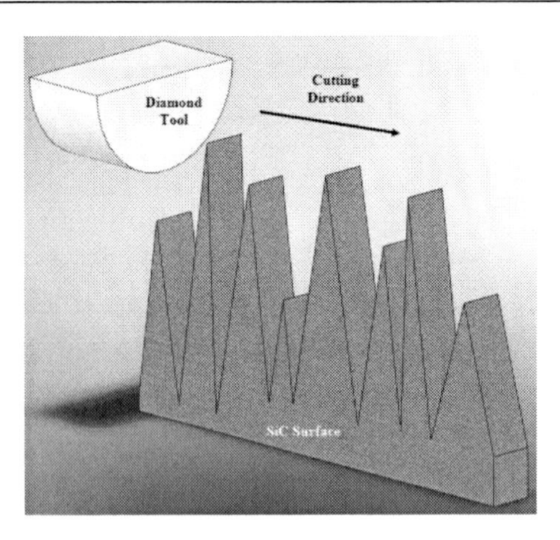

Figure 14. A model (not to scale) of the tool positioned before cutting through the peaks of the as-received SiC surface.

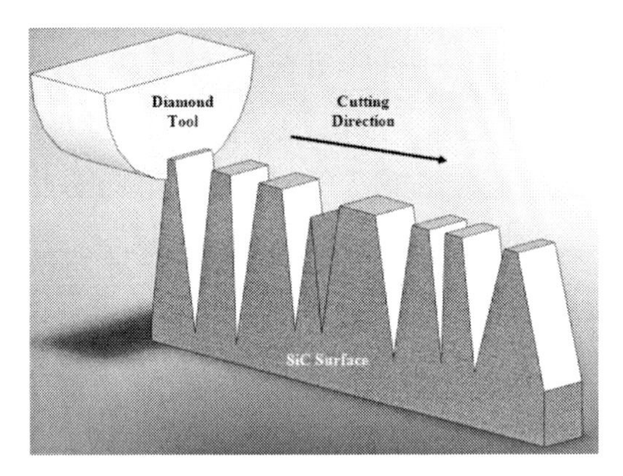

Figure 15. Tool positioned after the first pass and at the beginning of the second pass before cutting through the remaining peaks of the SiC surface.

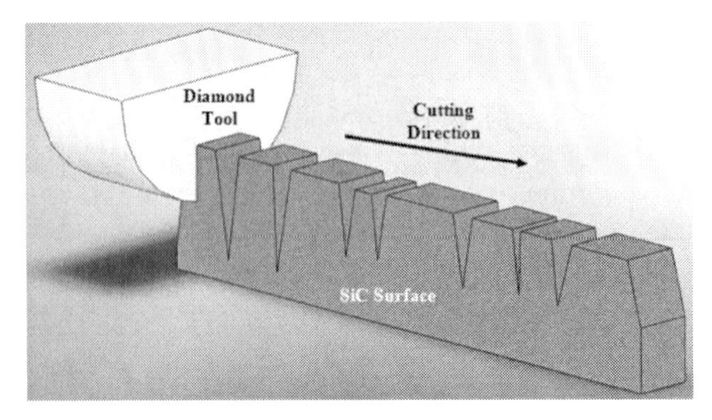

Figure 16. Tool positioned after the second pass and at the beginning of the third pass before cutting through the remaining blunt peaks.

Figures 14, 15 and *16* compare the surface topography (peaks and valleys on the surface) before and after machining. In general, the surface is seen to be improving as long as the material is removed in the ductile regime (plastic deformation). The initial pass/roughing pass concentrates more on removing the major peaks on the surface. The depth of cut of a smoother (flatter) surface as seen in *Figure 16* reduces due the greater contact area between the tool and workpiece surface (more material resistance to penetration of the tool).

There are several factors that could affect the surface quality of the workpiece while machining. These include tool geometry, tool wear, external vibration (from tool or spindle), cutting speed, feed, depth of cut and friction forces. Fritz Locke et al. in Jahanmir's book about machining of ceramics [29] describes the correlation between the machining parameters and the produced surface finish. It is stated that the surface quality of the work piece tends to worsen at higher feeds and this finding is consistent with the results obtained from this study. The tool nose radius plays an important role in the surface finish of the workpiece. In general, a larger tool nose radius and a smaller feed yields in a better surface. This relationship is given by the equation:

$$H_{max} \sim f^2 / 8R, \text{ for } f << R \quad (5)$$

where H_{max} is the predicted theoretical surface roughness (maximum peak to valley height), f is the feed and R is the tool nose radius. It is also important to realize that surface roughness degrades with machining time due to tool wear development [30]. The effect of SPDT on the surface finish of the workpiece will be discussed in the next section.

SPDT of CVD Coated SiC

Several preliminary tests were carried out using various depths of cuts and feed in order to find the optimum machining parameters. The preliminary single point diamond turning experiments were successful in reducing the surface roughness of a CVD coated silicon carbide disk [31]. The Ra was brought down by over one order of magnitude *(from 1158nm to 85nm)* in 16 passes. Since the primary goal of the preliminary tests was to develop machining parameters appropriate for ductile mode machining of CVD SiC, there were several additional steps (machining passes) that had to be carried out in order to confirm or verify the processing parameters. For the actual manufacturing process, many steps (passes) were eliminated to make the actual production process more cost and time efficient.

Table 1. Optimized machining parameters for improving surface quality of a CVD coated SiC disk by SPDT

Pass	Programmed Depth of Cut	Actual Depth of Cut	Feed (μm/rev)
1	2.0μm	1.3μm	30
2	2.0μm	1.2μm	30
3	2.0μm	845nm	5
4	500nm	255nm	1
5	500nm	210nm	1
6	500nm	160nm	1

A total of six passes have been suggested and carried out for the final manufacturing process to improve the surface roughness of silicon carbide. When an additional machining pass is found not to change/improve the surface roughness significantly, that pass is removed in the final recommendation as shown in Table 1. The actual depth of cut is always expected and measured to be less than the programmed depth of cut (in most cases for SiC, the actual depth of cut is about half the programmed depth of cut) due to the elastic properties of the material and tool, and the compliance of the equipment. In this study, all analysis was based upon the actual measured depth of cut. It is also important to note that some of the actual measured depths of cuts shown in Table 1 are greater than the DBT depth (~550nm) indicated (and yet no fracture was caused) is due to the different tool geometries used during the scratch tests and the machining passes. The DBT experiments were carried out using a 5μm tip diamond stylus that is known to have a much more aggressive cutting effect compared to a 3mm nose radius cutting tool.

Results and Discussion

All six machining passes carried out for the final machining experiment were successful. This section discusses the results for the final machining of the 6" CVD-SiC. *Figure 17* shows the surface roughness data for all passes.

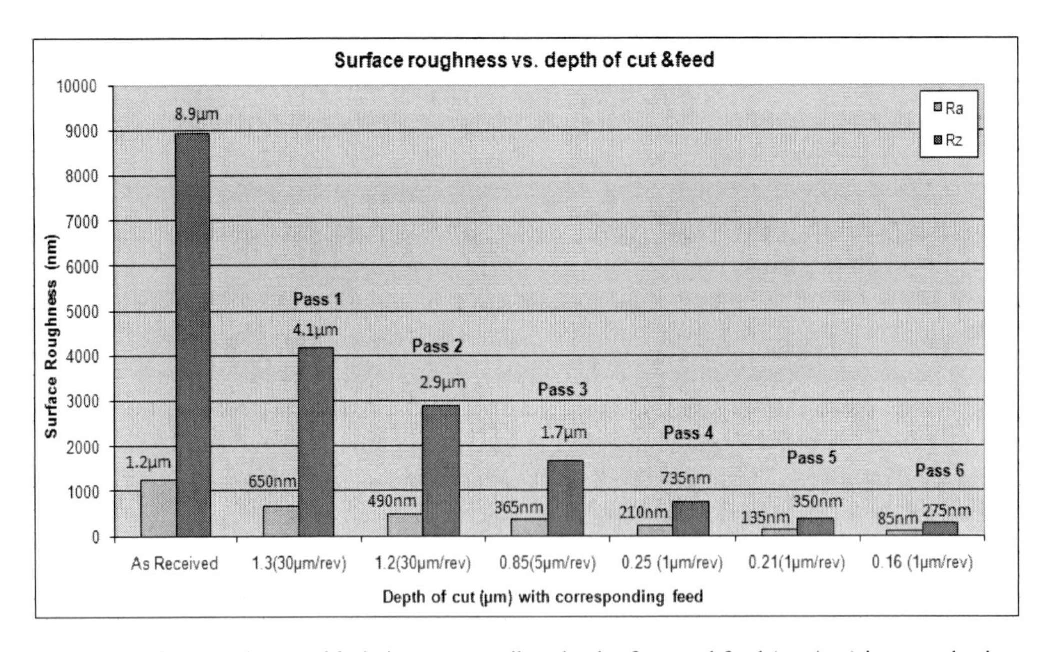

Figure 17. Surface roughness with their corresponding depth of cut and feed (μm/rev) in parenthesis.

The results suggest that the surface roughness improved after every machining pass. The trend was consistence with the results obtained in the preliminary machining experiment where the Ra value decreased as the peak-to-valley (Rz) value decreased. The surface roughness (Ra) was reduced from 1.23μm to 85nm in six passes. *Figure 18* shows the cutting force with the corresponding feed and depth of cut for all six passes.

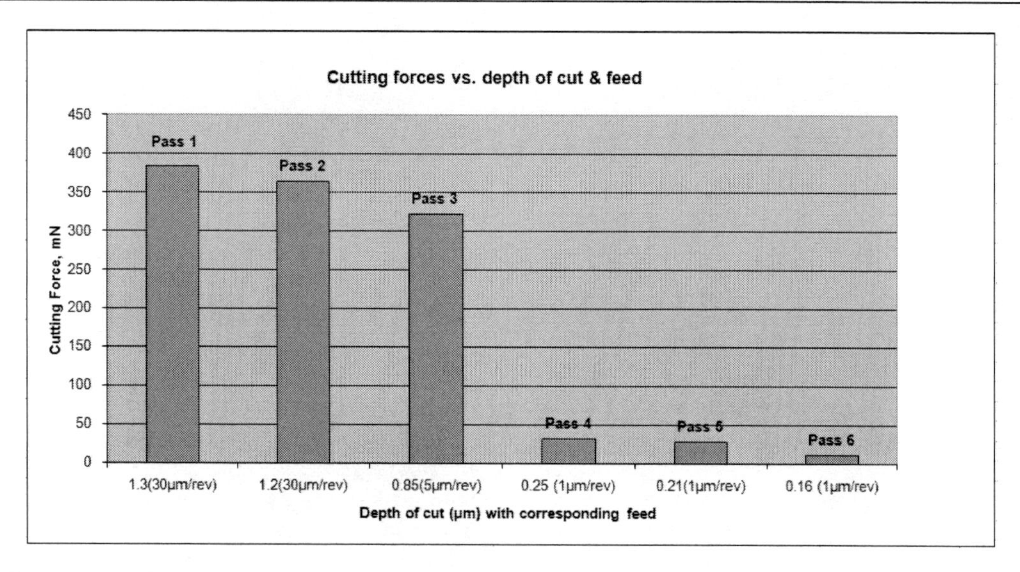

Figure 18. Cutting forces with their corresponding depth of cut and feed in parenthesis.

The cutting force constantly decreased after each pass. The main drop in cutting forces is observed at pass 4 due to the lower depth of cut. Cutting forces are a function of surface roughness (higher cutting forces due to rougher surfaces), depth of cut and feed. However, as seen in Figure 3, the most dominant parameter that influences the cutting force is the depth of cut (deeper cuts yield higher cutting forces).

An additional study was done for the final machining to evaluate tool wear. Due to the hardness and abrasiveness of ceramics (in this case SiC), tool wear becomes a major concern when attempting to SPDT, as diamond is only about four times 'harder' than SiC (100GPa vs. 26GPa). In precision machining, tool wear is usually observed at the micron level (at times even a few hundred nanometers) as even minor tool wear could play a huge role in changing the surface finish of the workpiece. *Table 2* shows the tool wear data for the respective machining conditions.

Table 2. Tool wear data for the machining passes carried out

Pass	*Actual DoC	Feed (µm/rev)	*Cutting Force, Fx (mN)*	***Wear Length* (µm)	Rake Wear (µm)	Flank Wear (µm)
1	1.3µm	30	385	370	6	36
2	1.2µm	30	365	360	4	29
3	845nm	5	320	345	3	27
4	255nm	1	33	220	2	14
5	210nm	1	28	215	2	11
6	160nm	1	11	155	2	9

Note: *Actual DoC refers to the actual measured depth of cut (not the programmed depth of cut).
**The wear length is the total wear measured along the cutting edge radius as seen in Figure 19(a).

A new tool was used for every pass and there were no tool failures reported for any of the machining passes. In general, the tool wear data shows that the wear length along the cutting edge radius is directly proportional to the depth of cut (the greater the depths of cut, the longer the measured wear along the cutting radius). The rake and flank wear are both a function of surface roughness, depth of cut and feed. In all six passes, the measured flank wear was more than the measured rake wear due to the larger contact area between the flank face of the tool and the workpiece.

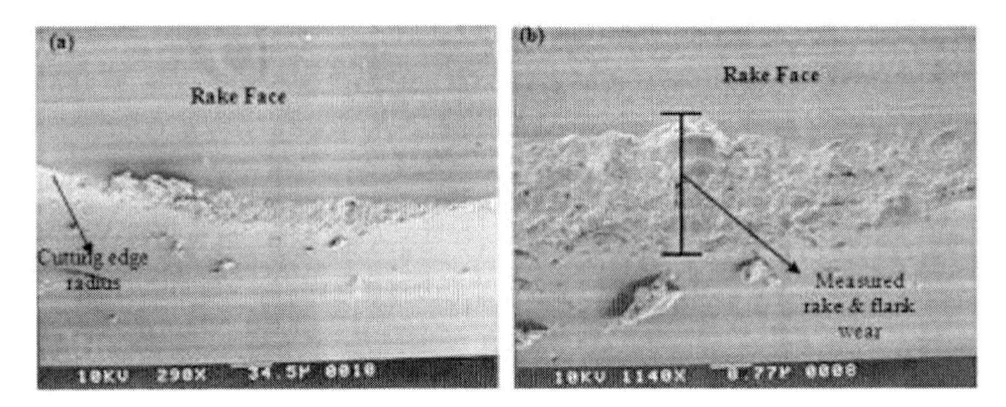

Figure 19. (a) and (b): SEM images of the cutting edge of the diamond tool after machining.

Figure 19(a) and *19(b)* are SEM micrographs taken at 290x and 1140x respectively to study different wear patterns on the tool tip. The tool shown in the micrograph was used in pass 1 (also known as the roughing pass) where the programmed depth of cut was 2μm with a 30μm/rev feed. The large measured wear on the tool is due to the rough surface of the as-received disk.

Subsurface Damage Analysis

Since SiC undergoes a high pressure phase transformation during machining, it is possible that subsurface damage occurs in this brittle material without any indication of surface damage. Two different non destructive techniques were used to investigate the subsurface damage of the machined SiC; laser micro-Raman spectroscopy and scanning acoustic microscopy (SAcM).

Laser Raman Spectroscopy

Laser Raman spectroscopy is a well known non destructive characterization technique often used for semiconductors. A 633nm wavelength He-Ne laser was used to study the sub-surface of the machined. The main purpose of the Raman spectroscopy in this study is to attempt to detect the amorphous layer beneath the machined surface.

In *Figure 20(a),* the spectrum shows sharp crystalline peaks of the unmachined SiC sample. Comparing these peaks with *Figure 20 (b),* it is seen that a combination of the crystalline peaks (sharp peaks) and amorphous broadening (broad peaks) are formed in the machined surface. The amorphous layer, represented by the amorphous broadening in *Figure 20(b),* is a good indication of a ductile material removal process (resulting from a back transformation from the HPPT). In general, the thickness of the amorphous layer increases as the depth of cut is increased.

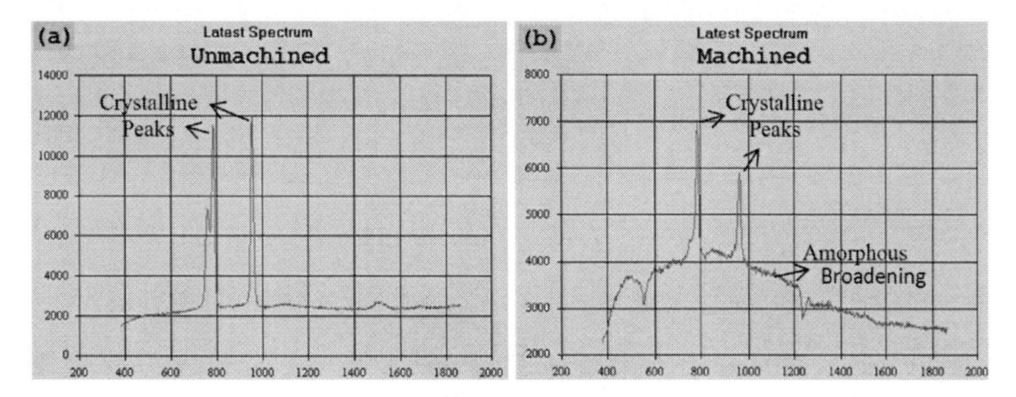

Figure 20. (a) and (b): Raman laser spectrum of the unmachined SiC (a) and the machined SiC (b) surfaces.

Scanning Acoustic Microscopy (SAcM)

SAcM is widely used in non-destructive evaluation (NDE) of materials utilizing high-frequency acoustic waves (60MHz to 2.0GHz) to reveal surface topography, subsurface features and elastic properties [32]. The concept of SAcM is schematically illustrated in *Figure 21*. Acoustic waves are produced by a transducer, pass through the coupling liquid (usually distilled water), and reflect from the focal plane (located at a distance 'z' below the specimen's surface). The reflected acoustic echoes from individual regions are detected during scanning and are used to assemble images [33].

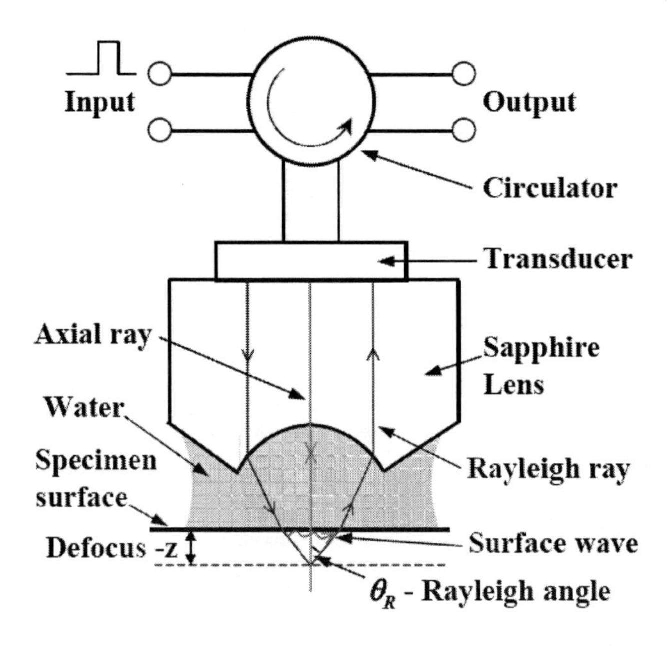

Figure 21. Schematic diagram of scanning acoustic microscopy.[23]

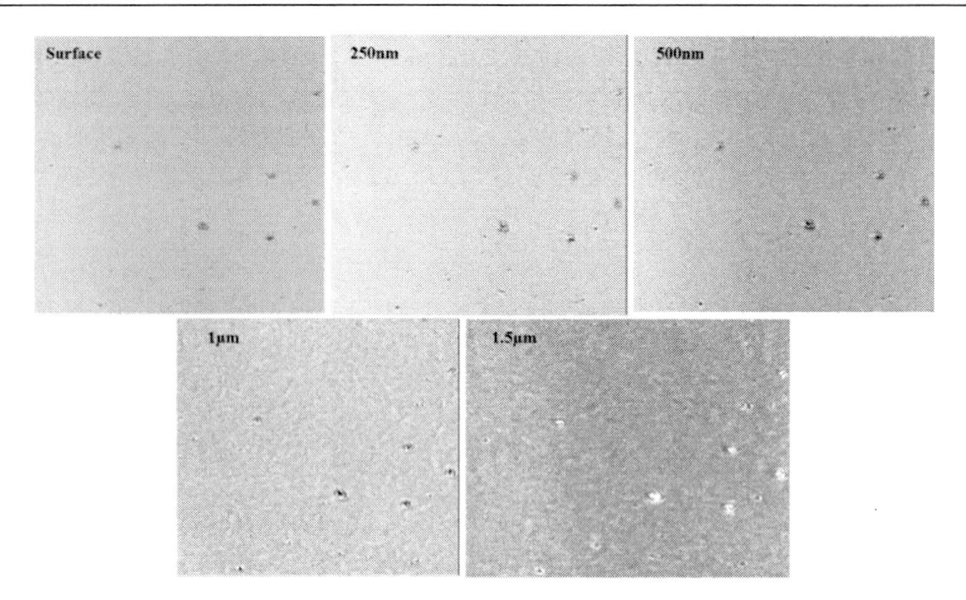

Figure 22. A sequence of images of the machined SiC sample (first image shows the surface and the numbers on the top left of each image represents the scanned depth beneath the machined surface).

The acoustic microscopy images show no signs of subsurface cracks or damage. The surface features seen in the images in *Figure 22* are pits and voids that existed in the as received material and were not caused by the SPDT operation.

Summary

The single point diamond turning experiments were successful in reducing the surface roughness of a CVD coated silicon carbide disk (mirror blank). The Ra was brought down by over one order of magnitude *(from 1.23μm to 85nm).* The most important consideration when machining in the ductile regime is not to exceed the critical depth or the DBT depth of the material, in order to avoid brittle fracture, which leads to higher surface roughness. It is possible to machine nominally brittle materials by plastic deformation at small scales i.e. below the critical depth or the DBT (the DBT for this material from a stylus scratch test was approximately 550nm) [18]. Note that some of the actual measured depths of cuts shown in *Table 1* are greater than the DBT depth (~550nm) indicated (and yet no fracture was caused) is due to the different tool geometries used during the scratch tests and the machining passes. The DBT experiments were carried out using a 5μm tip diamond stylus that is known to have a much more aggressive cutting effect compared to a 3mm nose radius cutting tool.

A total of six passes have been suggested and successfully carried out for the final manufacturing process to improve the surface roughness of silicon carbide. When an additional machining pass was found not to change/improve the surface roughness significantly, that pass was removed in the final machining experiment. The best surface finish was obtained with the lowest feed rate attempted (1μm/rev) but initially a higher feed rate was used (30μm/rev) to maximize the material removal rate and minimize tool wear. The trade off at lower feed rates is that the measured tool wear is much more than at the higher feeds as shown in the results (as seen in *Table 2*), but the resultant surface finish is improved.

The tool wear can be reduced by using a suitable cutting fluid to reduce frictional effects, and also by reducing the feed rates once the surface becomes reasonably smooth as suggested in *Table 2*. Both subsurface damage techniques showed complementing results, where no signs of brittle material removal were detected. SPDT was successful in improving the surface roughness of SiC without causing surface and subsurface damage. *Figure 23* compares the CVD-SiC surface before and after the machining was carried out. Note that the machined disk is significantly more reflective, indicating a better surface finish (lower Ra) than the as-received disk. The rings in the machined disk are artifacts due to the various feed rates uses (i.e. multiple non overlapping passes).

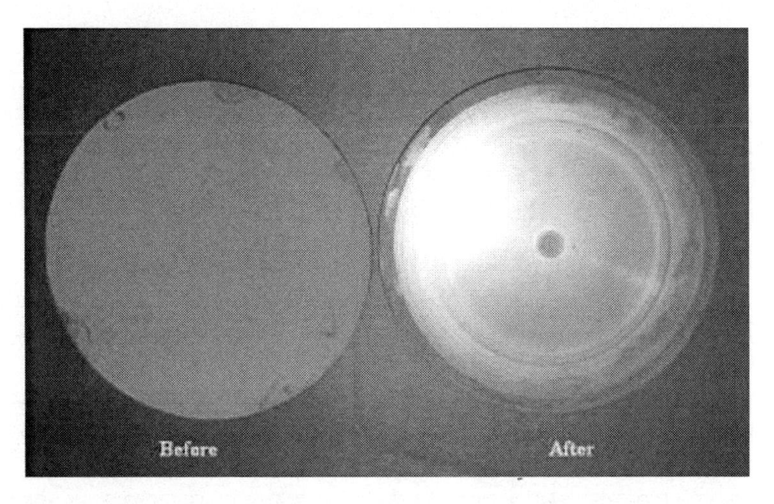

Figure 23. The figure above compares the 6" CVD-SiC before SPDT was carried out (left) and after the SPDT operation was carried out (right).

CONCLUSION

It is important to understand the fracture toughness/material limits (fracture tolerance) in ceramics when attempting to machine them. Ceramics (e.g., SiC) are extremely brittle materials but can be machined in the ductile regime at the nano/micro scale. The results of this research confirmed that brittle materials (i.e. SiC) can be machined in the "ductile regime".

It is also vital to establish ductile regime machining conditions for finishing and smoothing ceramics, as brittle mode behavior results in pitting, micro-cracks and uncontrolled subsurface damage. One of the first steps in attempting ductile mode machining is to establish/determine a critical depth of cut or the Ductile to Brittle Transition (DBT) of the material. The DBT of a material can be determined by performing cuts at the nano to micro scale on the material surface, which has been previously smoothed – generally by polishing, so as to be able to more readily determine the onset of fracture. One of the methodologies for determining the DBT of a single-crystal 4H-SiC wafer is demonstrated in this study [3]. The DBT was determined by performing nanometric cuts and then analyzing those cuts using an AFM. The results from the scratching tests were used as guidelines for the SPDT projects in this research [18]. The Blake and Scattergood [6] cutting model was used to estimate the calculated critical depth of cut as a function of tool geometry and feed.

The single point diamond turning (SPDT) study was carried out on the 6" CVD-SiC disk in an attempt to improve its surface roughness. SPDT was proven to be successful in smoothing CVD-SiC (Ra for SiC was reduced from 1.23μm to 85nm) to enhance its optical properties to be used as an optical mirror device. Six passes were carried out to achieve a final surface roughness (Ra) of 85nm. This is because of the maximum depth of cut limitation (at each pass) that cannot exceed the critical depth of cut (550 nm for CVD SiC) in order to achieve ductile-regime machining. All tools held up (no tool failure but tool wear was measured and reported) for all machining passes.

In general, there are several parameters that can be adjusted in order to increase the material removal rate (MRR). However, in precision engineering applications such as precision machining, there are often tradeoffs when increasing the MRR. An increase in feed can reduce the total machining time (increase material removal rate) and also reduce tool wear; however, a higher feed rate also results in higher surface roughness. For the SPDT experiments performed on the 6" CVD-SiC disk, a higher feed was used for the initial/roughing passes. The feeds and depth of cuts were decreased as the workpiece surface continued to improve (surface roughness reduced). The trade off at lower feeds is that the measured tool wear is much more than at higher feeds due to the longer track length covered by the tool at lower feeds, also, the MRR is less but the resultant surface finish is better compared to at higher feeds.

Results from this research also suggest that tool wear is a function of cutting force (generally lower forces result in less tool wear and better surface finish), depth of cut (the deeper the cut, the greater the cutting force and the larger the measured tool wear), surface roughness (the rougher the surface, the greater the wear), cutting speed (generally tool wear is less when machining closer to the center of the disk due to the lower cutting speed which is a function of the workpiece radius) and equipment stability (i.e., spindle vibration and workpiece runout). It is vital to eliminate tool chatter as it affects the productivity, surface finish, and tool life [34]. Cutting fluid/CMP slurry is used to reduce tool wear by minimizing the frictional force (there may also be some chemical activity or benefit associated with the cutting fluid). Although theoretically it is almost impossible for a layer of cutting fluid to remain at the cutting edge (due to the extremely high pressure generated at the cutting edge radius), it is believed that the slurry helps to remove the machining debris/chips from the cutting region and chemically reacts and changes (perhaps chemically softening the surface which would reduce the cutting force and friction) the workpiece surface during the machining process [35].

Acknowledgments

The authors would like to thank Dr. Jun Qu from the Material Science and Technology Division (Oak Ridge National Laboratories) and the High Temperature Materials Laboratory (Oak Ridge National Laboratories) for sharing their expertise and equipment (Kramer Scientific Instruments (KSI) SAM2000-Scanning Acoustic Microscope) in conducting the subsurface damage analysis. The authors would also like to thank Dr. Robert Hocken at the University of North Carolina at Charlotte for his contribution in the design and fabrication of the Nanocut II device.

REFERENCES

[1] D. Gao, M.B.J. Wijesundara, C. Carraro, C.W. Low, R.T. Howe, and R. Maboudian, 2003, "High Modulus Polycrystalline 3C-SiC Technology for RF MEMS", Proc. Transducers 12th Int. conf. *Solid-State Sensors and Actuators*, pp. 1160-1163

[2] W.R. Ashurst, M.B.J. Wijesundara, C. Carraro, and R. Maboudian, "Tribological Impact of SiC Coating on Released Polysilicon Structures", *Tribol. Lett.*, v.17, n.2, pp. 195-198, 2004.

[3] Deepak Ravindra, John Patten and Makoto Tano, 2007, "Ductile to Brittle Transition in a Single Crystal 4H SiC by Performing Nanometric Machining", ISAAT 2007 Precision Grinding and Abrasive Technology at SME International Grinding conference, *Advances in Abrasive Technology*, X, pp 459-465.

[4] Morris, J.C., Callaham, D.L. Kulik, J. Patten, and R.O. Scattergood, 1995, "Origins of the Ductile Regime in Single Point Diamond Turning of Semiconductors", *Journal of American Ceramic Society*, v78, no. 6, pp. 2015-2020

[5] T. G. Bifano, T. G. Dow and R. O. Scattergood, 1991, "Ductile Regime Grinding- A new technology for machining brittle materials", *Journal of Engineering for Industry*, 113, pp. 184-189.

[6] Blake and Scattergood, 1990, "Ductile-regime Machining of Germanium and Silicon", American Ceramic Society, *Journal of the America Ceramic Society*, Vol 73, Issue 4, pp 949-957.

[7] John A. Patten, W. Gao and K.Yasuto, 2005, Ductile Regime Nanomachining of Single-Crystal Silicon Carbide, *ASME*, v127, pp 522- 532.

[8] K. Kamitani, M. Grimsditch, J.C .Nipko, C.K .Leong, M. Okada, I. Kimura, 1997, The Elastic Constants of Silicon Carbide: A Brillouin-scattering Study of 4H and 6H SiC Single Crystals, *Journal of Applied Physics*, v82(6), pp 3152-3154

[9] J.C., Lovingood, 1999, Design, Optimization and Experiments with a Second Generation Nanometric Cutting Instrument, Master's Thesis, UNCC.

[10] Thimmaiah Ganapathi Kumbera, 2001, Study of Ductile Machining of Silicon Nitride, Master's Thesis, UNCC

[11] N.P.Hung, Y.Q.Fu, 2000, Effect of Crystalline in the Ductile-Regime Machining of Silicon, *The International Journal of Advance Manufacturing Technology*, 16, pp 871-876

[12] N.Axen, L. Kahlman, and Hutchings, I. M., 1997, "Correlations Between Tangential Force and Damage Mechanism in the Scratch Testing of Ceramics," *Tribol. Int.*, 30_7_, pp. 467–474.

[13] G. Pharr, 2003, Workshop on High Pressure Phase Transformations of Semiconductors and Ceramics, UNC Charlotte, Aug. 20–22.

[14] Ming Zhang, H.M. Hobgood, J.L. Demenet, P. Pirouz, 2003, Transition from Brittle Fracture to Ductile Behavior in 4H-SiC, *Journal of Materials Research*, v18(50, pp 1087-1095

[15] Deepak Ravindra, Bogac Poyraz and John Patten, "The Effect of Laser Heating on the Ductile to Brittle Transition of Silicon", The 5th International Conference on MicroManufacturing (ICOMM/4M), Wisconsin, USA, 2010.

[16] J.A.Patten, 2000, Advances in Abrasive Technology, ISAAT 2000, vol.III, Japan *Society of Abrasive Technology* pp. 87-98.

[17] http://www.poco.com/LinkClick.aspx?fileticket=l8QvibTuXcE%3dandtabid=194andmid=677.

[18] Biswarup Bhattacharya, 2005, "Ductile Regime Nano-Machining of Polycrystalline Silicon Carbide", Masters Theses, Western Michigan University.

[19] Y. Ishida, G. Ogawa, 1962, *Japanese Journal of Mechanical Engineering Laboratory*, v8, (1), pp.15-30.

[20] D.M. Marsh, 1964,"Plastic Flow and Fracture of Glass", Proc. R. A, v282, pp.33-43

[21] F.M. Ernsberger, 1968, "Glasses Under Point Loading", *Journal of the American Ceramic Society* , v51, pp.545-547

[22] R. Komanduri, 1996, "On Material Removal Mechanisms in Finishing of Advanced Ceramics and Glasses ", *Annals of the CIRP* (College International pour la Recherche en Productique), v45, p.509.

[23] M.A. Moore, F.S. King, 1980, "Abrasive wear of brittle solids ", *Wear*, v60, pp.123-130

[24] Chunhe Zhang, Teruko Kato, Wei Li and Hitoshi Ohmori, 2000, "A Comparative Study: Surface Characteristics of CVD-SiC Ground with Cast Iron Bond Diamond Wheel", *International Journal of Machine Tools and Manufacture*, Vol 40, pp 527-537

[25] FZ Fang, XD Liu, LC Lee, 2003, "Micro-machining of Optical Glasses- A Review of Diamond- Cutting Glasses", *Indian Academy of Sciences*, Vol 28, Part 5.

[26] http://www.buehler.com/productinfo/consumables/pdfs/FINAL_POLISHING.pdf.

[27] J. Wilks, 1980, "Performance of Diamonds as Cutting Tools for Precesion Machining", *Journal of Precision Engineering*, v 2 (2), pp. 57-72

[28] FZ Fang, XD Liu, LC Lee, 2003, "Micro-machining of Optical Glasses- A Review of Diamond- Cutting Glasses", *Indian Academy of Sciences*, Vol 28, Part 5.

[29] Fritz Klocke, Oliver Gerent and Christoph Schippers, 1999, "Green Machining of Advanced Ceramics, in Machining of Ceramics and Composites", Eds Jahanmir and Said, New York Mercel Dekker Inc., pp. 1-9

[30] Y. Kevin Chou and Hui Song, 2004, "Tool Nose Radius Effect on Finish Hard Turning", *Journal of Materials Processing Technology*, v148, pp. 259-268.

[31] Deepak Ravindra and John Patten, "Improving the Surface Roughness of a CVD Coated Silicon Carbide Disk by Performing Ductile Regime Single Point Diamond Turning", Proceedings of the 2008 *International Manufacturing Science and Engineering Conference*.

[32] Jun Qu, Peter J. Blau, Albert Shih, Samuel McSpadden, George Pharr and Jae-il Jang, 2004, "Scanning Acoustic Microscopy for Non-Destructive Evaluation of Subsurface Characteristics" Paper #435, Proceedings of the 6th *International Conference on Frontiers of Design and Manufacturing* (S.M. Wu Symposium on Manufacturing Sciences), Xi'an, China, Jun. 21-23, 2004.

[33] Jun Qu and Peter J. Blau, "Scanning Acoustic Microscopy for Characterization of Coatings and Near-Surface Features of Ceramics", 2006, Proceedings of the 30th *International Conference and Exposition on Advanced Ceramics and Composites*, Cocoa Beach, FL, Jan. 22-27, 2006, pp. 22-27.

[34] C. Mei, J.G. Cherng and Y. Wang, 2006, "Active Control of Regenerative Chatter During Metal Cutting Process", *Journal of Manufacturing Science and Engineering,* v128, Issue 1, pp. 346-349.

[35] Q.L. Zhao, D. Stephenson, J. Corbett, J. Hedge, J.H. Wang and Y.C. Liang, 2004, "Single Grit Diamond Grinding of Spectrosil 2000 Glass on Tetraform 'C'", *Advances in Abrasive Technology* VI, v257-258, pp.107-112.

In: Silicon Carbide: New Materials, Production ... ISBN: 978-1-61122-312-5
Editor: Sofia H. Vanger

Chapter 5

COMPUTER SIMULATION ON THE NANOMECHANICAL PROPERTIES OF SIC NANOWIRES

Zhiguo Wang[1] and Xiaotao Zu
[1]Department of Applied Physics, University of Electronic Science and Technology of China, Chengdu, P.R. China

ABSTRACT

One-dimensional (1D) SiC materials have drawn increasing attentions for their use in fundamental research, and for potential nano-technological applications in electronics and photonics operating in harsh environment. In the present chapter, we firstly report the fabrication methods and some novel properties of 1D SiC nanostructures. Then the results of computer simulation on the mechanical properties of SiC nanowires are presented. For perfect SiC nanowires, under axial tensile strain, the bonds of the nanowires are just stretched before the failure of nanowires by bond breakage. Under axial compressive strain, the collapse of the SiC nanowires by yielding or column buckling mode depends on the length and diameters of the nanowires. The nanowires collapse through a phase transformation from crystal to amorphous structure in several atomic layers under torsion strain. The effects of twinning and amorphous coating on the mechanical behavior are also presented. Amorphous layer coating can induce brittle to ductile transition in SiC nanowires. And the critical strain of the nanowires can be enhanced by the twin-stacking faults.

1. INTRODUCTION

Low-dimensional materials, particularly one-dimensional (1D), have stimulated great interest due to their importance in basic scientific research and potential technological applications. 1D materials (e.g. nanowires and nanotubes) can function both as active devices and interconnects. In addition, their low dimensionality associated with new quantum size effects has provided new ways to develop nanoscale electronics and optoelectronics. With high mechanical strength and thermal conductivity, excellent physical and chemical stability, silicon carbide (SiC) has found wide applications in biomedical materials, high-temperature semiconductor devices, synchrotron optical elements, and high strength structural materials [1]. After being fabricated into low-dimensional nanostructures, SiC nanostructures may exhibit unique properties due to their quantum size effect and may find applications in nanotechnology and nanoscale engineering. For example, bulk SiC shows weak emission at room temperature on account of its indirect band gap [2]. However, the emission intensity can be significantly enhanced in the nanometer materials [3,4]. In accordance with the quantum confinement effect, photoluminescence of the crystallites with diameters below the Bohr radius of bulk excitons is shifted to blue with decreasing the size [5]. SiC nanowires have outstanding mechanical property and high electrical conductance can be used to reinforce composites maters or as nanocontacts in a harsh environment. Moreover, the 1D SiC materials have good field electron emission properties and biocompatibility. In recent years, much effort has been paid to synthesis various SiC nanostructures. And their outstanding properties are also investigated. In the present chapter, we firstly report the fabrication methods and some novel properties of 1D SiC nanostructures. Then the results of computer simulation on the mechanical properties of SiC nanowires are discussed in detail.

2. Fabrication and Novel Properties of One-Dimentional SiC Structures

A variety of methods on the synthesis of SiC nanowires have been developed, including laser ablation, chemical vapor deposition via silicon precursor, physical evaporation, hydrothermal method and catalyst-assisted vapor liquid solid mechanism. The SiO vapor can be used to convert carbon fibers and nanotubes to SiC rods. Dai *et al.* [6] prepared SiC nanorods through a reaction between carbon nanotubes and SiO or SiI_2, the synthesized SiC nanorod with diameter (2-20 nm) similar to that of the starting nanotube reactants. A two-step reaction has also been developed to synthesize SiC nanorods at 1400℃ [7]. SiO vapor is first generated via reduction of silica, and then the SiO vapor reacts with the carbon nanotubes to form SiC nanorods. The TEM images reveal that the diameters of the SiC nanorods ranged from 3 to 40 nm. 3C-SiC nanorods with diameters of 5-20 nm and lengths of about 1 μm have been grown on porous silicon substrates by CVD using iron as the catalyst [8]. A tablet composed of pressed Si and SiO_2 powders is placed in the chamber in which the reaction takes place in the presence of CH_4 and H_2. 3C-SiC nanorods can also be synthesized by using a solid mixture of graphite, silicon, and silicon dioxide [9]. As is well known, SiC prossesses more than 250 crystal structures [10], 3C, 4H and 6H-SiC are the most important ones among them. Except for the 3C-SiC nanowires, 6H-SiC nanowires are also been fabricated. Li et al. [11] first reported the synthesis of cone shaped 6H-SiC nanorods by and arc-discharge process. Gao et al. [12] reported the synthesis of Al-doped 6H-SiC nanowires with diameter of 150-300 nm by catalyst assisted pyrolysis of polymer precursors. Wei et al. [13] reported the Al-doped 6H-SiC nanowires via the nano-Al assisted VLS process with diameter ranging from 5 to 200 nm. However, SiC nanowires reported are mostly in the 3C-SiC structure, which is quite different from their polytype bulk materials. This is because 3C-SiC is the most stable phase and its formation is more favorable in energy. Large areas of millimeter long 3C-SiC nanowires were also synthesized [14,15], The ultra long nanowires may open up new opportunities for integrated nanoelectronics and could server as unique building blocks linking integrated structures from the nanometer through macroscales.

Most of above methods are complicated processes with multi-step operations. Moreover, the metallic catalyst remaining in the SiC nanowires act as an impurity, which influences their properties and limit their further practical applications. New methods are developed to synthesized high purity SiC nanowires, such as microwave pyrolysis of methane [16]. The purity of 3C-SiC nanowires is raised obviously by using an ordered nanoporous anodic aluminum oxide template by the thermal evaporation method without any metal catalyst [17].

The photoluminescence position of 1D SiC nanomaterials varies in a wide range, which shows a blue-shift in comparison with that of bulk 3C-SiC. The SiC nanorods synthesized using carbon nanotubes as template show PL peaks at about 340 nm and SiC nanowires synthesized by heating a carbonaceous silica gel emit light peak at 400 nm [7,18]. Needle-shape SiC nanowires exhibit a strong broad PL peak at around 450 nm at room temperature [19]. SiC/SiO_x nanocalbes emit stable violet-blue light at wavelengths of about 315 and 360-400 nm [20]. Two broad emission peaks at 340 and 440 nm are also observed from SiC nanocables synthesized from activated carbon and silica gel mixture [21]. Li et al. found the SiC nanowires emit 290 and 396 nm PL peak prepared by sol-gel process [22]. The PL was assumed to be relevant with morphology and size as well as defects in the materials. Different shape and size of SiC nanowires can be used to realize wavelength-tunable and controllable luminescence.

Field emission is one of the most fascinating properties of one-dimensional nanostructured materials. For many practical applications, field emission from a cold-cathode is required to have low turn-on and threshold fields with uniform and high emission current. The field emission properties of SiC nanostructures showed that they are with low turn-on and low threshold electric field values, which indicate that they could have potential applications in electron field-emitting devices. The field emission properties of these well-oriented SiC nanowires were measured. Field emission current densities of 10 $\mu A/cm^2$ were observed at applied fields of 0.7-1.5 V/μm, and current densities of 10 mA/cm^2 were realized at applied fields as low as 2.5-3.5 V/μm [23]. The emission efficiency can be tuned though controlling the density of SiC nanowires. Lowest turn-on filed of 1.8 V/μm and highest field enhancement factor of 5.9×10^3 can be observed for medium density SiC nanowires [24].

Due to the large surface-to-volume ratio of nanowires, surface effects, such as surface energy, surface stress and surface elastic stiffness have an important effect on the mechanical behavior of nanowires. Wong et al. [25] using atomic force microscopy and lithography techniques showed that SiC nanowires and nanorods have yield strengths that can be over 50 GPa. This value is far larger than the corresponding values for microscale SiC whiskers and fibers. Therefore, SiC nanowires have great potential for use in composite materials as reinforcements with vary high strength and toughness. Measuring the mechanical properties of individual nanowires by conventional techniques is not trivial. Optical measurements used commonly in microelectromechanical systems are not readily applicable to nanowire resonators because the diameter is less than a visible wavelength. Atomistic molecular dynamic (MD) simulations can provide detailed information about deformation and fracture of nanowires. The computer simulation on the mechanical properties of SiC nanowires are discussed in detail in the following part.

3. Simulation Details

3.1. Atomic Potentials

The Tersoff bond-order potential [26,27] was used to describe the interactions among Si and C atoms, where the short-range interactions have been modified to match *ab initio* calculations [28], as described by Devannathem *et al.* [29]. The energy E as a function of distance r_{ij} between atom i and atom j, the angle θ_{ijk} between the ij and ik bonds is given by

$$E=\sum_{i} E_i=\frac{1}{2}\sum_{i\neq j} V_{ij}=\frac{1}{2}\sum_{i\neq j} f_c(r_{ij})[A_{ij}\exp(-\lambda_{ij}r_{ij})-b_{ij}B_{ij}\exp(-\mu i_j r_{ij}] \quad (1)$$

The cutoff function is

$$f_c(r_{ij})=\begin{cases} 1 & r_{ij}<R_{ij} \\ \frac{1}{2}+\frac{1}{2}\cos[\pi(r_{ij}-R_{ij})/(S_{ij}-R_{ij})] & R_{ij}<r_{ij}<S_{ij} \\ 0 & r_{ij}>S_{ij} \end{cases} \quad (2)$$

and the bond-order parameter is given by

$$bij=\chi i_j(1+\zeta_{ij}^{n_i})^{-1/2n_i} \quad (3)$$

where

$$\zeta ij=\sum_{k\neq i,j} f_c(r_{ik})\beta_i g(\theta_{ijk}) \quad (4)$$

and

$$g(\theta ijk)=1+\frac{c_i}{d_i}-\frac{c_i}{d_i^2+(h_i-\cos\theta_{ijk})^2} \quad (5)$$

λ_{ij} and μ_{ij} are arithmetic averages and A_{ij}, B_{ij}, R_{ij} and S_{ij} are geometric averages of the corresponding i and j values. The parameters and cutoff used can be found in Ref. [29].

The Tersoff potential is known to describe adequately the crystalline and amorphous phases of SiC and its nonstoichiometric alloys [30], and has been employed to study the ion irradiation induced defects [31-33], thermal properties [34,35] and surface reconstruction [36] of SiC. The modified potential has been used to study the melting temperature of β-SiC [37] using a "semiconstant volume simulations" [38]. The estimated melting temperature is 3050 K, which is reasonable agreement with the reported experimental value of 2800 K and recent MD simulation result of 3250K±50K [39]. All of these results have demonstrated that the

Tersoff potential is capable of describing the equilibrium and deformed structures of crystalline SiC nanowires.

3.2. Simulation Setup

For all the simulations, a time step of 0.5 *fs* is used and temperature was controlled by a scaling method [40], $v_i^{new} = v_i \sqrt{T_D / T_R}$, where v_i^{new} is the velocity of particle i after correction. T_D and T_R are the desired and actual temperatures of the system, respectively. A rigid boundary condition along the axial direction is used in this work. The tension and compressive strains were applied by displacing the top and bottom fixed four atomic layers in opposite directions at a constant strain rate, and the remaining atoms in the middle part were free to relax, as shown in Figure 1. The initial structures of the nanowires were equilibrated for 100 *ps* at a given temperature, which allows the nanowires to have stable configurations. The strain was then applied along the axial direction to study the tension and compression mechanical properties of the nanowires by imposing a displacement Δz. The atoms in the rigid borders were displaced by $\Delta z/2$, but the coordinates of the remaining atoms were scaled by a factor $(L+\Delta z)/L$ along the z direction. This deformed tube was relaxed for 10 *ps*, and then the relaxed structure was used as an initial configuration for the next strain simulation. The procedure was repeated until each nanowire failed. The stress during each strain increment was computed by averaging over the final 2000 relaxation steps. The torsion procedure was performed as follows: the edge atoms, residing within several atomic planes at both ends, are kept fixed in the MD relaxation procedure, and the remaining atoms in the middle part are free to relax. A fixed boundary condition is assumed at one end of the nanowires, while the torsion loads are applied to the other end of the nanowires.

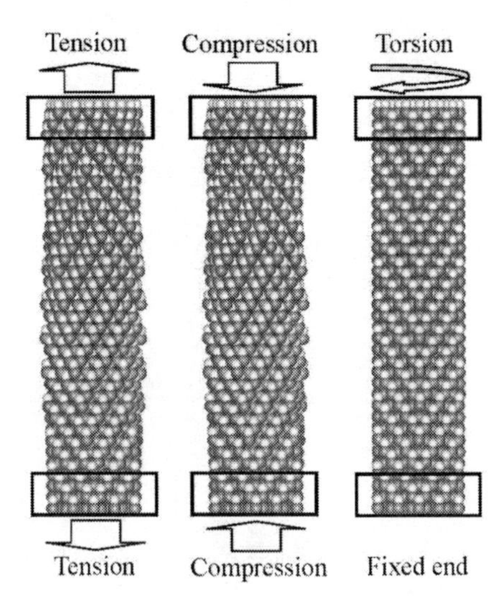

Figure 1. Simulation setup for a SiC nanowire under tensile, compressive and compressive strain.

The axial stress is taken as the arithmetic mean of the local stresses on all atoms, as follows [41]:

$$\sigma_z = \frac{1}{N}\sum_{i=1}^{N}\frac{1}{V_i}\left(m_i v_z^i v_z^i + \frac{1}{2}\sum_{\substack{j=1 \\ (j\neq i)}}^{N} F_z^{ij}(\varepsilon_z) r_z^{ij}(\varepsilon_z) \right) \tag{5}$$

where m_i is the mass of atom i, v_z^i is the velocity along the axis direction and F_z^{ij} refers to the component of the interatomic force along the axis direction between atoms i and j. r_z^{ij} is the interatomic distance along the axis direction between atoms i and j. V_i refers to the volume of atom i, which was assumed as a hard sphere in a closely packed crystal structure.

4. Results

4.1. Mechanical Properties of Perfect [111]-Oriented SiC Nanowires

Figure 2 shows the top views of the geometry of the four perfect SiC nanowires. The relaxed configurations show that the surface layer atoms undergo a bond-length contraction in which the C surface atoms only relax parallel to the surface, while the Si surface atoms relax parallel and perpendicular to the surface, moving radially inward by 0.012 nm. The relaxation of the atoms on the second outmost layers is very small. The relaxed Si-C bond length on the outmost surface layer is 0.186 nm, which contracts by 1.58% compared with that in the bulk SiC. The ripple surface structure obtained in this work is in agreement with other theoretical predictions [42,43].

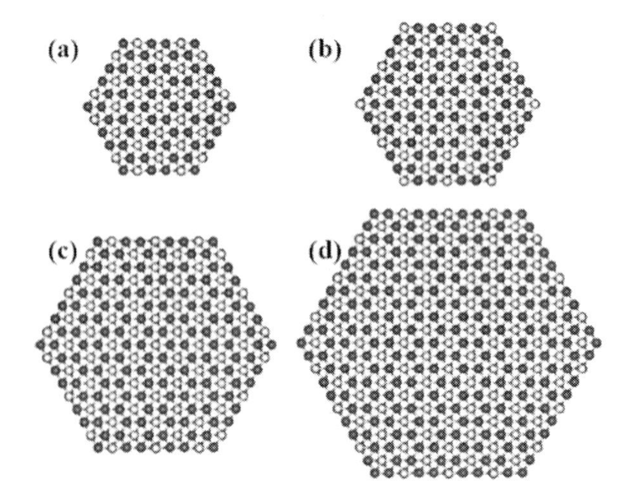

Figure 2. Cross section views of the [111]-oriented SiC nanowires. The nanowires diameters are: (a) 1.65 nm, (b) 2.00 nm, (c) 2.64 nm and (d) 3.31 nm.

Figure 3 shows the representative tensile stress-strain curves for the tensile process of SiC nanowires with a diameter of 2.00 nm at various temperatures. All the stress-strain curves are similar irrespective of temperatures. The stress-strain curves show the same characters as those of the [100]-oriented SiC rod under tensile strain with a extension rate of 2 m/s [30], which demonstrates the brittle failure. The stress increases with increasing strain up to a threshold value that is defined as the critical stress for yield, but the further increase in strain leads to abruptly dropping of stress. For example, the stress increases with increasing strain up to 111 GPa (at strain of 27.3%) and 47.1 GPa (at strain of 14.6%) at 300 K and 1800K, respectively, and further increases in strain lead to stress abruptly decreasing to ~ 0.0 GPa. Judging from the character of the stress-strain dependences, we can assume that the SiC nanowires will fail without exhibiting any plastic-like behavior.

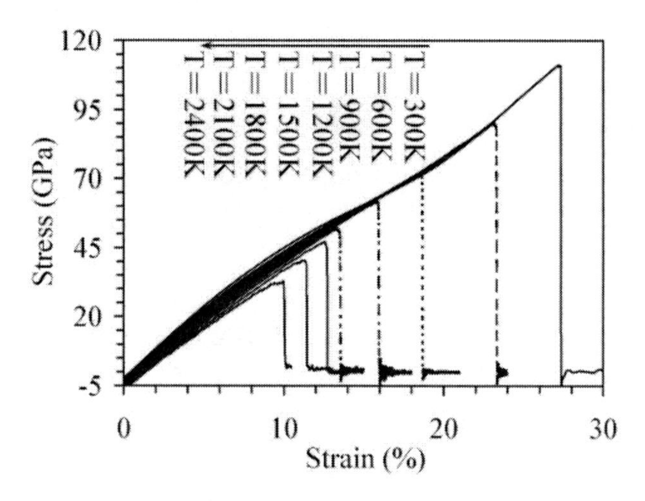

Figure 3. Stress-strain curves for the tensile process of SiC nanowires at various temperatures with diameters of 2.0 nm.

The evolution of atomic configurations at several stages under tension at 300 K and 1800K for the SiC nanowires with four different diameters is shown in Figure 4 and 5, respectively. Under the critical strain, the bonds of the nanowires are just stretched and preserve their fourfold coordination in the nanowires, and no structural defects appear at this stage. With further stretching, bond breakage in the outmost layer is observed, and rapidly spreads towards the center as the strain increases. Several atomic chains can be seen before the nanowires rupture.

Figure 6 shows the compressive stress-strain relationship of the SiC nanowires with diameter of 2.0 nm and lengths at 300K. The stress increases with the increase of strain and decreases after passing a threshold value. The atomic configurations show that short and long nanowires under compression exhibit different modes for the collapse of the SiC nanowires, one with yielding and another with column buckling, respectively. Figure 7 represents the compressive yielding process for the nanowires with a diameter of 2.00 nm at a length of 4.53 nm. It is clear that the SiC nanowires progressively shorten when compression increases, but remain straight up to a strain of 13.89 %; and the nanowires deform by yielding with further increase of strain. For the longer nanowires with the same diameter, the failure of the nanowires changes to a buckling mode due to the instability of the structure. The critical buckling stress is much smaller than the compressive yielding stress. As the nanowires buckle, instead of remaining straight, they become curved. Figure 8 shows the buckling configurations for the nanowires with a diameter of 2.00 nm at a length of 12.08 nm.

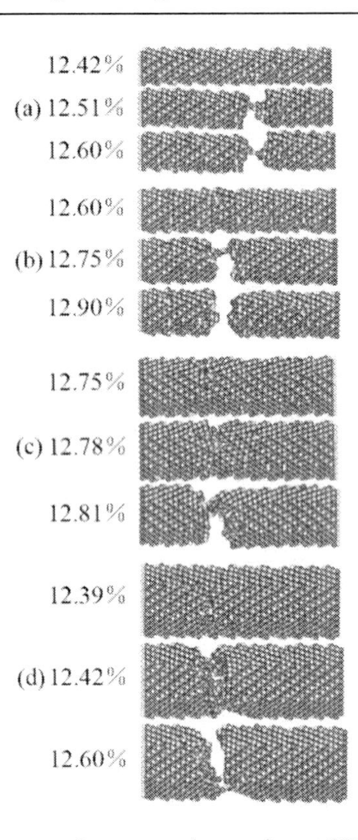

Figure 4. Atomic configurations at several stages under tension at 300 K for the SiC nanowires with four different diameters of: (a) 1.65 nm, (b) 2.00 nm, (c) 2.64 nm and (d) 3.31 nm.

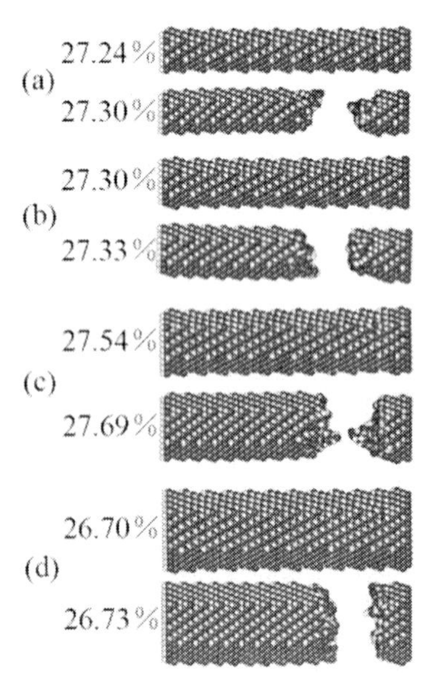

Figure 5. Atomic configurations at several stages under tension at 1800 K for the SiC nanowires with diameters of: (a) 1.65 nm, (b) 2.00 nm, (c) 2.64 nm and (d) 3.31 nm.

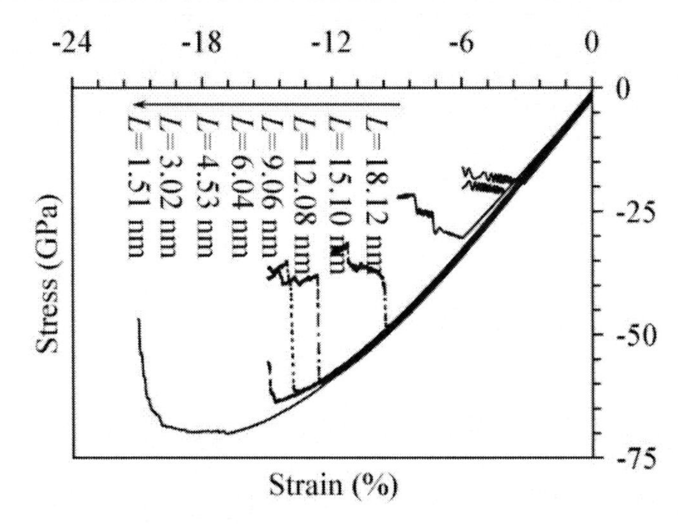

Figure 6. Compressive stress-strain relationship of the SiC nanowires with diameter of 2.0 nm and wire length, simulated at 300K.

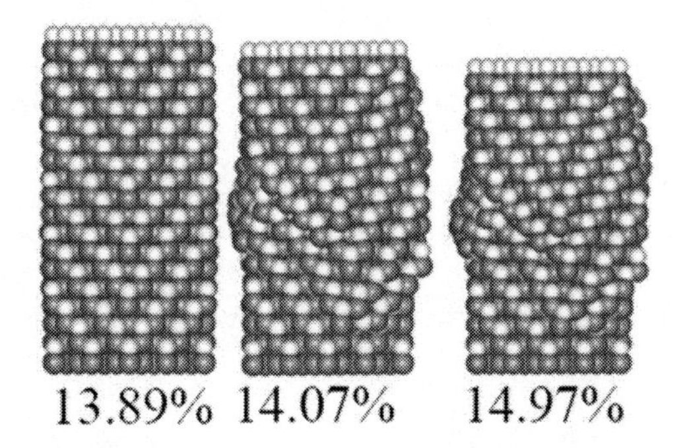

Figure 7. Atomic configurations show the compressive yielding process for the nanowires with diameter of 2.00 nm at a length of 4.53 nm.

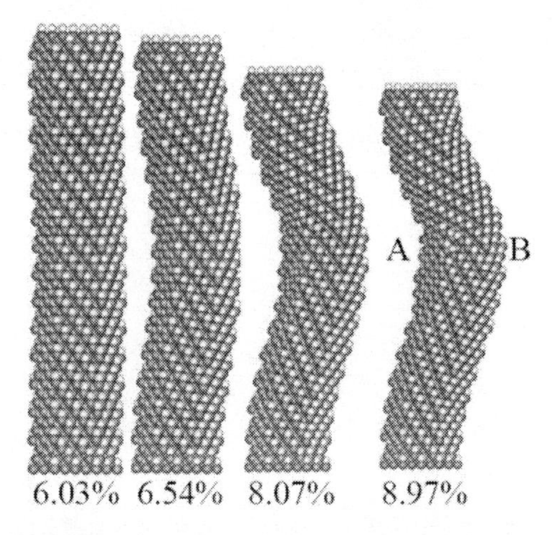

Figure 8. Atomic configurations show the column buckling process for the nanowires with diameter of 2.00 nm at a length of 12.08 nm.

The energy of the initial strain free configurations is denoted as E_0 and that of the strained system is denoted as E_l. The strain energy E_s=E_l-E_0 is accumulated in the nanowires due to torsion. Figure 9 (a) and (b) show the strain energy as a function of torsion angle for the nanowires with diameter of 2.0 nm and length ranged from 3.02 nm to 18.12 nm at 300 K and 1500 K, respectively. The torsion rates of 0.01° ps^{-1} and 0.15° ps^{-1} are used at 300 K and 1500 K, respectively. As seen from Figure 9, there is a critical value of torsional angle, beyond which plastic defects occur with a sudden drop in the accumulated torsional strain energy. This angle corresponds to the critical torque τc, below which the nanowires can sustain without any loss in structure integrity and the strain energy Es increases with the increasing of torsional angle θ, following a quadratic form: Es=kθ2. The evolution of the atomic configuration of the nanowire with a diameter of 2.0 nm and a length of 12.08 nm under different torsion angles at a simulation temperature of 300 K is shown in Figure 10. The whole structure retains hexagonal cross-section as the torsion angle is smaller than 221°. The atoms move only along the circumferential direction and the crystal lattices preserve their original arrangements. Once the torsion angle exceeds the critical value, a slight collapse of the structure takes place in the middle part of the nanowires. Crystal to amorphous structure transition can be observed in the collapse part. The configuration deforms more violently as the torsion angle increases.

The torsional buckling is also found to strongly depend on the length of the nanowire. The critical torsional angles as a function of wire length at 300 K and 1500K are shown in Figure 9 (c) and (d), respectively. It is clearly shown that the critical torsional angle increases with the increasing of wire length.

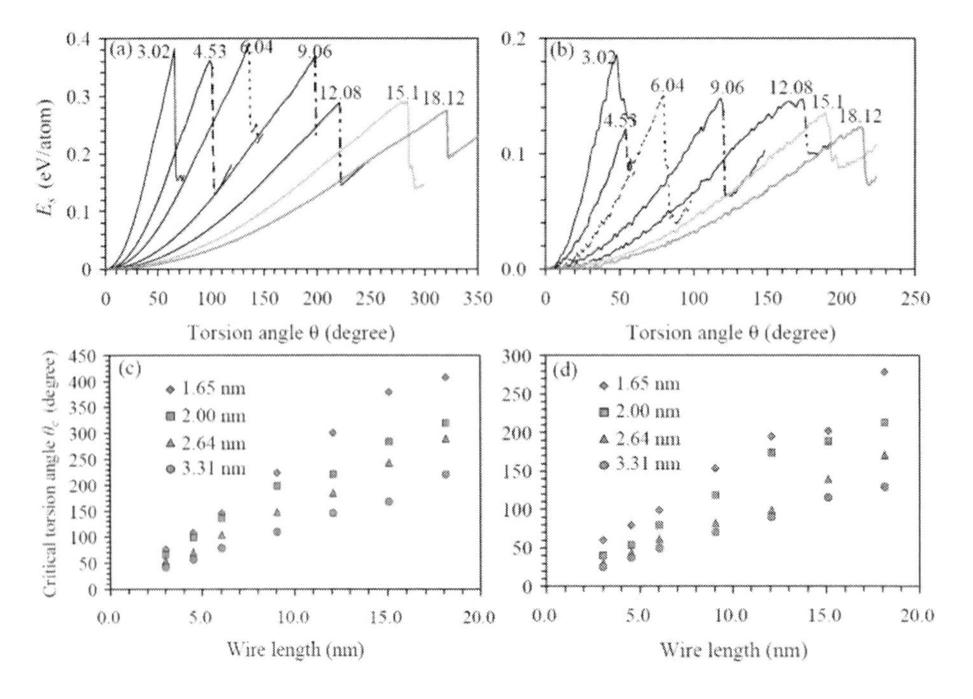

Figure 9. Strain energy as a function of torsion angle for the nanowires with diameter of 2.0 nm and length changed from 3.02 nm to 18.12 nm at (a) 300 K and (b) 1500 K. Critical torsional angles as a function of wire length at (c) 300 K and (d) 1500K.

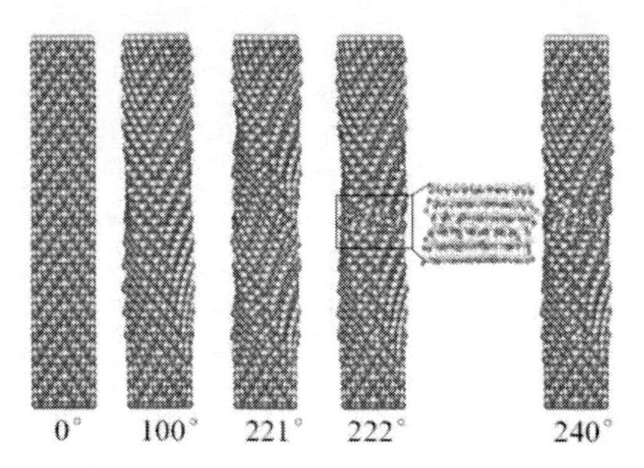

Figure 10. Evolution of the atomic configuration of nanowires with diameter of 2.0 nm and length of 12.08 nm under different torsion angles at a simulation temperature of 300 K.

4.2. Amorphous Coating Effect on the Mechanical Behavior of SiC Nanowires

Most of the experiment results show that SiC nanowires are of cubic zinc-blend structure (β-SiC) oriented along the [111] crystal directions with amorphous carbon [44-47] or SiO_2 coating [48-50]. SiC/SiO_x core-shell nanowires can emit stable and high intensity blue-green [19] or violet-blue light [20], so they would have great potential applications as light emitting devices. In addition, SiC/SiO_2 core-shell nanowires are expected to substitute for large-sized SiO_2 that act as the reinforcing phase in the polymer, which will improve the interface bonding intensity of the polymer. It is no doubt that characterizing and understanding the effect of amorphous coating on the mechanical behavior of nanowires are of both practical and academic importance.

The initial configuration was created by fixing the atoms of the core part of the nanowires, and melting the outer layer at a temperature of 6000 K. Then the system was quenched to 0 K. After the system was rescaled to a required temperature, the fixed atoms were released and the system was equilibrated for 100 *ps*. Top views of the geometry of the amorphous layer coated SiC nanowire is shown in Figure 11. Wire length (*L*) of 9.06 nm is used.

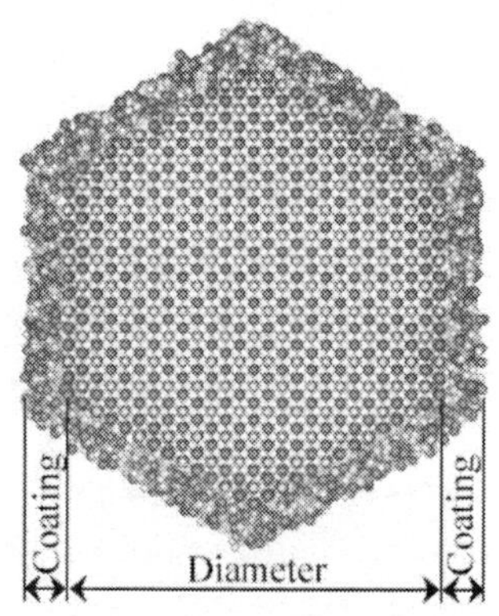

Figure 11. Cross section views of the [111]-oriented SiC nanowires coated with amorphous layer.

Figure 12 shows the tensile stress-strain relationship for the tensile process of SiC nanowires coated with different amorphous layers at various temperatures. The stress increases with increasing strain up to a threshold value that is defined as the critical stress for yield, but the further increase in strain leads to abruptly dropping of stress for the nanowire without amorphous coating. For example, the stress increases with increasing strain up to 113 GPa (at strain of 27.6%) and 53.7 GPa (at strain of 14.0%) at 300 K and 1500K, respectively, and further increases in strain lead to stress abruptly decreasing to ~ 0.0 GPa. Judging from the character of the stress-strain dependences, we can assume that the SiC nanowires will fail in a brittle manner. However, the stress-strain curves shows different behavior after the strain surpasses the critical strain for the nanowires with amorphous coating. For example, the stress increases with increasing strain up to 61.3 GPa (at strain of 22%) and 30.3 (at strain of 11.1%) at 300K and 1500K, respectively, and the stress does not decrease to 0 GPa with further increasing the strain. These results demonstrate that the amorphous coated nanowires fail in different manner compared with the uncoated ones.

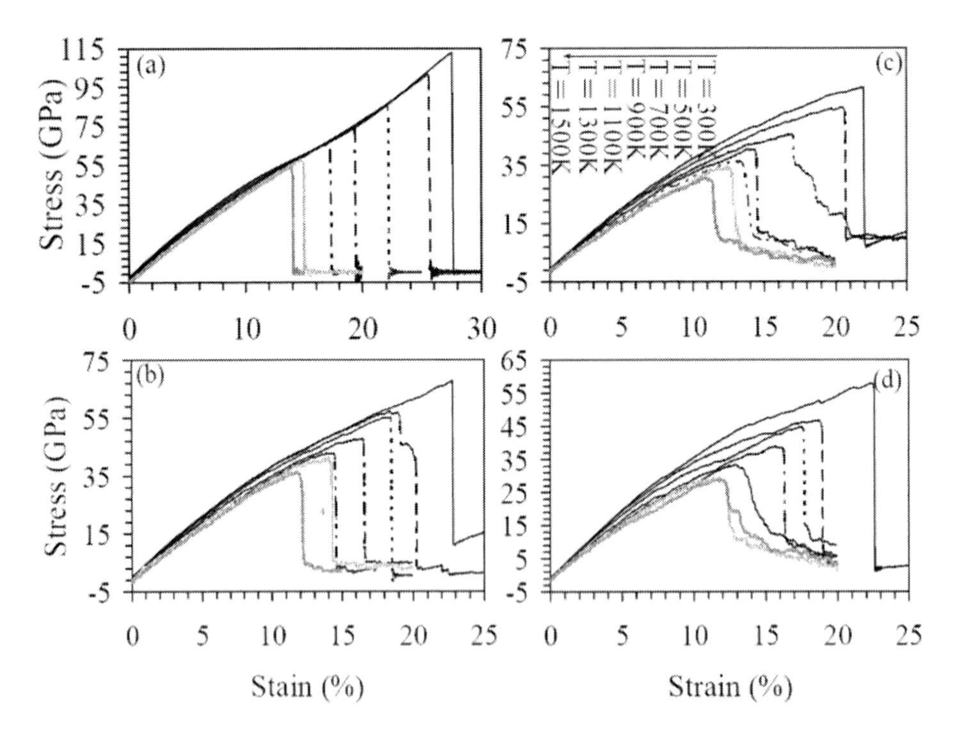

Figure 12. Tensile stress-strain curves for the SiC nanowires coated with amorphous layer of (a) 0 nm, (b) 0.15 nm, (c) 0.31 nm and (d) 0.46 nm at various temperatures.

The atomic configurations of the whole nanowires, core atoms and amorphous layer of the nanowires with amorphous layer of 0.47 nm at different states of the uniaxial tension at 300 K are shown in Figure 13. There are no significant changes of atomic configuration of the core atoms and amorphous layer up to a strain of 16.50%. Necking can be clearly seen with further extension for the core atoms and amorphous layer. The necking part develops into amorphous structure. These results show that the amorphous layer leads to the appearance of plastic deformation of SiC nanowires.

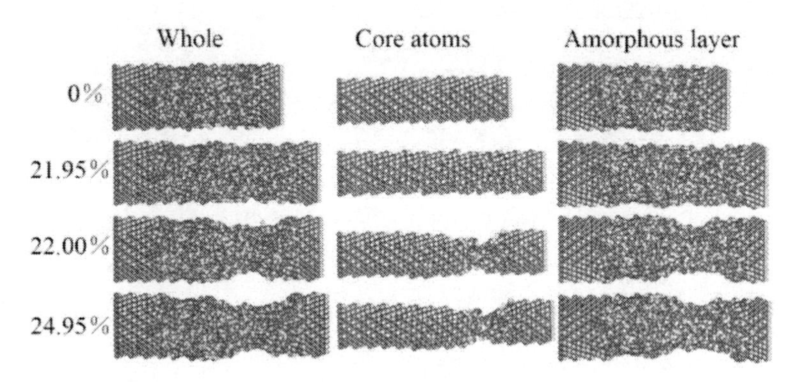

Figure 13. Atomic configurations of selected stages for the whole nanowires, core atoms and amorphous layer of the nanowires with amorphous layer of 0.47 nm at the temperature of 300K.

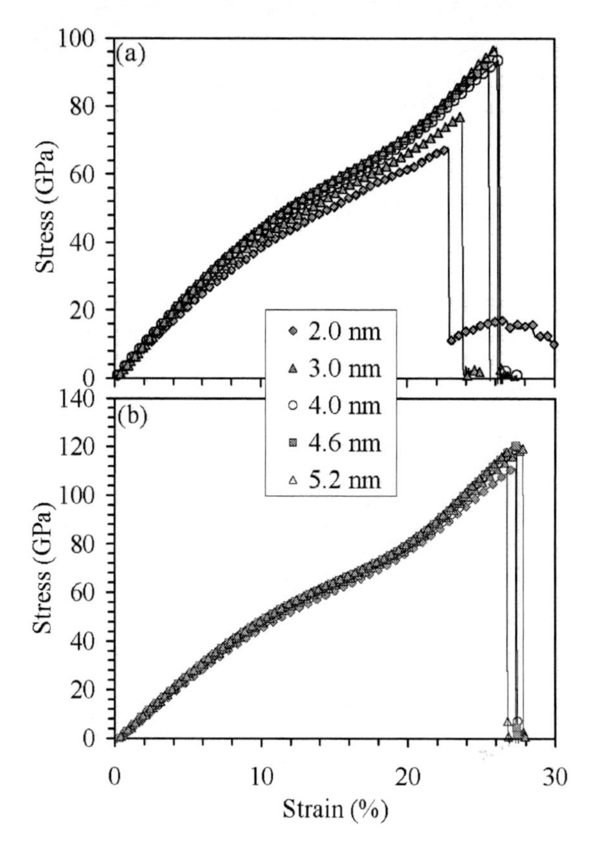

Figure 14. Stress-strain curves for the tensile process of SiC nanowires (a) without and (b) with amorphous coating layer.

The tensile stress-strain curves of the nanowires with and without amorphous coating are shown in Figure 14 (a) and (b), respectively. Before the dropping of the stress, all the stress-strain curves show the similar characters which is irrespective of diameters and amorphous coating. At the initial stage, the stress-strain relationship follows Hooke's law: the stress increases almost linearly with increasing strain. After passing the linear relationship region, the stress increases with increasing strain up to a threshold value that is defined as the critical stress for yield, but the further increase in strain leads to abruptly dropping of stress. For the uncoated nanowires, the stress drops to about zero GPa after passing the critical strain. Judging from the character of the stress-strain dependences, it can be assume that the nanowires without amorphous coating fail without exhibiting any plastic-like behaviour. As for the amorphous layer coated SiC nanowires with diameter of 2.0 nm, the stress does not decrease to zero GPa with further increasing the strain, but vibrates at about 10 GPa, the stress decrease to about zero GPa for the coated nanowires with larger diameters. This indicates that the crystalline SiC nanowires under high-rate uniaxial strains mostly behave elastically until they fail in a brittle manner, whereas the amorphous layer coated nanowires with thin diameters and large diameters fails in a ductile and brittle manner, respectively.

The critical strain as a function of nanowire diameters is shown in Figure 15 (a). For the amorphous layer coated nanowires, the critical stress increases with the increasing of the diameter of SiC nanowires, i.e. the critical stress of the thin one is lower than that of the large one. While for the uncoated nanowires, the critical stress increases to a saturation value of about 120 GPa after the diameter is larger than 3.0 nm. The critical stress of the coated nanowire is smaller than that of the uncoated one with the same diameter, and the difference of the critical stress increases with the decrease of the diameters. For example, the differences are 44 and 22 GPa for the nanowires with diameter of 2.0 and 5.2 nm, respectively. The Young's moduli determined from the stress-strain curves with the strain $<3.0\%$ using linear regression are shown in Figure 15 (b). The Young's moduli increase with increasing the diameter of SiC nanowires. But the Young's modulus of coated nanowires is smaller than that of the uncoated one with the same diameters. For example, the Young's moduli are 564 and 498 GPa for the nanowires with and without amorphous coating layer with diameter of 4.0 nm, respectively. From the above results we can see that the amorphous coating lead to the decrease of critical stress and Young's modulus, especially for the thin nanowires. The decrease of critical stress and Young's modulus is attributed to the weaking of the Si-C bonds in the amorphous coating layers [30]. The ratio of atoms in the amorphous layer and core part decreases from 0.657 to 0.207 as the diameter of nanowire increases from 2.0 to 5.2 nm. The weaking of the Si-C bonds becomes more evident in the thin nanowire, so the decrease of the critical stress and Young's modulus is more prominent.

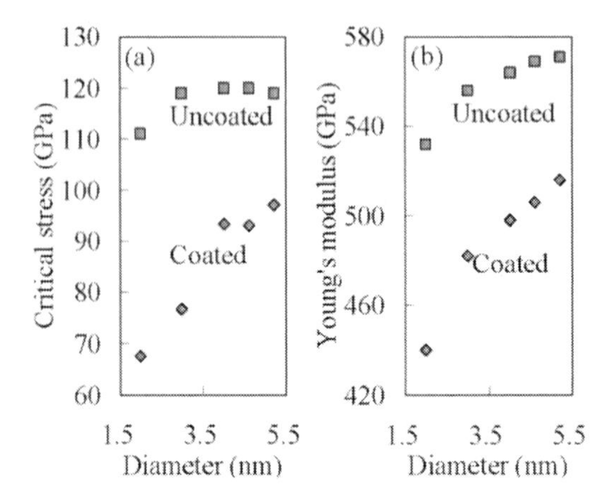

Figure 15. (a) Critical stress and (b) Young's modulus as function of diameters of SiC nanowires with and without amorphous coating layer.

When a single crystal is stretched, the fundamental deformation mechanism is a shearing action based on the resolved shear stress on an active slip system. The β-SiC is a diamond structure and its slip plane is {111} [51,52], whereas the load is along the [111] orientation in the present simulation, the resolved shear stress on the (111) plane is zero, which implies that no shearing takes place. The deformation will proceed by bond stretching and breakage without local necking for the uncoated nanowires. And the abrupt rupture of the bond leads to the stress abruptly dropping. As amorphous phase is homogeneous and plastic deformation occurs easily for the atoms move easily to release the external strain. So the amorphous layer coated nanowires show ductile behavior.

4.3. Mechanical Behavior of Twinned SiC Nanowires

The twined SiC nanowires consist of a mixture of hexagonal and cubic stackings of Si-C layers as shown in Figure16. Hexagonal (2H) and cubic (3C) stackings are represented by ABAB... and ABCABC..., respectively. One interesting observation is that the twin boundaries of SiC nanowires are well separated with almost equally spaced, and the nanowires have <111> axes.

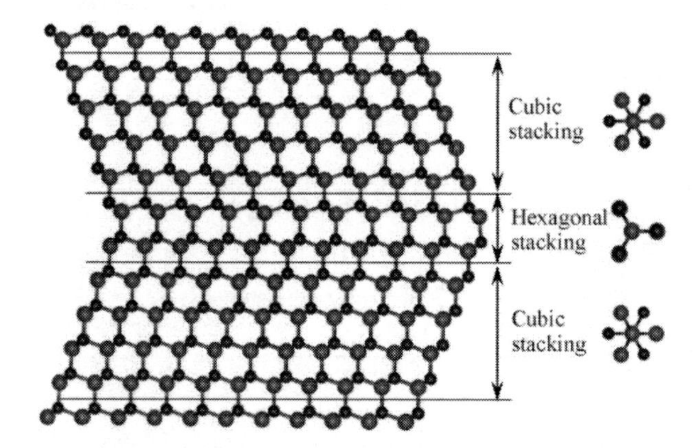

Figure 16. Atomic configuration of a twinned nanowire consists of the hexagonal and cubic stacking of Si-C layers.

The twinned SiC nanowire with its axis along the [111] direction is surrounded by {111}A and {111}B side facets which are tilted in opposite directions (by $\theta \approx 19.5°$) with respect to the nanowire axis. The A and B side facets are terminated with silicon and carbon atoms, respectively. Along the [111] direction, the {111}A edges move inward and their length increases, while the {111}B edges move outward and their length decreases. The nanowire can be constructed by rotating the twin segment through 60° or 180° and shifting about one twin segment thickness. In Figure 17(a)-(c), the models of the twinned SiC nanowires are shown in various projections with twin segment thickness (*t*) of 0.5, 1.5 and 2.5 nm. A zigzag appearance can be obviously seen from the figures. The relative stability of twinned nanowire with different segment thicknesses and diameters was investigated. Figure 17(d) shows the relative potential energies per atom of the twinned nanowires calculated at 0

K. The relative potential energy is defined as the difference the potential energy per atom in a nanowire and corresponding bulk value.

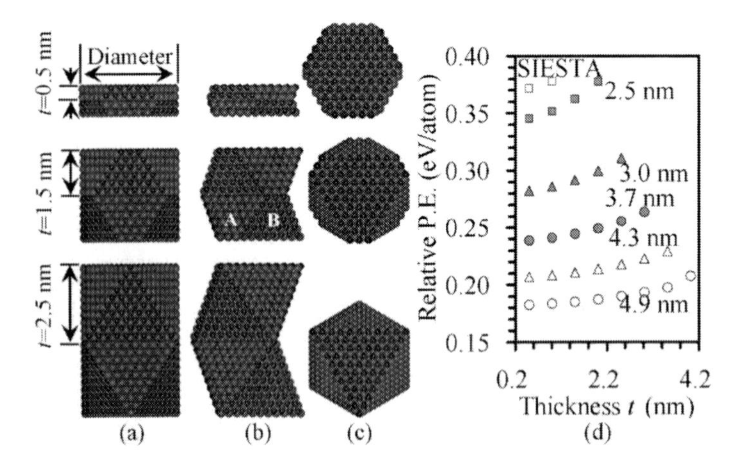

Figure 17. Atomic configurations of the twined SiC nanowires in a cubic crystal structure with non-parallel {111} side facets viewed from (a) the <110> direction, (b) the <112> direction and (c) the <111> direction. Red and blue balls represent the Si and C atoms, respectively. (d) Potential energies of the twinned nanowires with various diameters and different segment thicknesses calculated at 0 K.

The relationships between tensile stress and strain for the nanowires with diameter of 3.0 nm and different segment thicknesses are shown in Figure 18 with a strain rate of 0.005% ps^{-1} at 300 K. All the stress-strain curves are similar irrespective of the segment thickness. For low strains the tension-strain relationship follows Hooke's law, and the stress increases linearly with the increase of strain. The stress-strain curves are very smooth before the strain increases to a threshold value (namely critical strain, and the corresponding stress corresponds to critical stress).

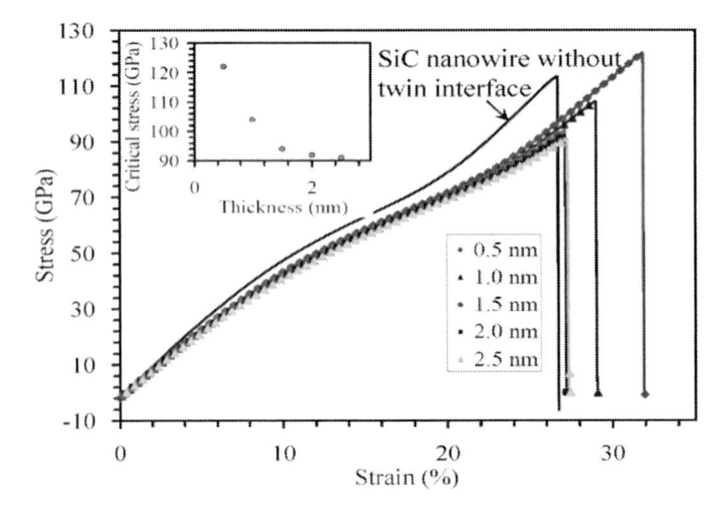

Figure 18. Stress-strain curves of the twinned SiC nanowires with a diameter of 3.0 nm and different segment thicknesses under applied tensile loading at 300 K. The inset shows the critical stress as a function of segment thickness. The stress-strain curve of a (111)-oriented nanowire with a diameter of 3.31 nm without twin stacking fault is also imposed for comparison.

The stress drops abruptly when the strain is larger than the critical strain. For example, the stress increases with increasing strain up to 121 GPa (at strain of 31.85%) and 91 GPa (at strain of 27.3%) for the nanowires with segment thickness of 0.5 and 2.5 nm, respectively, and further increases in strain lead to stress abruptly decreasing to 1.6 and 0 GPa. The critical stress as a function of segment thickness is shown in the inset of Figure 18. It is clear that the critical stress decreases with increasing the segment thickness. The evolution of atomic configurations before and after failure for the SiC nanowires with the segment thicknesses of 0.5 and 2.5 nm is shown in Figure 19 (a) and (b), respectively. Under the critical strain, the bonds of the nanowires are just stretched and preserve their original coordination in the nanowires, and no structural defects appear at this stage. Upon passing the critical strain, the atomic chain is observed for the nanowire with the segment thickness of 0.5 nm, while the nanowire with the segment thickness of 2.5 nm is ruptured almost with a clean cut. The stress-strain curve of a (111)-oriented nanowire with a diameter of 3.31 nm without twin stacking fault is also shown. It is clear that the critical strain of the twinned nanowire is larger than that of the nanowire without stacking fault.

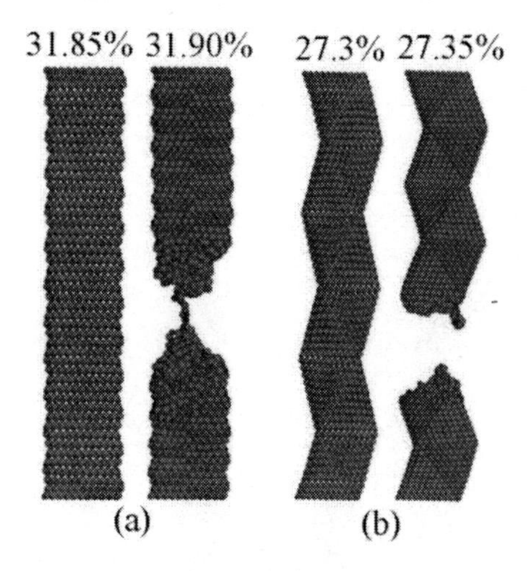

Figure 19. Atomic configurations before and after failure for the twinned SiC nanowires with a diameter of 3.0 nm and segment thickness of (a) 0.5 and (b) 2.5 nm. An atom chain is observed for the nanowire with a segment thickness of 0.5 nm.

Figure 20 shows the compressive stress-strain relationship under the applied compress for the twined SiC nanowires with a diameter of 3.0 nm and five different segment thicknesses at temperature of 300 K. All the stress-strain curves are similar irrespective of segment thickness except for the critical strain value, after which the compressive stress decreases. A linear relationship between the stress and strain can be clearly seen. The inset shows the value of critical compressive stress as a function of segment thickness. The stress decreases with the increase of thickness, which may attribute to the strength effect of twin faults.

Figure 21 shows the stress-strain curves of the nanowires with different lengths, which were simulated at 300 K with a strain rate of 0.005%/ps. For all the wire lengths employed, the stresses increase linearly with strain initially. Below the elastic limit, the stress-strain curves for the all the nanowires are almost completely overlapped, which implies that the wire length has no effect on the elastic properties of the twinned nanowires. However, the length has an effect on the buckling stress as shown in the inset of Figure 21.

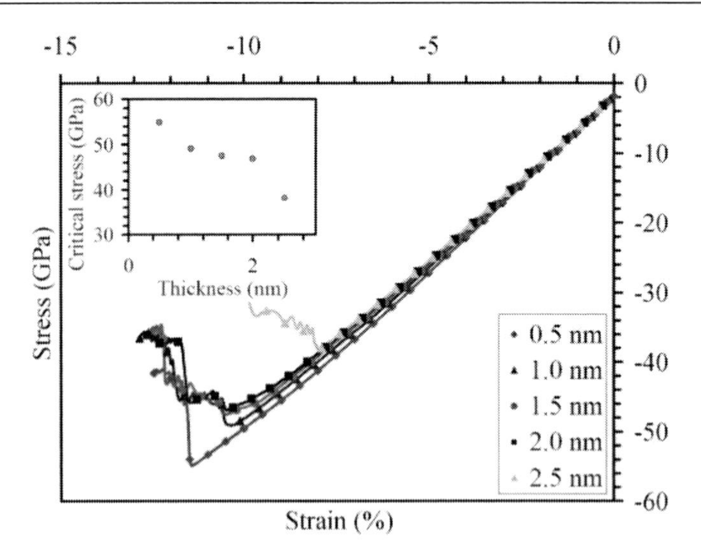

Figure 20. Compressive stress-strain curves for the twinned SiC nanowires with a diameter of 3.0 nm and different segment thicknesses at 300 K. The inset shows the critical stress as a function of segment thickness.

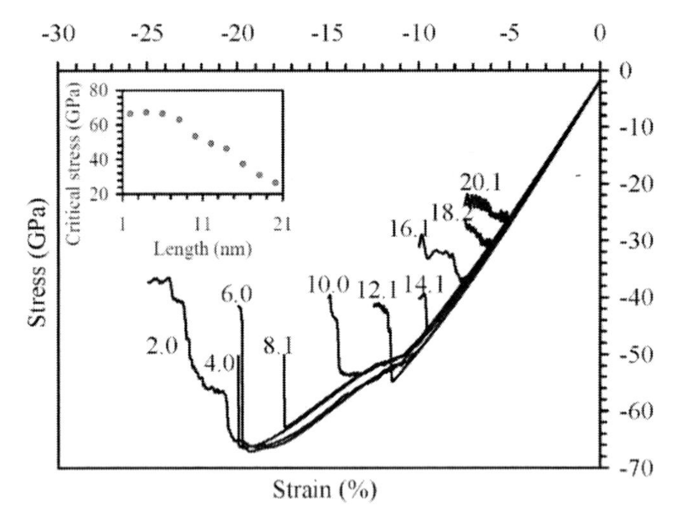

Figure 21. Compressive stress as a function of the compressive strain for the nanowires with a diameter of 3.1 nm, a segment length 1.0 nm and the lengths ranging from 2.0 to 20.1 nm at 300 K. The inset shows the effect of the wire length on critical buckling stress.

The buckling stress shows no dependence on the length that is smaller than 6.0 nm, and decreases with increasing the length, as it is larger than 6.0 nm. The buckling stress decreases from 66.3 GPa to 26.5 GPa, as the length increases from 6.0 nm to 20.1 nm. This behavior is due to the fact that there exit two different buckling modes, i.e. shell buckling mode for short nanowire lengths and columnar buckling for longer nanowire lengths. Shell buckling means local buckling with lobes and half waves along the wire, but the wire's axis remains straight [53]. On the other hand, columnar buckling means global buckling, where the nanowire remains it original cross section and buckles sideways as a whole similar to beam bending [53]. As for columnar buckling, the critical buckling load can be predicted using Euler theory

[53]: $P_{cr} = \frac{\pi^2 EI}{L_e^2}$, where E is the Young's modulus, L_e is the effective length of the wire, and I is the moment of inertia. The critical load is inversely proportion to the square of the length, and thus, the critical stress decreases with increasing length. Representative atom configurations of the twined nanowires with the length of 6.0, 8.0 and 14 nm under compression are shown in Figure 22. As for the nanowire with a length of 6.0 nm, the nanowire retains its hexagonal symmetry until the strain is reached 19.95%, after which the nanowire begins to buckle. The region near point "A" is highly compressed and significant rearrangement and even randomization of the atoms occur to release the built up stain in this region. As for the nanowires with the lengths of 8.0 and 14 nm, their lengths decrease with increasing the compression, but remain straight up to the critical strain; As the compressive stress further increases, the nanowires start to buckle, and become curved, instead of remaining straight. Under uniaxial compressive strain, the SiC nanowires buckle with two modes, shell buckling and column buckling, which depend on the length of the nanowires, i.e. shell buckling for short nanowire lengths and columnar buckling for longer nanowire lengths.

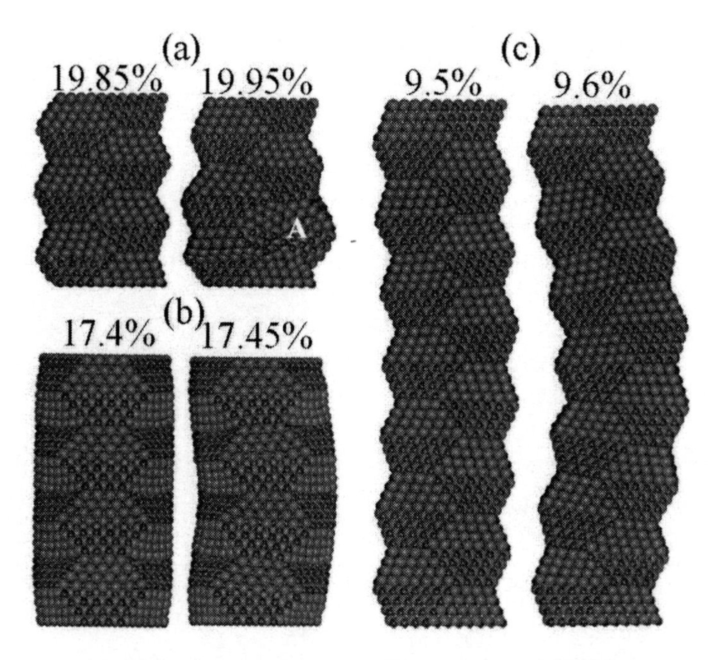

Figure 22. Atomic configurations show the two types of buckling modes for the nanowires with a diameter of 3.1 nm, a segment length 1.0 nm and the lengths of (a) 6.0, (b) 8.0 and (c) 14 nm.

The surface atoms and stacking fault all can affect the mechanical properties of the twinned SiC nanowires. The critical strain of the twinned nanowire is larger than that of the perfect one with the same diameter either under tensile or compressive loading. Under tensile loading the critical strain decreases from 31.85 to 27.3% for the nanowires with a diameter of 3.0 nm and segment thicknesses increasing from 0.5 to 2.5 nm. However, the ratio of the surface atoms increases from 0.18 to 0.20, the difference between them is very small, thus stacking fault is the key factor to affect the critical strain. As the twined SiC nanowires consist of a mixture of hexagonal and cubic stacking of Si-C layers, the percentage of hexagonal stacking is called the hexagonality [54]. The critical strain of the twinned nanowire is larger than that of the perfect one should be related to the hexagonality. The hexagonality for a perfect 3C-SiC nanowire is zero, and the hexagonality decreases from 49.7% to 8% for the nanowires with a diameter of 3.0 nm and segment thicknesses increasing from 0.5 to 2.5 nm. The mechanical behavior can be strengthened by the hexagonal stacking. The hexagonalities are 50%, 33% and 0 and the Young's moduli are 448.2, 442.6 and 425.3 GPa for the 4H-, 6H- and 3C-SiC polytypes [54], respectively. The Young's modulus increases with increasing the hexagonalities. Our simulation Young's moduli decrease from 499 to 469 GPa as the segment thickness is increased from 0.5 to 2.5 nm of the nanowires with a diameter of 3.0 nm. The stacking fault strengthens the mechanical behavior of twinned SiC nanowires.

Conclusion

The perfect SiC nanowires deform through bond-stretching and breaking in response to the axial tension. Under axial compressive strain, the collapse of SiC nanowires through yielding or column buckling mode depends on the length and diameter of the nanowires. The nanowire collapses through a phase transformation from crystal to amorphous structure in several atomic layers under torsion strain.

Amorphous coating leads an appearance of plastic deformation. The amorphous coating also leads to the decrease of Young's modulus. The critical stress and Young's modulus are decreased by the occurrence of the amorphous layer; the decreasing can be attributed to the weakening of the Si-C bonds in the amorphous coating layers. The amorphous layer does not affect the fracture mode of nanowires with large diameter and thin coating layer.

For the twinned nanowires, the segment thickness and diameters have an important effect on the mechanical behavior of the SiC nanowires. The critical strain of the nanowire can be enhanced by the twin stacking fault. The critical strains of the twinned nanowires are larger than those of the corresponding perfect nanowires with the same diameters. An atomic chain is observed for the thin nanowires with the small segment thicknesses. Under axial compressive strain, the deformation of the twinned SiC nanowires depends on the length and diameter of the nanowires, and exhibits two differently failure modes (shell buckling for short length nanowires and columnar buckling for longer length nanowires).

Acknowledgments

Z. Wang was financially supported by the National Natural Science Foundation of China (10704014) and the Young Scientists Foundation of Sichuan (09ZQ026-029) and UESTC (JX0731).

References

[1] Fissel, A.; Schroter, B.; Richter, W. *Appl. Phys. Lett.* 1995, 66, 3182-3184.
[2] Devaty, R. P.; Choyke, W. *J. Phys. Stat. Sol.* (a) 1997, 162, 5-38.
[3] Canham, L. T. *Appl. Phys. Lett.* 1990, 57, 1406-1408.

[4] Cullis, A. G.; Canham, L. T.; Calcott, D. J. J. *Appl. Phys*. 1997, 82, 909-965.
[5] Brus, L. E. *J. Chem. Phys*. 1983, 79, 5566-5571.
[6] Dai, H. J.; Wong, E. W.; Lu, Y. Z.; Fang, S. S.; Lieber, C. M. *Nature* 1995, 375, 769-772.
[7] Han, W.; Fan, S.; Li, Q.; Liang, W.; Gu, B.; Yu, D. *Chem. Phys. Lett*. 1997, 265, 374-378.
[8] Zhou, X. T.; Lai, H. L.; Peng, H. Y.; Au, F. C. K.; Liao, L. S.; Wang, N. *Chem. Phys. Lett*. 2000, 318, 58-62.
[9] Lai, H. L.; Wong, N. B.; Zhou, X. T.; Peng, H. Y.; Au, F. C. K.; Wang, N. *Appl. Phys. Lett*. 2000, 76, 294-296.
[10] Han, R. J.; Xu, X. G.; Hu, X. B.; Yu, N. S.; Wang, J. Y.; Tian, Y. L.; Huang, W. X. *Opt. Mater*. 2003, 23, 415-420.
[11] Li, Y. B.; Xie, S. S.; Zhou, W. Y.; Ci, L. J.; Bando, Y. *Chem. Phys. Lett*. 2002, 356, 325-330.
[12] Gao, F. M.; Yang, W. Y.; Wang, H. T.; Fan, Y.; Xie, Z. P.; An, L. A. *Cryst. Growth Des*. 2008, 8, 1461-1464.
[13] Wei, G. D.; Qin, W. P.; Wang, G. F.; Sun, J. B.; Lin, J. J.; Kim, R.; Zhang, D. S.; Zheng, K .Z. J. Phys. D: *Appl. Phys*. 2008, 41, 235102.
[14] Cai, K. F.; Lei, Q.; Zhang, A. X. J. *Nanosci. Nanotechnol.* 2007, 7, 580-586.
[15] Li, G. Y.; Li, X. D.; Wang, H.; Liu, L. *Solid State Sci*. 2009, 11, 2167-2172.
[16] Fu, D. J.; Zeng, X. R.; Zou, J. Z.; Li, L.; Li, X. H.; Deng, F. J. *Alloys. Comp*. 2009, 486, 406-409.
[17] Zhang, E. L.; Tang, Y. H.; Zhang, Y.; Guo, C. *Physica E* 2009, 41, 655-659.
[18] Gundiah, G.; Madhav, G. V.; Govindaraj, A.; Seikh, M. M.; Rao, C. N. R. *J. Mater. Chem*. 2002, 12, 1606-1611.
[19] Feng, D. H.; Jia, T. Q.; Li, X. X.; Xu, Z. Z.; Chen, J.; Deng, S. Z.; Wu, Z. S.; Xu, N. S. *Sold. State. Commun*. 2003, 128, 295-297.
[20] Liu, X. M.; Yao, K. F. *Nanotechnology* 2005, 16, 2932-2935.
[21] Liang, C. H.; Meng, G. W.; Zhang, L. D.; Wu, Y. C.; Cui, Z. *Chem. Phys. Lett*. 2000, 329, 323-328.
[22] Li, K. Z.; Wei, J.; Li, H. J.; Li, Z. J.; Hou, D. S.; Zhang, Y. L. *Mater. Sci. Eng*. A 2007, 460-461, 233-237.
[23] Pan, Z. W.; Lai, H. L.; Au, F. C. K.; Duan, X. F.; Zhou, W. Y.; Shi, W. S.; Wang, N.; Lee, C. S.; Wong, N. B.; Lee, S. T.; Xie, S. S. *Adv. Mater*. 2000, 12, 1186-1190.
[24] Senthil, K.; Yong, K. *J. Mater. Chem. Phys*. 2008, 112, 88-93.
[25] Wong, E. W.; Sheehan, P. E.; Lieber, C. M. *Science* 1997, 277, 1971-1975.
[26] Tersoff, J. *Phys. Rev. Lett*. 1986, 56, 632-635.
[27] Tersoff, J. *Phys. Rev*. B 1989, 39, 5566-5568.
[28] Nordlund, K.; Keinonen, J; Mattila, T. *Phys. Rev. Lett*. 1996, 77, 699-702.
[29] Devanathan, R.; de la Rubia, T. D.; Weber, W. J. *J. Nucl. Mater*. 1998, 253, 47-52.
[30] Ivashchenko, V. I.; Turchi, P. E. A.; Shevchenko, V. I. *Phys. Rev*. B 2007, 75, 085209.
[31] Gao, F.; Weber, W. J. *Appl. Phys. Lett*. 2003, 82, 913-915.
[32] Gao, F.; Weber, W. J. *Phys. Rev*. B 2006, 66, 024106.
[33] Devanathan, R.; Weber, W. J.; Gao, F. J. *Appl. Phys*. 2001, 90, 2303-2309.
[34] Porter, L. J.; Li, J.; Yip. S. J. *Nucl. Mater*. 1997, 246, 53-59.
[35] Tang, M. J.; Yip, S. *Phys. Rev*. B, 1995, 52, 15150-15159.

[36] Luo, X.; Qian, G. F.; Fei, W. D.; Wang, E. G.; Chen, C. F. *Phys. Rev*. B 1998, 57, 9234-9240.

[37] Gao, F.; Weber, W. J. *Phys. Rev*. B 2000, 63, 054101.

[38] Osetsky, Y. N.; Serra, A. *Phys. Rev*. B 1998, 57, 755-763.

[39] Vashishta, P.; Kalia, R. K.; Nakano, A.; Rino, J. P. *J. Appl. Phys*. 2007, 101, 103515.

[40] Haile, J. M. *Molecular Dynamics Simulation* (Wiley, New York, 1992).

[41] Ju, S. P.; Lin, J. S.; Lee, W. J. *Nanotechnology* 2004,15: 1221-1225.

[42] Wenzien, B.; Käckell, P.; Bechstedt, F. *Surf. Sci*. 1994, 307, 989-994.

[43] Sabisch, M.; Kürger, P.; Pollmann, J. *Phys. Rev*. B 1995, 51, 13367-13380.

[44] Yang, W.; Araki, H.; Tang, C. C.; Hu, Q. L.; Suzuki, H.; Noda, T. J. *Nanosci. Nanotechn*. 2005, 5, 255-258.

[45] Yang, W.; Araki, H.; Tang, C. C.; Thaveethavorn, S.; Kohyama, A.; Suzuki, H.; Noda, T. *Adv. Mater*. 2005, 17, 1519-1523.

[46] Kang, B. C.; Lee, S. B.; Boo, J. H. *J. Vac. Sci. Technol*. B 2005, 23, 1722-1725.

[47] Ryu Y.; Yong, K. *J. Vac. Sci. Technol*. B 2005, 23, 2069-2072.

[48] Baek, Y.; Ryu, Y. H.; Yong, K. J. *Mater. Sci. Eng*. C 2006, 26, 805-808.

[49] Ye, H. H.; Titchenal, N.; Gogotsi, Y.; Ko, F. *Adv. Mater*. 2005, 17, 1531-1535.

[50] Kholmanov, I. N.; Kharlamov, A.; Barborini, E.; Lenardi, C.; Bassi, A. L.; Bottani, C. E.; Ducati, C.; Maffi, S.; Kirillova, N. V.; Milani, P. J. *Nanosci. Nanotechnol*. 2002, 2, 453-456.

[51] Stevens, R. J. *Mater. Sci*. 1970, 5, 474-478.

[52] Sitch, P. K.; Jones, R.; Oberg, S.; Heggie, M. I. *Phys. Rev*. B 1995, 52, 4951-4955.

[53] Timoshenko, S. P.; Gere, J. M. *Theory of elastic stability* New York, McGraw-Hill. 2nd edition, 1961, pp541

[54] Nakashima, S.; Hangyo, M. *Solid State Commun*. 1991, 80, 21-24.

In: Silicon Carbide: New Materials, Production ... ISBN: 978-1-61122-312-5
Editor: Sofia H. Vanger

Chapter 6

POTENTIALITIES AND LIMITATIONS OF SIC IN THE LIQUID-STATE PROCESSING OF AL-MMCS

M. I. Pech-Canul[1]
Cinvestav Saltillo, Saltillo Coahuila, México

ABSTRACT

By virtue of its physicochemical, thermal and mechanical properties, silicon carbide has been widely recognized as a leading candidate for the manufacture of aluminum matrix composites. However, when designing Al composites reinforced with SiC, consideration of the processing route is of prime importance. Particularly, when Al/SiC composites are processed by the liquid state route, there always exists the potential of affecting the SiC reinforcements, going from a mild sensitizing to a severe degradation and eventually to a reactive dissolution, accompanied by the formation of the deleterious aluminum carbide (Al_4C_3) phase. And even though in many cases the composite may exhibit a good appearance, the presence of Al_4C_3 might sabotage its performance in terms of time, stress level and atmospheric conditions. The latter is on account of the susceptibility of Al_4C_3 of reacting with humidity in the atmosphere. In the worst case scenario the composite collapses. Certainly, a number of processing factors determine whether the reinforcements will be threatened or whether they will ultimately keep their properties and exhibit their full potential. Based on the current literature and on the processing→microstructure→properties paradigm, in this contribution, an analysis of the potentialities and limitations of SiC in liquid aluminum, and their consequences, is made. Likewise, suggestions for preventing SiC degradation are outlined.

1. INTRODUCTION

In spite of the promises of silicon carbide (SiC) to act as a reinforcing material in Al/SiC composites, processing by the liquid state route can lead to a number of concerns that deserve careful attention. One of the main issues is the lack of wetting; in general, liquid metals do

[1]Phone: +52 844 438 9600, Exts. 9678 and 9671; Fax: +52 844 438 9600. E-mail: martin.pech@cinvestav.edu.mx, martpech@gmail.com, martpech@hotmail.com.

not wet ceramic materials. The second subject matter to put into consideration is the development of negative reactions at the metal/ceramic interface that give place to unwanted reaction products, of which aluminum carbide (Al_4C_3) is the most deleterious one. Another (ternary) compound, Al_4SiC_4 can also be formed under non-optimal conditions. This doesn't mean, however, that SiC has to be discarded as a reinforcing phase. On the contrary, the cumulus of knowledge generated at least during the last three decades, has paved the way for an optimum production at a cost-effective and massive scale.

It is well known that in ambient air at room temperature SiC is highly stable, it is not corroded and conserves its physical and mechanical properties. Nonetheless, owing to thermodynamic considerations, in the presence of liquid aluminum SiC becomes unstable, and the tendency for its dissolution comes into existence. The effect can go from a partial to a severe or total dissolution, and the consequences can be dramatic, thus sabotaging the effectiveness of SiC within the composite. In chronological order, the factors influencing Al_4C_3 formation can viewed from two distinct perspectives, namely, during-processing, and post-processing.

Several factors are known to influence the tendency for SiC dissolution during processing: temperature, time, atmosphere type, quasi-static/dynamic condition, SiC composition and Al-alloy composition. In some cases, in terms of experimental design, the effect of the interaction of certain factors might even be more significant than the individual factors themselves; for instance, the interaction of the contact time and temperature. Besides, the shape and size of SiC also play a crucial role. On the other hand, the post-processing behavior of SiC – as part of the composite – is dictated by such factors as heat treatments (homogenization, aging), service temperature, strain level, environment and thermal cycle in service: If Al_4C_3 is formed during processing, then the most critical post-processing factor is humidity in the atmosphere or the contact with liquid water.

If formation of aluminum carbide is inevitable, then, a paradigm shift will have to be put forward, essentially one oriented toward the protection of the composites, such as surface treatments. Possible solutions to mitigate the effect of Al_4C_3 include: (i) formation of coatings on the Al/SiC composites and, (ii) stimulation of the reaction of superficial Al_4C_3 with reactants in gas or liquid state to produce more stable and water resistant phases.

Certainly, silicon carbide has a remarkable role to play as a reinforcing phase in Al/SiC composites – fabricated by the liquid state route –, and for it to achieve its full potential, specific optimization of the processing conditions still remains to be completed. In this contribution, a discussion of the advantages and limitations of using SiC for the manufacture (with liquid aluminum) of Al/SiC composites is presented. This chapter does not pretend to cover all aspects, but some of them linked to the formation of aluminum carbide; how is it formed? , thermodynamics and kinetics of formation, microstructure related aspects (crystallography, morphology, etc.), its interaction with humidity or water, and how to prevent it or mitigate its effect.

2. The Liquid State Processing of Al/SiC Composites

Although Al/SiC composites can be processed by the solid and semi-solid state routes, in addition to the liquid-state route, this particular contribution will focus exclusively in the latter. An attempt is made to present the discussion on the light of the materials science and engineering´s central paradigm, i.e., processing→ microstructure → properties → performance. Moreover, the attention is centered on the behavior of silicon carbide when it is exposed to liquid aluminum, because of the potential reactions that can take place.

According to I. A. Ibrahim et al. [1], the liquid metal-ceramic particulate mixing category includes: (a) injection of powders entrained in an inert carrier gas into the melt using an injection gun; (b) addition of particulates into the molten stream as it fills the mould; (c) addition of particulates into the melt via a vortex introduced by mechanical agitation; (d) addition of small briquettes into the melt followed by stirring (the briquettes are made from co-pressed aggregates of the base alloy powder and the solid particulates); (e) dispersion of the particulates in the melt by using centrifugal acceleration; (f) pushing of the particulates in the melt by using reciprocating rods; (g) injection of the particulates in the melt while the melt is being irradiated with ultrasound; and (h) zero gravity processing. The latter involves utilizing ultra-high vacuum and high temperatures for long periods of time [1].

Accordingly, the processes for manufacturing MMCs with the *metal as a liquid phase* can be categorized into two main groups: (i) dynamic processes and, (ii) quasi-static processes. All the above mentioned techniques from (a)-(h) fall within the first category, while those in which the ceramic preform, compact or powder bed remain static pertain to the second group. In the first group either one of the constituents or both, the ceramic and the liquid metal, are put in motion to combine with the ceramic phase, which could be in the form of continuous (fibers) or as discontinuous (particulates) reinforcements. In the quasi-static processes – with or without the aid of mechanical devices –, some sort of transport mechanism, like liquid metal movement through the channels in a porous body, or diffusion, exists. The two variants of infiltration – non-assisted and assisted – are considered within the second category.

2.1. Assisted Infiltration: Pressure Die Casting and Squeeze Casting Techniques

There is a variety of techniques by which the liquid metal is forced to fill the voids in a porous body. It can be either by the application of pressure or vacuum to the system, though some attempts have been made using electromagnetic fields or ultrasonic waves. In the case of Al/SiC composites – regarding the conventional "assisted" infiltration – most of the work done up to now has centered in the use of pressure. By this route, a preform or powder bed of SiC is placed below the liquid metal and the application of pressure forces the liquid to penetrate the interstices in the porous body. Two important variants of pressure assisted infiltration are *pressure die casting and squeeze casting*. The most important parameters in the former are the speed and pressure of the plunger, and the die temperature. Low die filling rates avoid the deformation of the preform and reduce turbulence and gas pickup. A low filling rate is obtained by combining a low plunger speed with a suitable size of the ingate. In pressure die casting, the compression of the preform is higher than in squeeze casting owing to much higher melt velocity and infiltration rates [2]. The squeeze casting method has been associated with a homogeneous dispersion of the reinforcement particles in the metal matrix, which is almost pore free, showing complete infiltration of the aluminum melt into the SiC preform [3].

The main advantages of pressure assisted infiltration are that the speed of the liquid incorporation can be controlled and that the threat for the occurrence of negative reactions is minimized. However – considering that both, preforms or powder beds can be used –, there are some inherent drawbacks, because the preform can be deformed or even worse, broken; in the case of a powder bed, particle segregation can also be induced.

2.2. The Pressureless, Non-Assisted or Spontaneous Infiltration Technique

To reduce to its most elemental terms, *pressureless*, *non-assisted* or *spontaneous* infiltration refers to the process by which the interstices in a porous body are filled by the liquid metal without the aid of pressure, vacuum, electromagnetic field or any kind force, except for the gravitational one. When the body is on top of the liquid metal, incorporation occurs by capillary action [4, 5]. If the liquid is on top of the solid, gravity plays a dominant role, and then the most appropriate term in this case is "percolation". Despite this difference, in both cases, the existence of wetting of the solid by the liquid is paramount. And even when it could be thought that gravity is assisting to a certain extent, that is true, but it is still considered as non-assisted because "gravity does not imply any cost". It should be made clear however, that the term "spontaneous" does not refer to the kinetics or that incorporation of the liquid occurs suddenly or instantaneously, but that the thermodynamic conditions are favorable for it to occur. The incorporation depends upon a number of factors, including the nature (chemical composition, non-oxide or oxide ceramic, etc.) of the metal/ceramic system itself, temperature, time, atmosphere and ceramic surface condition, amongst others. A schematic of the pressureless infiltration process is shown in Figure 1. It is well established that the processing parameters impact two fundamental wettability parameters, the contact angle (θ) and liquid-vapor surface tension (γ_{lv}).

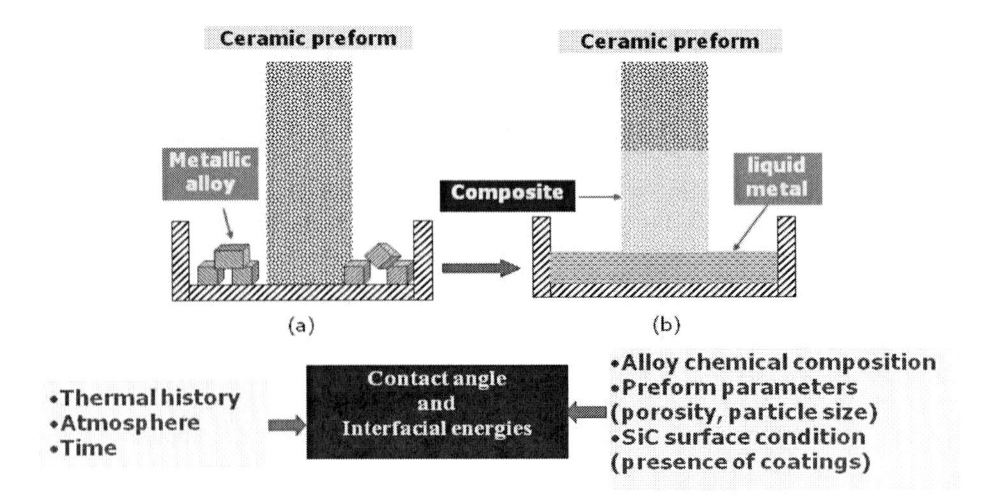

Figure 1. Schematic representation of the pressureless infiltration process.

It is evident that in the absence of an applied force, the ceramic preform is less prone to deformation or in the worst case scenario, to fracture. However, the infiltration kinetics can be negatively affected and the development of deleterious reactions at the ceramic/metal interface is more likely to occur. A deep understanding of wettability and more specifically, of reactive wetting – of the specific system – is required in order to engineer the metal/ceramic interface composition and strength. A study based on the thermodynamics and kinetics, and on the correlation of the processing parameters with composite microstructure and properties, is considered to be necessary.

It should be recognized that both, the quasi-static and the dynamic processes posses their own pros and cons for the Al/SiC system with aluminum in liquid phase. As for dynamic processes, due to the friction effect, if SiC is used in the form of fibers or fine particles, then, they will tend to dissolve and trigger Al_4C_3 formation. The assisted infiltration variants will also have a similar effect. On the other hand, due to the longer processing times, in the case of pressureless infiltration – if the processing conditions are not optimized –, the tendency for SiC dissolution also there exists, and it will be more intense with decrease in SiC particle size. Nonetheless, under optimum processing parameters, non-assisted infiltration allows avoiding preform deformation and/or fracture, and particle segregation. Consequently, a more uniform reinforcement distribution and homogeneous aluminum matrix composition are more likely to be produced.

Either activated in quasi-static or dynamic processes, it is of prime importance to analyze the SiC behavior when under inappropriate processing parameters it is put in contact with liquid aluminum. Two possibilities are pictured, a *partial* or *total dissolution*. The next section is devoted precisely to such analysis.

3. Potential Interfacial Reactions between SiC and Aluminum

Compared with interfaces in conventional monolithic alloys, in discontinuously reinforced MMCs the interfaces play a more dominant role. The interfacial area is higher and the interfaces are generally of high energy and chemically unstable. And contrasting with multiphase alloys, MMCs are generally thermodynamically unstable systems, so there is a tendency for reactions to occur between the matrix and the reinforcements during high temperature exposure. High temperature exposures can occur during fabrication, heat treatment or in service. However, particularly with the metal in liquid state, the fabrication stage is usually the most influential, because for most composites this involves the highest temperature [6].

The reaction of SiC with molten aluminum at temperatures above 700 °C is perhaps the most prominent example of these types of reactions. It is associated with the dissolution of SiC by liquid aluminum, according to the following reaction equation [6, 7]:

$$3\ SiC_{(s)} + 4\ Al_{(l)} \leftrightarrow Al_4C_{3(s)} + 3\ Si_{(\mathrm{in}\ l\ \mathrm{Al})} \quad (1)$$

It has been reported that between 800 and 1000 °C, the free energy change for reaction (1) is positive and does not spontaneously proceed from the thermodynamic point of view [7]. On the other hand, investigations of phase equilibria in the ternary system Al-C-Si have shown that silicon carbide is attacked by pure aluminum at temperatures higher or equal to 650 ± 3 °C and up to about 1327 °C [8].

Although the vast majority of investigations report Al_4C_3 as the only carbide compound produced by the reaction between SiC and aluminum, a ternary Al_4SiC_4 compound might also be formed in accordance to [1].

$$4\ SiC_{(s)} + 4\ Al_{(l)} \leftrightarrow Al_4SiC_{4(s)} + 3\ Si_{(in\ l\ Al)} \quad (2)$$

The detrimental effect associated to the formation of binary and ternary aluminum carbides can be better understood *chronologically* from two different perspectives, namely, during processing and post-processing. During processing, Al_4C_3 formation has been associated with a negative impact on the fluidity of the melt [9]. This is because of the increase in the viscosity of the melt. It's easy to recognize that in the case of infiltration, it blocks the porous bodies' channels.

Because of these deleterious effects, fabrication of SiC/Al composites devoid of Al_4C_3 has become a major challenge. The post-processing effects of reaction (1) in its turn can be separated into two main groups, mechanical and corrosion behavior: (i) the mechanical properties of SiC will be degraded due to the formation of Al_4C_3, which on account of its brittleness, degrades the interfacial strength and consequently, the composite strength; (ii) Si formation as an interfacial reaction product will produce Al-Si eutectic at the interface and the grain boundary regions, resulting in poor mechanical properties of the composite, and (iii) the composite can be susceptible to corrosive environments because the reaction product Al_4C_3 is unstable in some environments (water, methanol, HCl, etc.) [10].

Hydrolysis of Al_4C_3 – or its reaction with water – can occur according to the following possible reactions [11, 12]:

$$Al_4C_{3(s)} + 12\ H_2O_{(g)} \rightarrow 4\ Al(OH)_{3(s)} + 3\ CH_{4(g)} \quad (3)$$

$$Al_4C_{3(s)} + 18\ H_2O_{(l)} \rightarrow 4\ Al(OH)_{3(s)} + 3\ CO_{2(g)} + 12\ H_{2(g)} \quad (4)$$

Apparently, either methane or carbon dioxide is formed as gas reaction product, depending on whether water is in liquid state or as vapor. An in situ analysis of the gas reaction products is required because the mere formation of bubbles allowed speculating that reaction (4) took place in an experiment of 120 h exposure in water of SiC/6061 composites containing Al_4C_3 [12]. It is clear from reactions (3) and (4) that the effect of reaction (1) goes beyond the simple generation of the unwanted aluminum-carbide reaction products, because it is not only the presence of aluminum carbide that weakens the composite, but also the further formation of aluminum hydroxide. Ultimately and dramatically, the composite can disintegrate back into powders.

As observed earlier [7], it is not difficult to understand that reaction (1) proceeds faster when the grain size of SiC is smaller. Associated to this aspect is the morphology of the SiC reinforcements because, just to illustrate, the kinetics and mechanism of dissolution will not be the same for fine fibers and for large particles. This will also influence the way in which the reaction products are formed, specifically, as continuous layers surrounding the SiC reinforcements or as isolated precipitates [1]. Fine fibers will tend to dissolve forming isolated precipitates, the severity of which will depend on one additional but not less important factor, the existence of a quasi-static or dynamic condition. In all likelihood, the latter – for instance, in the presence of agitation – will favor the formation of isolated fine precipitates.

Several investigations have been devoted to study the kinetics and mechanism of SiC dissolution reaction or the chemical interaction between SiC and liquid aluminum. A study of the interfacial reaction between pure Al and SiC allowed determining that *the reaction is not controlled by the diffusion of Si atoms in aluminum; the chemical dissolution of the solid SiC in molten aluminum appears to control the Al/SiC interaction*. The rate of the SiC dissolution was determined for two temperatures and expressed in distinctive equations. Certainly, for composites with Al alloys as the matrix, the equations do not apply, because alloying elements will affect the interfacial reactions. The activation energy of the SiC dissolution was also evaluated [13].

A more recent study on the mechanism and kinetics of reaction (1) was conducted using 6H silicon carbide platelets with broad Si (0001) and C ($000\bar{1}$) faces. The specimens were isothermally heated at 727 °C in a large excess of liquid aluminum. Characterization revealed that the reaction proceeds in both faces via a dissolution-precipitation mechanism. However, the polarity of the substrate surface strikingly influences the rate at which silicon carbide decomposed: dissolution starts much more rapidly on the Si face than on the carbon face, but, while a barrier layer of aluminum carbide is formed on the Si face protecting it against further attack, the major part of the C face remains exposed to liquid aluminum and thus, may continue to dissolve at low but constant rate up to complete decomposition of the α-SiC crystal [8]. The mechanism involves the migration of carbon atoms by liquid phase diffusion from places where the SiC surface is in direct contact with the metal matrix to the growing faces of Al_4C_3 crystals located at or near the metal/carbide interface. The silicon liberated in this reaction dissolves in aluminum in excess, forming a liquid Al-Si alloy. The rate at which such a decomposition reaction proceeds, as well as the morphology of the resulting reaction zone, greatly depends, however, on the polarity of the substrate surface exposed to aluminum attack. The (0001) Si face and the randomly oriented faces behave in a similar manner. These faces dissolve at a rather fast rate in aluminum and, correlatively, Al_4C_3 crystallites nucleate onto the SiC surface with the *c*-axis of the substrate. Lateral extension of these crystallites results in the formation of an adherent and continuous layer of Al_4C_3 crystals that very efficiently protect the underlying substrate from further decomposition. Due to these preferential germination and growth orientations, passivation takes place sooner at the Si face than at the randomly oriented faces. As for the C ($000\bar{1}$) C face, it exhibits a very particular behavior: it dissolves at a much slower rate than any other face (six to ten times slower), but as Al_4C_3 crystals cannot nucleate or remain fixed onto this face, passivation never occurs. Consequently, if exposed for a very long time to aluminum attack, the C face will appear more damaged than any other face [8].

The chemical interaction between aluminum and SiC was experimentally studied under atmospheric pressure over a wide temperature range (27 to 1627 °C). A thermodynamic model based on a stable and metastable phase equilibria in the Al-C-Si ternary system was set up in order to provide a general description of the chemical interaction between aluminum and SiC. According to the model, aluminum and SiC are in thermodynamic equilibrium at every temperature lower than 650 °C; at 10 °C below the melting point of pure aluminum, a quasiperitectic invariant transformation occurs in the Al-C-Si system. In this transformation, solid aluminum reacts with SiC to give Al_4C_3 and a ternary (Al-C-Si) liquid phase. The carbon content of this liquid phase is very low; its silicon content is 1.5 ± 0.4 at%. From 650 to about 1347 °C, aluminum partially reacts with an excess of SiC, leading to a metastable monovariant equilibrium involving SiC, Al_4C_3 and an aluminum-rich (Al-C-Si) ternary liquid phase, L. The carbon content of this phase, L, remains very low, whereas its silicon content increases with temperature from 1.5 ± 0.4 at% at 650 °C to 16.5 ± 1 at% at 1347 °C. In the temperature range 1397 to 1627 °C, two other three-phased monovariant equilibria can be reached by reacting aluminum and SiC. These equilibria involve on the one hand SiC, Al_4SiC_4 and a liquid phase, L', and on the other hand, Al_4SiC_4, Al_4C_3 and a liquid phase, L". The former is a stable equilibrium; the latter is a metastable one [14].

But despite various efforts to understand the chemical interaction between SiC and liquid aluminum, there is one essential fact that cannot be ignored; due to thermodynamic grounds, most of the ceramics tend to form an oxide outward layer. Thus, liquid aluminum is not in direct contact with SiC, but with a SiO_2 layer. Moreover, liquid aluminum is not bare either; it is covered by a layer of aluminum oxide.

Using a variety of metals other than aluminum, a study of the interaction with SiC allowed concluding that the reaction is governed by the SiO_2 layer on the ceramic. This particular study was conducted in air at temperatures near 1000 °C. In exposure times up to 100 h, the reaction products were silicides, silicates and carbides. The severity of the interaction was found to depend on the temperature and the ease of migration of free silicon from the ceramic to the metallic phase. The occurrence of these reactions may be deleterious in applications in which silicon based ceramics and alloys are in contact for extended periods at high temperatures [15].

The presence of the oxide layers, both on the ceramic and the metal, has implications on the wetting behavior, and usually the effect is negative. In fact, the stable silica layer obscures the true wetting behavior of the metal [16]. In the case of Al/SiC composites, because of the great affinity of the silica reduction reaction by aluminum, the SiO_2 coating of SiC particles or fibers would have positive repercussions upon wetting, and consequently on infiltration or incorporation of the liquid aluminum into a porous body of SiC. Nonetheless, experimentally it is found that this energy is dissipated before the two phases are in contact and the reaction product does not favor wetting. And since the reduction reaction does not take place during the spreading of aluminum on SiO_2, the event cannot be considered reactive wetting [17]. Particularly in the field of processing MMCs by the liquid state route, wettability is of prime importance. And given the limitations by the reactivity of SiC and liquid aluminum, a common practice has been to induce reactive wetting using coatings on SiC or modifying the Al alloy composition.

Even though some observations are encouraging, there is always a possibility of finding Al_4C_3, yet in an insidious mode. For instance in a study of the wetting of SiC by liquid aluminum using the micro-droplet technique, low contact angles were found at the SiC surface, indicating good wetting. However, around the droplet a contour of reaction products, presumably of Al_4C_3 was also observed [16].

With regard to the role of aluminum oxide, in a study evaluating wettability by means of pressure infiltration it was found that no significant differences in the threshold pressures between the different particulates (SiC, TiC, and Al_2O_3) were observed, presumably because of the oxide layer that covers aluminum [18]. Thus, the problem of SiC/Al interfaces becomes a matter of interfaces between oxides.

A detailed analysis of possible interfaces in cast aluminum-silicon based reinforced composites containing SiC indicates that several different kinds of interfaces can form. The reinforcement may be totally surrounded by primary-phase, or primary silicon, or by the eutectic between Al and Si. In addition, some of the original coatings or their reaction products in the case of coated particles (such as nickel or nickel-aluminum intermetallics in nickel coated reinforcements and Cu or Cu aluminum intermetallics in Cu coated reinforcements) may also form the interface. The reaction between the dispersoids and the alloy itself can form complex interfaces [19].

Several approaches have been put forward in order to prevent the attack of SiC by liquid aluminum, namely: a) the use of alloying elements, like silicon, which in addition, improves fluidity [20]; b) coating the SiC reinforcements with materials that protect them and that at the same time are better wetted; c) mixing the SiC reinforcements with compounds that avert the attack of SiC but promote their interaction with liquid aluminum, by some kind of sacrifice material – this is the case of SiO_2 powders, which induce formation of spinel ($MgAl_2O_4$) in the case of Al-Mg or Al-Si-Mg alloys [21]. Approach a) is based on Le Chateleir´s principle because a Si excess in the system would tend to reverse equation (1). However, as some investigations have proven, even with high levels of Si in the system Al_4C_3 cannot be totally eliminated.

From earlier studies it was understood that the reaction between SiC and aluminum occurred at the melting point of aluminum, and that the saturated value of the extent of reaction depends on the activity of silicon in liquid aluminum. With the findings that a decrease of the extent of the reaction – for Al_4C_3 formation – is obtained by addition of silicon, it was recommended to use sintered SiC, which contains free silicon. This conclusion was drawn from a comparison with pressureless sintered SiC and aluminum, where aluminum carbide was formed [7].

Several conventional and nonconventional experimental techniques have been applied for the study of the chemical interaction between SiC and aluminum. For instance, the liquid-metal X-ray diffraction, which provides insight into the melt structural characteristics relevant to the interfacial reaction above liquidus in $SiC_{(p)}$/Al composites. It allowed finding that yet with addition of 25 wt. % Si in the aluminum alloy matrix, Al_4C_3 was present, indicating that the interfacial reaction is still active. At elevated temperatures, Si, either formed from the interfacial reaction or added into the matrix, distributes as short-range order and segregates around the SiC particles [22]. Al_4C_3 formation has also been studied by x-ray and ultraviolet photoelectron spectroscopies as a function of annealing temperature. At 600 or 800 °C the disappearance of the electron density of states at the Fermi level and the Al core binding energy shifts indicates that aluminum has reacted with SiC, presumably to form aluminum carbide [23].

In an attempt to study the morphology of Al_4C_3, several techniques have been put into practice, like the electrochemical reaction, which was used to extract interfacial reaction products by a preferential dissolution technique [24]. And using this technique it was confirmed that regarding morphology, Al_4C_3 forms as hexagonal-shaped crystalline platelets [12, 24], and not as needles, as it had been suggested earlier [25]. Moreover, it was found that the interfacial reaction products (Al_4C_3 + silicon in a dendritic pattern) are formed even at 450

°C [24]. Due to the presence of Si reaction product, which will later form eutectic Al-Si (with lower melting point), heat treatments at elevated temperature will lead to an alteration of the mechanical properties of the SiC_p/Al composites. Hexagonal-platelet shaped Al_4C_3 and Si were the two major interfacial reaction products found in a study of the effect of various processing methods on the interfacial reactions in $SiC_p/2024$ Al composite prepared [10]. Remarkably, significant amounts of interfacial reactions were observed in composites prepared by the powder metallurgical hot pressing, thixoforming, and compocasting techniques, while almost no interfacial reactions occurred in composites produced by spray forming [10]. Moreover, the extent of interfacial reactions was observed to be dependent on the processing temperature and holding time.

The problem of the presence of Al_4C_3 is magnified by its subsequent hydrolysis and aluminum hydroxide formation. This is perhaps the most dramatic manifestation of the detrimental effect of reaction (1), because the composite eventually collapses. It has been found that Al_4C_3 dissolution can occur in less than 120 h when exposed to a wet environment and that failure occurs predominantly at the particle/matrix interface laden with Al_4C_3 [12]. However, we should recognize that the rate of interaction with humidity or liquid water may vary, depending on the environment, Al_4C_3content and alloy matrix composition, and it can be manifested either as a gradual or rapid degradation of the composite.

The detrimental effect of the attack of silicon carbide by liquid aluminum can be so dramatic, as illustrated in the following photomicrograph showing how a SiC particle collapses under indentation tests (see Figure 2). The composite was prepared by pressureless infiltration using an Al-Si-Mg alloy, purposely designed for enhancing wettability in non-assisted infiltration. Part of the experiment consisted in studying the effect of Si_2O particles in the SiC particulate preforms on the microstructure and properties of the composites [26].

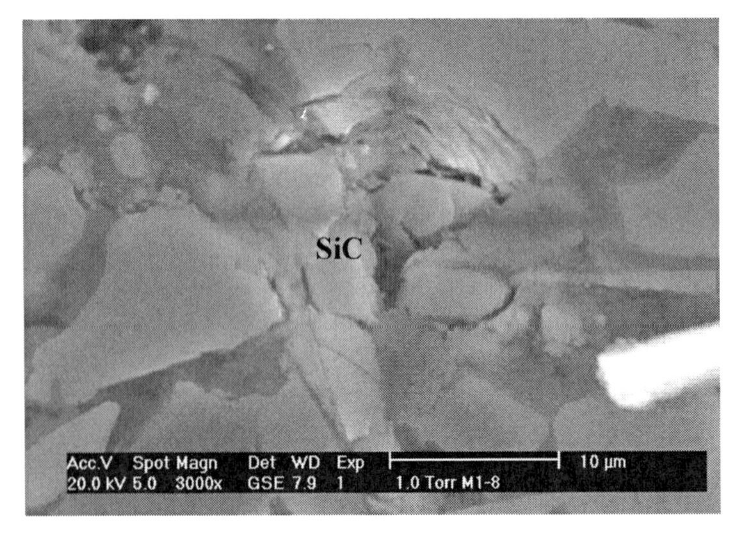

Figure 2. Photomicrograph illustrating the damage produced on SiC particles by treating in liquid aluminum under non-optimal conditions. The SiC particle cannot withstand a Vickers indentation [26].

When SiO_2 was used in the system, the final condition of the composite was ameliorated and the condition of the SiC particles enabled to perform indentation tests (see Figure 3) [26].

Figure 3. Photomicrograph illustrating the improvement produced on SiC particles by using SiO_2 powders in the system [26].

Even under the conditions used in well-known patented processes, the likelihood for Al_4C_3 formation there always exists. This is the case of Al_4C_3 and Si formed as reaction products in Al/SiC composites processed under N_2 atmosphere by the spontaneous infiltration of molten Al alloys into a bed of powder mixture of Al, Mg and SiC in the range of 700 to 1000 °C [27]. In a different work, it was recognized that in Al-Zn-Mg/SiC_p composites processed by a liquid metal processing (stir casting) followed by extrusion, the reaction between SiC and molten aluminum occurred in selective locations [28]. This confirms the difficulty to avoid the formation of Al_4C_3 completely.

In order to provide protection to the reinforcements in the presence of liquid aluminum, the oxidation of SiC has become a common practice [28]. However, since Si is always rejected in the matrix the interfacial reactions may influence the age hardening of the composites [29].

In summary, as it has been illustrated, a great deal of knowledge has been generated regarding the behavior of SiC in liquid aluminum. Although many of the investigations have been conducted with pure aluminum, it is expected that in the presence of alloying elements, the story will be different, not only with respect to the SiC behavior, but also from that of the composite behavior, especially in the presence of water or humid environments. For instance, when Si and Mg are used as alloying elements, there is a great tendency for magnesium silicide (Mg_2Si) formation during the solidification event. This intermetallic has been associated to the corrosion of Al/SiC composites, because it is electrochemically active and plays an anodic role in galvanic couples [30, 31]. On the other hand, this intermetallic has been attributed a beneficial effect on the corrosion resistance of Al-4Mg/SiC composites produced by pressure infiltration technique [32]. What is for sure is that in the presence of Al_4C_3 and Mg_2Si, the degradation process becomes complex.

4. Advantages And Limitations of the Liquid State Processing of Al/SiC Composites

The promise of SiC as an outstanding reinforcement for Al/SiC composites is unquestionable. From design, practicability and mechanical behavior viewpoints, there are numerous advantages that motivate to make these composites reach their full potential. However – due to thermodynamic grounds –, in contact with liquid aluminum, SiC may undergo decomposition via a dissolution reaction recognized as a dissolution-precipitation mechanism.

As mentioned previously, depending on the processing conditions – including particle size –, in liquid aluminum, the SiC reinforcements may undergo partial or total dissolution. The detrimental effect can be viewed from the three sides, namely, that of the reinforcement, the alloy matrix and the overall composite. It can also be analyzed considering whether the composite is in the absence or presence of water (liquid or vapor).

Assuming the absence of water, if the reinforcement is partially dissolved, then the reaction products, either as a continuous layer or isolated precipitates, will dominate the metal/ceramic interface, causing a decrease in both, the reinforcement and interface strength. In the presence of the brittle Al_4C_3 phase as an interface reaction product, failure of the load transmission from the matrix to the reinforcement is imminent. If total dissolution has occurred, then, the total failure of the composite is also impending because the SiC reinforcements will be totally replaced by Al_4C_3. Considering that there is always a particle size distribution – even when utilizing a monomodal distribution –, it can be recognized that the extent of dissolution will not be the same for all particles; some of them will be more attacked than others. An immediate consequence of all this is the alteration of the alloy composition, resulting in an unplanned microstructure. Eventually, from the mechanical behavior point of view, under an applied load, the composite will collapse. It is evident that the resulting composite will exhibit poor properties, lower than those intended or planned in the early design stage. It is even possible, that the composite exhibits a good appearance, physically, but internally, or microstructurally, it is an unsound material.

In the presence of water, the story will be different, because the corrosion products from the hydrolysis of Al_4C_3 will play the most dominant role. The failure of the composite will depend on the kinetics of hydrolysis of the aluminum carbide phase and on certain parameters, being, temperature one of the most important ones. It has been proved that the corrosion kinetics is accelerated significantly when the temperature of composites exposed to a controlled humid ambient is increased from 25 to 50 °C [33].

It should be recalled that it is the thermodynamic nature of the Al/SiC system which above all, accounts for the tendency of SiC dissolution by liquid aluminum. This is a central fact that cannot be ignored. Various processing parameters are known to influence this tendency. These include: a) time; b) temperature; c) atmosphere type; d) aluminum alloy composition; e) SiC composition, shape, and particle size; f) the presence of coatings on the SiC reinforcements; g) the existence of agitation or some sort of forced motion.

The first manifestation of inadequate processing parameters will be in the magnitude of the contact angle and liquid-vapor surface tension. It is believed that even with a careful control of the aforesaid processing parameters, in lesser or greater degree, aluminum carbide will be present. Perhaps the most it can be done is to help minimize the occurrence of such an undesirable process. In this context, the new approaches will have to consider either coating the Al/SiC composites with or inducing the reaction of Al_4C_3 in order to form more stable phases.

CONCLUDING REMARKS

The liquid state route for the processing of Al/SiC composites can be at the forefront only if a systematic correlation of the processing parameters with the microstructure, properties and performance is established. Even those variants like the pressureless infiltration technique, are promising ones. The following steps are recommended in order to undertake such a systematic investigation: a) A wettability study, based on an experimental design, and aimed at optimizing the processing parameters for obtaining both, the lowest contact angle (θ) and liquid-vapor surface tension (γ_{lv}); b) Processing of the composites using any variant with liquid aluminum; c) Interface characterization study, in order to estimate the extent of the reaction – phase identification, composition, etc. –, or even better, to make sure that the reaction products are absent; d) Mechanical property evaluation, followed by a meticulous characterization of the fracture surface – the soundness of the Al/SiC interface is of vital importance –; e) Humidity tests at various temperatures – more detailed investigations should include corrosion tests, using electrochemical techniques. Some important processing parameters to consider in the design of the experiment include: 1) SiC composition and particle size distribution, 2) Al alloy composition, 3) Processing atmosphere, 4) Presence of coatings on the SiC reinforcements, 5) Temperature, and 6) Time.

With all the valuable research work conducted until now in the field, nothing else can be predicted for SiC but the best to achieve its full potential as a reinforcing phase in Al/SiC composites produced by the liquid state route in a safe and cost-effective way. In all likelihood, Al_4C_3 will always be present, even in the minutest amount. Nevertheless, it is expected that the development of ground-breaking techniques to overcome the problem will emerge opportunely. This is supported by the continuous advances in characterization techniques and the ever growing interest and interaction of the research groups. Promising approaches include coatings on the Al/SiC composites and reactions with Al_4C_3 in order to produce stable phases, especially in the presence of water or humid environments. The most appealing approaches are those in which the surface appearance does not change dramatically or better yet, when they confer the composites an aesthetic surface finish.

REFERENCES

[1] A. Ibrahim, F. A. Mohamed, E. J. Lavernia, "Particulate reinforced metal matrix composites – a review", *J. Mater. Sci.* Vol. 26, pp. 1137-1156 (1991).

[2] R. Asthana, "Cast Metal-Matrix Composites: I Fabrication Techniques", *J. Mater. Synthesis and Processing,* Vol. 4, pp. 251-278 (1997).

[3] S. M. Seyed Reihani, Processing of squeeze cast Al6061-30v% SiC composites and their characterization, *Materials and Design*, Vol. 27, pp. 216-222 (2006).

[4] M. K. Aghajanian, M. A. Rocazella, J. T. Burke, S. D. Deck, "The Fabrication of Metal Matrix Composites by a Pressureless Infiltration Technique", *J. Mater. Sci.,* Vol. 26, pp. 447-454 (1991).

[5] M.I. Pech-Canul, R. N. Katz, M.M. Makhlouf; " Optimum Parameters for Wetting Silicon Carbide by Aluminum Alloys; *J. of Metallurgical and Materials Transactions* A, Vol, 31 A, No. 2, pp. 565-573, (2000).

[6] J. E. King and D. Bhattacharjee, "Interfacial Effects on Fatigue and Fracture in Discontinuously Reinforced Metal Matrix Composites", *Materials Science Forum,* Vols. 189-190, pp. 43-56 (1995).

[7] T. Iseki, T. Kameda, T. Maruyama, "Interfacial reactions between SiC and aluminum during joining", *J. Mater. Sci.,* Vol.19, pp. 1692-1698 (1984).

[8] J. C. Viala, F. Bosselet, V. Laurent, Y. Lepetitcorps, "Mechanism and kinetics of the chemical interaction between liquid aluminum and silicon-carbide single crystals", *J. Mater. Sci.*, Vol. 28, pp. 5301-5312 (1993).

[9] T. F. Klimowicz, "The Large-Scale Commercialization of Aluminum-Matrix Composites", *JOM,* November, pp. 49-53 (1994).

[10] J.-C. Lee, J.-Y. Byun, C.-S. Oh, H.-K. Seok and H.-O. Lee, "Effect of various processing methods on the interfacial reactions in SiCp/2024 Al composites", *Acta Mater.* Vol. 45, No. 12, pp. 5303-5315 (1997).

[11] T. Ya Kosolapova, "Carbides, Properties, Production and Applications", Plenum Press, New York, 1971.

[12] J. C. Park and J. P. Lucas, "Moisture Effect on SiCp/6061 Al MMC: Dissolution of Interfacial Al_4C_3", *Scripta Materialia*, Vo. 37, No. 4, pp. 511-516 (1997).

[13] R. Y. Lin and K. Kannikeswaran, "Interfacial reaction kinetics of Al/SiC composite during casting", in Interfaces in Metal-Ceramic Composites, Edited by R. Y. Lin, R. J. Arsenault, G. P. Martins, and S. G. Fishman, *The Minerals, Metals and Materials Society*, 1989.

[14] J. C. Viala, P. Fortier, J. Bouix, "Stable and metastable phase equilibria in the chemical interaction between aluminum and silicon carbide", *J. Mater. Sci.,* Vol. 25, pp. 1842-1850 (1990).

[15] R. L. Mehan, D. W. Mackee, "Interaction of metals and alloys with silicon-based ceramics", *J. Mater. Sci.*, Vol. 11, pp. 1009-1018 (1976).

[16] A. C. Ferro and B. Derby, "Development of a micro-droplet technique for wettability studies: application to the Al-Si/SiC system", *Scripta Metall. et Materialia,* Vol. 33, No. 5, pp. 837-842 (1995).

[17] V. Laurent, D. Chatain and N. Eustathopoulos, "Wettability of SiO_2 and oxidized SiC by aluminum", *Mater. Sci. and Eng.* A135, pp. 89-94 (1991).

[18] A. Alonso, A. Pamies, J. Narciso, C. García-Cordovilla, and E. Louis, "Evaluation of the wettability of liquid aluminum with ceramic particulates (SiC, TiC, Al_2O_3) by means of pressure infiltration", *Metallurgical Transactions* A, Vol. 24A, pp. 1423-1432 (1993).

[19] P. K. Rohatgi, S. Ray, R. Asthana and C. S. Narendranath, "Interfaces in cast metal-matrix composites", *Materials Science and Eng.* A, Vol. 162, pp. 163-174 (1993).

[20] M.I. Pech-Canul, R. N. Katz, M.M. Makhlouf, S. Pickard; "The Role of Silicon in Wetting and Pressureless Infiltration of SiCp Preforms by Aluminum Alloys; *J. of Materials Science,* Vol. 35, No. 9, pp. 2167-2173, (2000).

[21] M. Rodríguez-Reyes, M. I. Pech-Canul, J. C. Rendón-Angeles and J. López-Cuevas, "Limiting the Development of Al_4C_3 to Prevent Degradation of Al/SiCP Composites Processed by Pressureless Infiltration", *Composites Science and Technology*, Vol. 66, pp. 1056-1062 (2006).
[22] Z. Shi, J. M. Yang, T. Fan, D. Zhang, R. Wu, "The melt structural characteristics concerning the interfacial reaction in SiC(p)/Al composites", *Applied Physics A*, Vol. 71 pp. 203-209 (2000).
[23] L. Porte, "Photoemission spectroscopy study of the Al/SiC interface", *J. Appl. Phys.*, Vol. 60 No. 2, pp. 635-638 (1986).
[24] J. C. Lee, J.-I. Lee, H.-I. Lee, "Observation of three-dimensional interfacial morphologies in SiCp/Al composites and its characterization", *J. Mater. Sci. Lett.*, Vol. 15, pp. 1539-152 (1996).
[25] D. J. Lloyd, H. Lagace, A. McLeod and P. L. Morris, "Microstructural aspects of aluminum-silicon carbide particulate composites produced by a casting method", *Mater. Sci. and Eng.* A107, pp. 73-80 (1989).
[26] M. Rodríguez-Reyes, Ph. D. Thesis "Phenomenology associated to the incorporation of SiO_{2p} into SiCp preforms and its effect on the microstructure and mechanical properties of Al/SiCp composites", Cinvestav-Saltillo, Saltillo Coah. México, 2005.
[27] K. B. Lee, H. Kwon, "Interfacial reactions in SiCp/Al composite fabricated by pressureless infiltration", *Scripta Materialia*, Vol. 36, No. 8, pp. 847-852 (1997).
[28] N. V. Ravi Kumar, B. C. Pai and E. Dwarakadasa, "Microstructural evolution in liquid metal processed Al-alloy/SiC_p composites", *Int. J. Cast Metal Res.*, Vol. 15, pp. 573-579 (2003).
[29] J. G. Legoux, L. Salvo, H. Ribes, G. L. 'Esperance and M. Suery, Microstructural characterization of the interfacial region in SiC reinforced Al-Mg alloys", in Interfaces in Metal-Ceramic Composites, Edited by R. Y. Lin, R. J. Arsenault, G. P. Martins and S. G. Fish, *The Minerals, Metals and Materials Society*, 1989.pp. 187-196.
[30] N. Birbilis, R. G. Buchheit, "Electrochemical characteristics of intermetallic phases in aluminum alloys", *J. Electrochem. Soc.* Vol. No. 152, No. 4, pp. B140-B151 (2005).
[31] R. Escalera-Lozano, C. A. Gutiérrez, M. A. Pech-Canul, M. I. Pech-Canul, "Degradation of Al/SiC_p composites produced with rice-hull ash and aluminum cans", *Waste Management*, Vol. 28, pp. 389-395 (2008).
[32] S. Candan, E. Bilgic, "Corrosion behavior of Al-60 vol. % SiC_p composites in NaCl solution", *Materials Letters*, Vol. 58, pp. 2787-2790 (2004).
[33] R. Escalera-Lozano, M. A. Pech-Canul, M. I. Pech-Canul, "Degradación de compósitos Al-Si-Mg/SiCp en atmósferas húmedas", *Suplemento de la Revista Latinoamericana de Metalurgia y Materiales*, SI (3), pp. 1117-1123 (2009).

In: Silicon Carbide: New Materials, Production ... ISBN: 978-1-61122-312-5
Editor: Sofia H. Vanger

Chapter 7

EFFECTS OF ION IMPLANTATION IN SILICON CARBIDE

S. Leclerc, M. F. Beaufort, A. Declémy, and J. F. Barbot
CEMHTI, CNRS, Université d'Orléans, Orléans, France

ABSTRACT

The current status of the effects of ion implantation into SiC is reviewed with a focus on the implantation of helium ions and its consequences on the microstructure of SiC. The helium implantation results in a strain profile build-up as a consequence of the generation and accumulation of both point defects and helium-related complexes. At high fluence, the damage accumulation leads to amorphization that can be avoided by performing implantation at elevated temperature. Upon post-implantation annealing, the point defect recovery leads to the relaxation of the elastic strain and the helium-related complexes evolve into extended defects. The evolution of the strain and defects as a function of implantation temperature and annealing temperature is discussed with regard to the mobility of involved species. Moreover, all the defects are associated with a change in volume resulting in surface expansion or swelling of the implanted material that must be taken into account for the applications of SiC. Finally, the effects of implantation on the mechanical properties of SiC are discussed with regard to the microstructure changes.

INTRODUCTION

The unique features of SiC make it suitable for many electronic applications aimed at working under extreme conditions. The conventional thermal diffusion techniques used in device processing are inefficient in SiC because of low diffusivity of dopants. Ion implantation is therefore the preferred technique to embed dopants as it is compatible with usual industrial technologies and provides spatially well-defined dopant profiles that are often required to separate regions of p- and n-type conductivity in advanced devices. The implantation process however induces a massive production of defects which, although sometimes helpful (creation of semi-insulating areas, recombination centers reducing the carrier lifetime in high frequency devices [1]...), affects the microstructural and physical properties of the material. The induced defects are especially known to electrically compensate the implanted dopants therefore hindering the performance of SiC-based devices. In silicon, a non negligible part of the as-created defects is annihilated during implantation. In comparison, silicon carbide is much more efficient in creating stable defects that survive the implantation [2, 3]. Thus, post-implantation treatments such as thermal annealing are required that nevertheless show a poor efficiency in removing the unwanted defects and in activating the dopants (i. e. move the dopants onto substitutional sites). The heat treatment is moreover likely to induce the formation of stable defects and/or trigger the dopant diffusion consequently leading to the broadening of the dopant profile. The knowledge of implantation-induced defects and their behaviour under post-implantation treatments is thus essential for the success of SiC technology as well as any semiconductor material technology. Understanding the role of implantation-induced defects and accumulation of damage on the microstructural and mechanical properties of SiC is moreover crucial to predict the behaviour of the material under radiation for nuclear applications. SiC has for example been proposed as a structural component for the first wall of fusion reactors [4, 5]. An important question with respect to nuclear applications of SiC is its resistance to radiation damage that results not only from atomic displacements but also from nuclear reactions that create rare gas atoms within the material. Both potential electronic and nuclear applications have motivated a wide research on implantation-induced defects and damage in silicon carbide. It is now well-attested that the damage resulting from implantation and the follow-on properties are highly dependent on the parameters of implantation, namely particle, temperature, energy, and fluence. As a function of these parameters, implantation can lead to the mere formation of isolated point defects or to the formation of more complex defects, up to the amorphization of the material. Light ion implantations are of fundamental interest since they enable the study of the whole damage range. Among light particles, helium is particularly studied as it is produced in large amount from nuclear reactions. It is a noble gas with extremely low solubility so that the formation of bubbles, i. e., gas-filled cavities, is strongly expected. When formed at the surface, they lead to a large erosion of the material by the fracture of blisters upon annealing [6]. Such an erosion reduces the lifetime of plasma facing material in fusion devices whilst affecting the purity of the plasma. In another context, hydrogen implantation is used to create extended defects named platelets which induce upon subsequent annealing large microcracks in a narrow depth around the projected range of the hydrogen ions. The propagation and interactions of these microcracks can lead to the full layer transfer of the implanted layer when bonded to a stiffener (Smart Cut™ technology). The co-implantation of helium and hydrogen results in a significant reduction of the fluence required for the layer transfer. This process is also investigated to form high-quality and low cost SiCOI (Silicon Carbide On Insulator) structures. The control of damage introduced by light ion implantation is thus of crucial importance to improve such technological solutions.

In the following chapter, experimental results based on a Transmission Electron Microscopy (TEM) and X-Ray Diffraction (XRD) study will be provided on the microstructural changes induced by helium implantation in 4H-SiC. These changes will be reported as a function of fluence and temperature of implantation, from room temperature (RT) to 800°C. The energies of implantation range from a few tens of keV to a few MeV leading to a mean projected range of ions from a few hundred of nm to a few μm. The recovery of damage will then be considered through a thermal annealing study. The defects induced by implantation and the resulting microstructure lead to a swelling of the material increased by the presence of helium which may cause damage as severe as cracks and therefore, has to be avoided in nuclear systems. Damage also leads to changes in the mechanical properties of the material that have to be anticipated for its potential applications. Both aspects of swelling and mechanical properties will be finally reviewed. In the whole chapter, only single crystalline SiC will be regarded.

ROOM TEMPERATURE IMPLANTATION

Ion implantation into silicon carbide produces a great variety of intrinsic point defects such as silicon and carbon vacancies, related interstitials, and antisites. The latter, formed by the exchange between a silicon and a carbon atom in substitutional site have the lowest formation energy of all the intrinsic point defects [7, 8, 9]. During implantation, the point defects may diffuse and either agglomerate to form more stable defects or recombine, which is usually referred to as dynamic annealing. The concentration profile of vacancies created during room temperature implantation, as well as the concentration profile of implanted species, can be calculated using programs such as SRIM [10] that however does not take into account the effects of dynamic annealing. The calculated profiles are reported in figure 1 for three energies of implantation. Whilst the profiles of implanted atoms are roughly Gaussian, the vacancy profiles are asymmetric with a larger tail towards the surface: a non negligible concentration of defects is created in the near surface region.

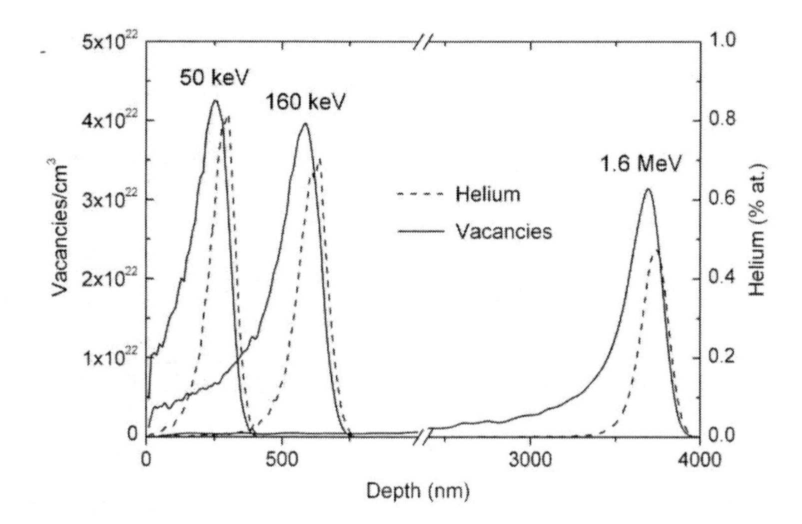

Figure 1. Vacancy and helium concentration profiles calculated by SRIM for three energies of implantation and for a fluence of $1x10^{16}$ cm^{-2}.

Each defect induces a local volume change that contributes to the overall strain of the lattice together with the implanted atoms. Unlike silicon [11], silicon carbide has vacancies and interstitials whose relaxation volumes do not cancel [12] which may lead to high strain values. Calculations made on 3C-SiC have even shown that except for the C_{Si} antisite, all the defects including vacancies induce a positive lattice volume change [13]. In thick SiC single crystals, the resultant of local volume changes induced by the defects and implanted atoms is a strong dilatation gradient localized along the direction perpendicular to the surface.

By performing XRD ω-2θ scans along this direction, the strain gradient can be fully characterized. The relation between the angular deviation $\Delta\theta$ from the Bragg angle θ_{Bragg} and the normal strain $(\Delta d/d)_N$ is given by the derivation of the Bragg's law. When the diffracting planes are not parallel with the surface and for small strains, i. e., $(\Delta d/d)_N << 1$, this relation can be written as : $(\Delta d/d)_N \approx - \Delta\theta / \cos^2\alpha \tan \theta_{Bragg}$ with α the angle between the surface normal direction and the normal to the diffracting planes [14]. The strain can thus be easily extracted from the spreading of scattered intensity on XRD curves. An illustration is given in Figure 2 with the XRD curves of samples implanted with helium ions at 50 keV. Regardless of the fluence and the energy of helium implantation in the keV-MeV range, the curves show a main sharp Bragg peak at zero strain (i. e., θ_{Bragg}), resulting from the diffraction of the unperturbed bulk. At its low angle side, a large tail of scattered intensity is observed that results from the dilatation gradient of the lattice along the surface normal direction. At the lowest fluences, this scattering tail is made up of interference fringes arising from the coherent diffraction of two damaged zones of a same level of strain on either sides of the maximum damage area. The position of the last fringe away from the Bragg peak gives the maximum strain value that is reached deep in the sample. The diffraction of the near surface zone, if crystalline, results in an intense satellite peak at the left of the main Bragg peak.

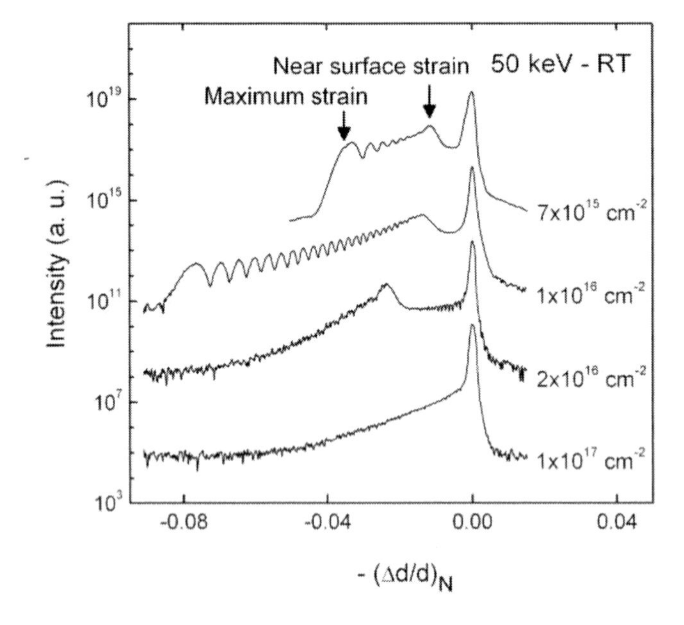

Figure 2. X-ray scattered intensity distribution along the direction normal to the sample surface close to the (0004) reflection in 4H-SiC implanted with helium ions at RT-50 keV to different fluences. $(\Delta d/d)_N$ is the normal strain.

As seen in Figure 2 and regardless of the energy of implantation, the normal strain induced by implantation in SiC single crystals can reach values of several %, including in the near surface region where the concentration of defects is the smallest encountered. At fluences for which the helium concentration does not exceed 0.5%, the strain mainly results from intrinsic point defects: The strain profiles obtained from the simulation of the curves [15] are proportional to the vacancy profile calculated by SRIM [16]. Likewise, damage profiles obtained from RBS/C experiments performed on ion implanted SiC have been shown to coincide well with the SRIM calculated profiles when implantation is performed at low temperature [17, 18]. This indicates that there is no long-range migration of the defects created during low temperature implantation as well as during room temperature implantation of helium ions. In this latter case, helium ions have also been shown to be frozen in [19]. This does not exclude short-range rearrangements and the formation of helium-vacancy complexes at helium concentration lower than 0.1% [20]. These defects are "extremely" stable and constitute the precursors for the formation of bubbles upon post-implantation annealing, as will be seen in a following subsection. These helium-vacancy defects are assumed to grow and become more stable with increasing fluence of implantation as they need higher temperature to dissociate [21, 22]. When the concentration of helium exceeds 0.5%, they have a dominant effect on the lattice expansion as shown by the concordance of the strain profile and SRIM calculated helium profile in the zones where the concentration of helium exceeds 0.5% [16].

Roughly, when the fluence increases, the concentration of defects increases inducing an increase of strain in both the near surface region and the highly damaged region where helium atoms are implanted. When a critical defect concentration is reached that corresponds to a strain value higher than 9 %, amorphization begins to take place. At these damage levels, TEM micrographs show a continuous buried amorphous layer where a dark contrasted layer containing defects not resolvable by conventional TEM was observed at lower fluences (see Figure 3). The amorphous layer is always surrounded by two dark-contrasted crystalline damaged layers containing a high density of defects as shown in Figure 3 (b). At sufficient fluence, a uniform distribution of small bubbles, 1-2 nm in diameter, is observed in the deeper part of the amorphous layer [23]. In the keV-MeV energy range, such bubbles exclusively appear in an amorphous matrix containing helium atoms at a minimum concentration of about 3% [24].

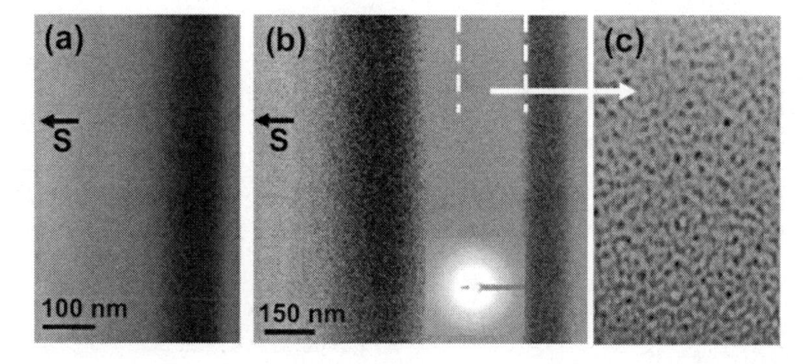

Figure 3. Cross-sectional TEM micrographs of 4H-SiC implanted with helium ions at room temperature. S denotes the surface. (a) Crystalline damaged layer as observed at fluences avoiding amorphization. Here, after implantation at 160 keV to $1x10^{16}$ cm^{-2}. (b) Damaged layer beyond the amorphization threshold (1.6 MeV to $1x10^{17}$ cm^{-2}). The damaged layer is made of a central amorphous layer surrounded by two highly defective layers. In the inset: diffraction pattern characteristic of amorphous material. (c) Small bubbles observed in the deeper part of the amorphous layer where the concentration of helium exceeds ~3%.

IMPLANTATION AT ELEVATED TEMPERATURE

During implantation at elevated temperature, simultaneous dynamic effects occur that involve a competition between the recombination and agglomeration of point defects. This may lead to the formation of complex defects that are more difficult to anneal out than those produced at room temperature [18, 22]. It also implies that a higher fluence is needed to induce the crystalline to amorphous transition. Above a critical temperature, amorphization is avoided regardless of the fluence. Reported critical temperatures vary between a few tens of degrees for electron irradiation [25] to ~300-400°C for gallium and antimony implantation [26].

When performing helium implantation at temperatures higher than 200°C, amorphization does not take place. The comparison of strain values obtained at elevated temperature with those obtained at RT shows an enhancement of the recovery of damage with increasing temperature that results from an increasing rate of recombination of point defects [19]. In particular, a strong enhancement occurs between RT and 200°C as previously observed in SiC implanted with heavier ions [27, 28]. In addition, the recovery rate increases with defect density thus being more important deeper in the sample than in the near surface region. At low fluence, this results in a smooth and soft strain gradient over the whole implanted region. At high fluence, a different behaviour is observed: the strain is strongly magnified over a thin region (~ 100 nm) deep in the sample (see Figure 4).

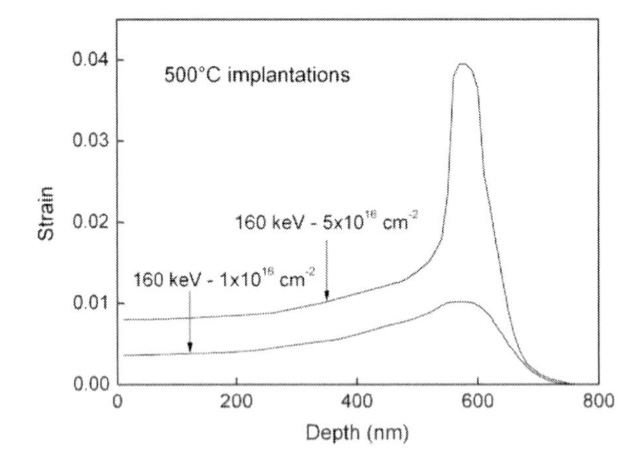

Figure 4. Strain profiles obtained from the simulation of the XRD curves obtained after helium implantation at a fixed temperature of 500°C to fluences of 1×10^{16} and 5×10^{16} cm^{-2} (160 keV). At 1×10^{16} cm^{-2}, the strain gradient is smooth and soft over the whole implanted region whilst at 5×10^{16} cm^{-2}, an abrupt variation of strain is observed in its deeper part. [19].

This magnification of strain is the consequence of the migration and accumulation of interstitial-type defects driven from the near surface region to the deep highly damaged region under the combined effects of temperature and strain gradient [19]. With increasing fluence and temperature, a further increase of strain in the deep highly damaged region may also result from the formation of helium bubbles. A layer of small bubbles (1-2 nm) with high density is indeed observed when helium implantation is performed at sufficient temperature and fluence (see Figure 5) [22, 29, 30]. In the near surface region, the outflow of interstitials and the recombination of point defects lead to the saturation of the strain beyond a threshold fluence. The strain at saturation strongly depends on the temperature of implantation; it amounts to 0.8% when implantation is performed at 50 keV to 600°C (see Figure 6) [31].

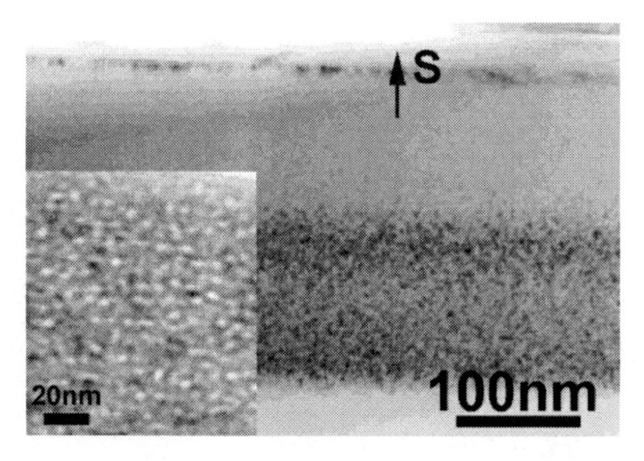

Figure 5. XTEM image of 4H-SiC implanted at elevated temperature (750°C) to $1x10^{17}$ cm^{-2} (50 keV). Inset: enlarged view of the damaged layer showing bubbles. [29]

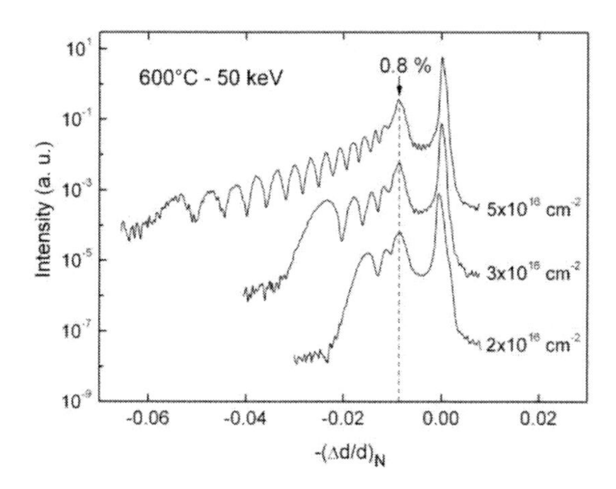

Figure 6. XRD curves obtained after helium implantation at 600°C – 50 keV to different fluences. Whilst the maximum strain increases with increasing fluence, no increase of strain is observed in the near surface region beyond a fluence of $2x10^{16}$ cm^{-2}.

ANNEALING

To anneal defects out of a wafer and to activate the dopants introduced by implantation, post-implant annealing at high temperature is required. In silicon carbide, some defects (including close Frenkel pairs) anneal out at relatively low temperature and in most cases, an annealing at temperature slightly higher than the temperature of implantation already induces a change of physical properties [32, 33]. Nevertheless, some residual damage induced by irradiation or implantation persists at temperatures higher than 1500°C as detected by TEM or techniques sensitive to point defects such as positron annihilation spectroscopy (PAS) and electron paramagnetic resonance spectroscopy (EPR) [34, 35, 36]. The presence of helium close to the defects modifies the overall behaviour of the material upon annealing since helium stabilizes vacancy-type defects via the formation of helium-vacancy complexes thus lowering the efficiency of annealing in recombining point defects. As observed in a RBS/C study, only partial recovery of the helium implantation damage is possible even when amorphization does not occur whilst complete recovery is observed for implantation with ions such as silicon ions [37].

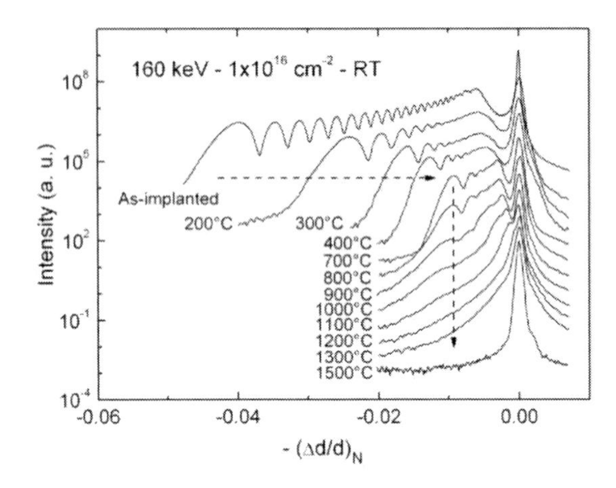

Figure 7. X-ray scattered intensity measurements along the surface normal direction close to the (0004) reflection for various annealing temperatures in 4H-SiC implanted with 160 keV helium ions to $1x10^{16}$ cm^{-2}. [16]

As a function of depth location, i. e., depending on the proximity of the helium profile, different microstructural evolutions are observed [16]. The evolution with annealing temperature of the XRD curves at fluences for which amorphization is avoided is exemplified in Figure 7. With increasing annealing temperature, the peak ascribed to the diffraction of the near surface zone continuously shifts towards the main Bragg peak. Accordingly, in the near surface region where mainly point defects are created, the strain relaxes progressively over the whole temperature range, from 100°C (not shown) to 1500°C. The near surface satellite peak and the main Bragg peak finally overlap showing that the near surface strain has completely relaxed. The last fringe away from the Bragg peak that results from the diffraction of the deep highly damaged region also shifts towards the main Bragg peak from 100°C. The relaxation of strain in both the near surface region and the highly damaged region at this low temperature concords with previous studies and is ascribed to the recombination of close Frenkel pairs [38, 39]. With further increase of temperature, interstitials migrate favouring their annihilation with vacancies [40] and a further relaxation of strain. From 600-700°C, the shift of the last fringes is stopped whilst the scattered intensity decreases. Thus, in the highly damaged region, the relaxation of strain stops at a temperature where clustering of defects occurs from stable defects, in agreement with the observation of complex helium-vacancy defects in the as-implanted state. The Figure 8 summarizes the strong microstructural evolution arising from the helium-vacancy defects that occurs in this region during the annealing process. Up to 600°C as well as in the as-implanted state, only a dark-contrasted layer is observed (figure 8(a)). At 700°C, platelets, i. e., lenticular cavities, form along (0001) planes. As seen in figure 8(b), their length reaches 10-20 nm and the strain contrasts that surround them are indicative of a high inner pressure. Upon annealing at 800°C, the platelets evolve into two-dimensional clusters of bubbles of nanometer size. This transformation starts at a temperature where the migration of vacancies takes place as observed by PAS [41]. With further increase in annealing temperature, the clusters, and the bubbles they are composed of, grow with further supply of mobile species including divacancies [41, 42]. The Figure 8(d) shows the large clusters obtained after annealing at 1400°C. Their length reaches 500 nm with bubbles a few tens of nanometers. Moreover, small dislocation loops (5-10 nm) have been punched out by the overpressurized bubbles. The weak-beam image in Figure 8(e) shows the distribution of these loops: whilst they migrate through the front strained material, they are stopped at the rear interface between the implanted region and the unimplanted bulk. The migration of loops is thus promoted by the implantation-induced strain. By migrating, the loops induce plastic deformation which results in the decrease of the scattered intensity on XRD curves. The decrease of scattered intensity is associated with an activation energy of 1.6 eV [16] which is lower than the previous values of a few eV reported for the motion of loops in 4H-SiC [43, 44] but supports the idea that the strain enhances the migration.

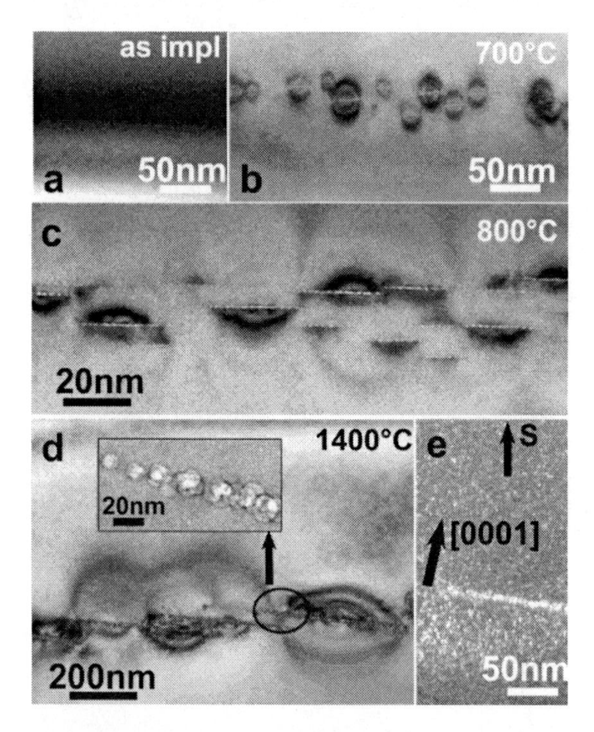

Figure 8. Cross-sectional TEM images of helium implanted 4H-SiC at 160 keV to a fluence of $1x10^{16}$ cm^{-2}: (a) as-implanted state, (b) annealed at 700°C, (c) 800°C, (d) 1400°C, (e) 1400°C weak-beam image. [16].

Annealing processes occurring in samples implanted at high fluence and the resulting microstructure are quite different due to the presence of a buried amorphous layer. In such samples, temperatures higher than 800°C are required to induce a notable change on the microstructure observed by TEM, although lower temperatures have been shown to induce an increase of the density of the amorphous network [45]. After a 800°C annealing, only minor changes are observed on the width and appearance of the damaged layers. The recrystallization occurs at a temperature between 800 and 1000°C [46-49] and results in a complex crystalline structure with inclusions of 3C-SiC and many stacking faults as seen in Figure 9. With increasing annealing temperature, the bubbles grow and reach different sizes as a function of their location within or at the periphery of the recrystallized layer [50]. In contrast to the above mentioned clusters of bubbles observed at lower fluences, their distribution is quite uniform in a defined layer.

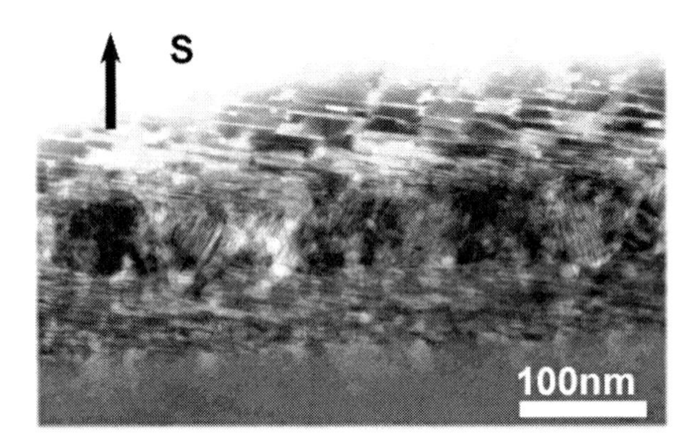

Figure 9. Bright field image of a recrystallized layer of 4H-SiC submitted to high fluence implantation of helium and subsequent annealing at 1500°C. Polytypism, stacking faults and large bubbles are observed.

Like the annealing processes and the resulting microstructures, the annealing kinetics and recovery rates depend on the as-implanted state which makes higher temperatures necessary to anneal out larger defects. It results that the damage produced by implantation at high temperature, which involves agglomeration of point defects, is more difficult to anneal out than the damage produced at room temperature [18]. Annealing of layers of bubbles produced by implantation at elevated temperature, similarly to annealing of those produced at room temperature, may lead to a coarsening of the bubbles depending on the temperature of annealing. As a function of implantation and annealing conditions, different sizes and distributions of the bubbles can be obtained. Layers of fine or large bubbles with high density (see Figure 10) and layers of planar clusters of bubbles have been observed [22, 29, 30]. In any case, it seems that the bubbles align along basal (0001) planes.

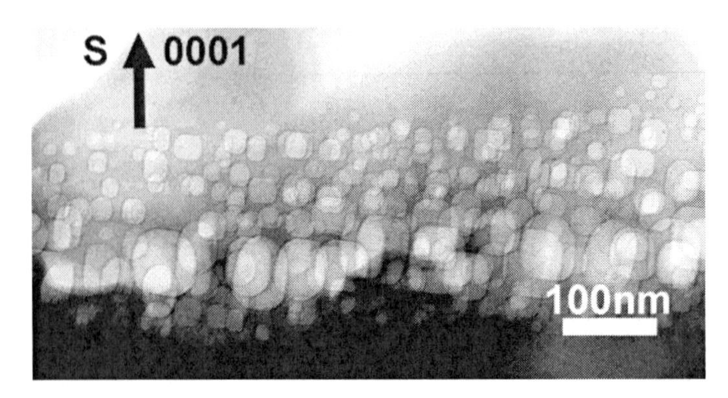

Figure 10. XTEM image of 4H-SiC implanted at elevated temperature (750°C) to $1x10^{17}$ cm^{-2} and annealed at 1500°C for 30 mn. [29]

SWELLING

When a part of the sample is masked during implantation, a step is formed at the interface between the unimplanted and the implanted areas due to the surface swelling of the material. This swelling results from different contributions whose relative weights depend on the conditions of implantation and subsequent treatments. These contributions include the contribution of point defects and small defects (including small helium-vacancy complexes) taken into account in the strain measured by XRD. Larger defects like bubbles or cavities and extended defects such as dislocations of interstitial type, which do not participate in the XRD strain, also induce a swelling of the material. As previously mentioned, the cavities grow with thermal annealing at sufficient temperature which may induce a drastic increase of the swelling [29] even though most of the point defects have recombined. During implantation at elevated temperature, dynamic annealing also results in the decrease of the swelling induced by point defects. The cavities and extended defects however induce a significant swelling under specific conditions of implantation. Recent densitometry and TEM experiments performed on neutron irradiated polycrystalline SiC have shown that the cavity swelling increases significantly with increasing temperature and fluence of irradiation above ~1100°C [51, 52, 53]. A significant volume expansion is also observed when amorphization takes place. Reported values in the literature range from 12% to 23% [54-58]. The discrepancy of the values can be partly explained by the variety of the techniques used. Annealing studies on amorphous SiC have also shown a decrease of the swelling upon annealing at temperature lower than the recrystallization temperature showing that different density states exist for the amorphous material [45].

CHANGE OF MECHANICAL PROPERTIES

There has been great interest in determining the mechanical property changes of SiC under irradiation or ion implantation to predict its behaviour in nuclear applications or to tailor the near-surface properties. In case of ion implantation, the implanted layer is thin resulting in a composite system and the accurate determination of its mechanical properties is difficult. Nanoindentation tests are therefore well adapted to study the mechanical properties (hardness and elastic modulus) of implanted materials as well as their variation with depth. Moreover the use of Atomic Force Microscopy (AFM) for nano-scale observations of surface around the indent imprints can provide additional information concerning the deformation mechanisms on the as-modified materials.

Several studies of the mechanical properties of neutron irradiated SiC have been reported in the literature as a function of irradiation temperature [59]. In particular, it has been shown that the damage increases the fracture toughness of SiC in the intermediate irradiation temperature range (<800°C) [60]. However, the microstructural evaluation of SiC for fusion application requires simultaneous displacement damage and helium generation that can be studied through helium implantation. In this section, we discuss the mechanical properties of helium implanted SiC.

The typical continuous load-displacement curves of as-grown SiC shows an abrupt increase of the penetration depth at a given load. This discontinuity (or "pop-in"), generally observed in ceramics and semiconductors shows the transition from purely elastic to elastoplastic behaviour and is ascribed to the homogeneous nucleation of dislocations [61]. The as-deformed volumes are small and considered as dislocation free. The absence of pop-in in strained SiC (implanted at low fluence) shows that the point defects generated by the implanted ions along their path promote the heterogeneous dislocation nucleation [62]. Figure 11 shows typical AFM images of the imprints in 4H-SiC. Radial cracks are observed at the corners of the imprint only in the as-grown state (Figure 11a). The high residual compressive in-plane stress induced by the ion implantation impedes thus the formation of cracks. Moreover, unevenness associated to slip lines observed inside the imprints corroborate that the plastic deformation of SiC is mainly governed by dislocation nucleation and slip [63]. At high fluence of implantation where the amorphization takes place from the surface, neither slip lines nor cracks are observed as expected for deformation behaviour of amorphous solids. Similar results have been reported after micro-indentation [64] scratch [56] and wear tests [65] on implanted SiC.

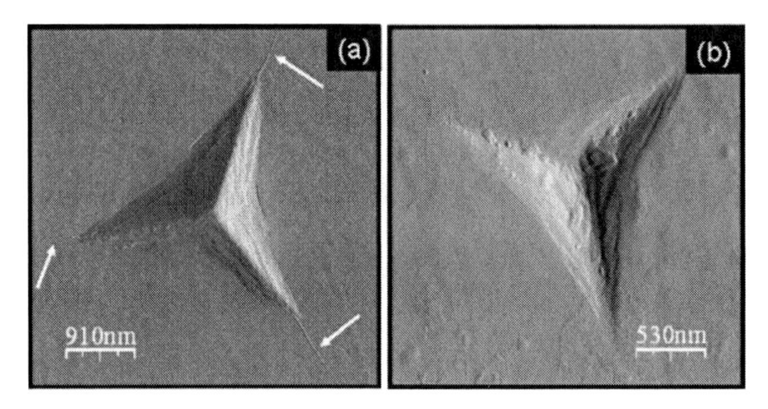

Figure 11. AFM image of the imprints of (a) as-grown and (b) "helium-damaged" (RT, 50 keV, $7x10^{15}$ $he.cm^{-2}$) 4H-SiC. Cracks on the corner of the imprint are only observed in bulk SiC (arrows). [62].

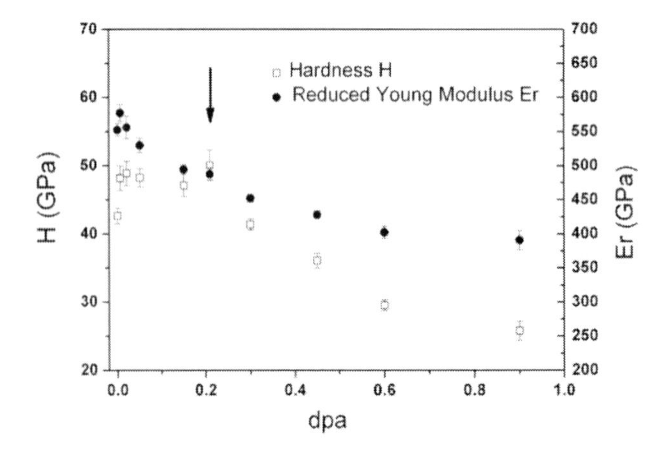

Figure 12. Hardness and reduced elastic modulus versus dpa (displacement per atom: $1dpa=1x10^{23}$ $vacancies.cm^{-3}$) for multi-energy implantation of helium at RT into 4H-SiC (maximum load 15 nN). The arrow shows the improvement of mechanical properties when bubbles form. [70]

Figure 12 shows the variations of the hardness and the reduced elastic modulus according to the helium fluence in dpa at room temperature in 4H-SiC. To obtain a number of dpa roughly constant over a wide depth, samples were prepared by multiple-energy implantation. In the unperturbed bulk, a value of 42±1 GPa is found for the hardness. This value is higher than the one generally determined for bulk 4H-SiC (37±1 GPa) due to the indentation size effect (ISE) [66]. This effect of hardness increase with decreasing penetration depth is often observed in crystalline materials and is ascribed to the increasing contribution of dislocations at small scale. The value of the elastic reduced modulus 510±10 GPa is relatively constant for the whole indenter penetration depth (not affected by slip) [65]. As seen on the curve of hardness versus dpa, a slight increase of hardness is observed at a low dpa level, up to 15% of increase at 0.021 dpa. This increase was also observed on Au irradiated 6H-SiC at RT up to a fluence of 0.03 dpa [67]. This effect of near-surface hardness as function of ion fluence was already reported for a wide range of ceramics or insulating materials [68] due to radiation or solution hardening. In SiC this hardening effect may be related to the point defect generation and accumulation along the ion path. Then, with increasing dpa, a monotonous decrease of the hardness (softening) is expected because of the increasing density of defects up to the amorphisation. This is observed in case of implantation of Au or Cr ions where the hardness and the elastic modulus are found to drop with increasing fluence up to the whole amorphisation. The degradation of the mechanical properties is also observed after neutron irradiation of SiC [69]. In case of helium implantation, an unexpected abrupt increase of the hardness is clearly observed at 0.21 dpa (see the arrow on Figure 12) [70]. A less marked discontinuity is also observed on the elastic modulus curve. For such level of dpa, the concentration of helium is close to 0.5 % at which complex defects are formed (see previous section). These complex defects which are supposed to be vacancy-helium clusters could act as small pinning obstacles for gliding dislocations which is the physical mechanism for plastic deformation. Moreover the as-produced out-of-plane strain (1.8%) and thus the in-plane stress are no more negligible and could participate to the improvement of these mechanical properties. With increasing fluence, the accumulation of damage (lattice disruption) leads to the dramatic degradation of mechanical properties. At 0.9 dpa, the implanted SiC is amorphous; the obtained values of hardness and elastic reduced modulus of 25±1 GPa and 400 GPa respectively are in good agreement with those reported for amorphous SiC [56, 69].

Finally, SiC being efficient in dynamical annealing of defects, the mechanical properties changes are shifted toward higher levels of dpa with the increase of the temperature of implantation [64, 67]. Moreover, in case of gas implantation at high temperature, the as-formed nanometric bubbles localized in a well defined buried layer should interact with the gliding dislocations and therefore may also induce changes in mechanical properties of SiC.

CONCLUSION

The effects of ion implantation and more specifically helium ion implantation into SiC have been reviewed. Significant progress has been achieved for the past decade in the understanding of the different processes occurring during the ion implantation and the subsequent annealing of SiC. However, compared to other semiconductors such as Si, the understanding of the implantation damage in SiC in which defects are more stable is still unsatisfactory. More work is needed to clearly understand the implantation damage and its influence on the physical and mechanical properties of SiC. Even for low fluence implantation the electrical properties of SiC are strongly modified and post-implantation annealing at high temperature is required to remove the as-produced defects. When helium is implanted and regardless of the fluence of implantation, the formation of stable helium-vacancy defects in the as-implanted state leads to a strong microstructural evolution upon annealing. These complex defects evolve in large cavities and at fluences avoiding amorphization, large clusters of bubbles are produced that punch out dislocation loops in the material. At elevated temperature of implantation, dynamic annealing takes place resulting in the recombination of point defects but also in the migration and accumulation of interstitial-type defects. All these defects contribute to the surface swelling and also modify the mechanical properties of SiC. Thus, a deeper understanding of the formation and interaction of the defects produced during implantation and post-implant treatments is necessary for the use of SiC in SiC-based devices as well as in advanced nuclear power plants. We hope that this chapter mainly related to helium implantation will stimulate additional works on the ion damage and its influence on the physical and mechanical properties of SiC, which is of technological importance for the SiC applications.

References

[1] Edwards, A.; Dwight, D. N.; Rao, M. V.; Ridgway, M. C.; Kelner, G.; Papanicolaou, N. *J. of Appl. Phys*. 1997, *82,* 4223.

[2] Posselt, M.; Bischoff, L.; Grambole, D.; Herrmann, F. *Appl. Phys. Lett*. 2006, 89, 151918-1-15918-3.

[3] Hallen, A.; Henry, A.; Pellegrino, P.; Svensson, B. G.; Åberg, D. *Mat. Sci. And Eng.* B 1999, 61-62, 378-381.

[4] Hopkins, G. R.; Chin, J. *J. of Nucl. Mater*. 1986, 141-143, 148-151.

[5] Raj. B.; Vijayalakshmi, M.; Vasudeva Rao, P. R.; Rao, K. B. S. *MRS Bulletin*, 2008, 33, 327-337.

[6] Yamauchi, Y.; Hirohata, Y.; Hino, T. *J. of Nucl. Mater*. 2003, 313-316, 408-412.

[7] Gali, A.; Deák, P. ; Rauls, E. ; Son, N. T.; Ivanov, I.G. ; Carlsson, F.H.C. ; Janzén, E.; Choyke, W.J. *Phys. Rev. B* 2003, 67, 155203-1-1055203-5.

[8] Salvador, M.; Perlado, J. M.; Mattoni, A.; Berbardini, F.; Colombo, L. *J. of Nucl. Mater.* 2004, 329-333, 1219-1222.

[9] Lucas, G.; Pizzagalli, L. *Nucl. Instr. And Meth. In Phys. Res. B* 2007, 255, 124-129.

[10] Ziegler, J. F.; Biersack, J. P. SRIM-2010.01 computer code. http://www.srim.org.

[11] Nordlund, K.; Partika, P.; Averback, R. S.; Robinson, I. K.; Erhart, P. *J. of Appl. Phys.* 2000, 88, 2278-2288.

[12] Miyasaki, H.; Suzuki, T.; Yano, T.; Iseki, T. *J. of Nucl. Sci. and Tech*. 1992, 29, 656-663.

[13] Li, J.; Porter, L.; Yip, S. *J. of Nucl. Mater*. 1998, 255, 139-152.

[14] Rao, S. I.; Houska, C. *J. of Appl. Phys*. 1991, 69, 8096-8103.

[15] Stepanov, S. GID_sl (1997). http://sergey.gmca.aps.anl.gov/gid_sl.html.

[16] Leclerc, S.; Beaufort, M. F.; Declémy, A.; Barbot, J. F. *Appl. Phys. Lett*. 2008, 93, 122101-1-122101-3.

[17] Grimaldi, M. G.; Calcagno, L.; Musumeci, P.; Frangis, N.; Van Landuyt, J. *J. of Appl. Phys.* 1997, 81, 7181-7185.

[18] Wendler, E.; Heft, A.; Wesch, W. *Nucl. Instr. and Meth. in Phys. Res. B* 1998, 141, 105-117.

[19] Leclerc, S.; Beaufort, M. F.; Declémy, A.; Barbot, J. F. *J. of Nucl. Mater.* 2010, 397, 132-134.

[20] Oliviero, E.; Van Veen, A.; Fedorov, A. V.; Beaufort, M. F.; Barbot, J. F. *Nucl. Instr. And Meth. In Phys. Res. B* 2002, 186, 223-228.

[21] Oliviero, E.; Beaufort, M. F.; Barbot, J. F.; Van Veen, A.; Fedorov, A. V. *J. of Appl. Phys.* 2003, 93, 231-238.

[22] Zhang, C. H.; Donnelly, S. E.; Vishnyakov, V. M.; Evans, J. H. *J. of Appl. Phys.* 2003, 94, 6017-6022.

[23] Oliviero, E.; Beaufort, M.F.; Pailloux, F.; Barbot, J.F. *Nucl. Instr. And Meth. In Phys. Res. B* 2004, 218, 391-395.

[24] Leclerc, S. Damage induced by helium implantation in silicon carbide; Ph.D. Thesis, University of Poitiers, 2007.

[25] Inui, H.; Mori, H.; Fujita, H. *Phil. Mag. B* 1990, 61, 107-124.

[26] Heft, A.; Wendler, E.; Heindl, J.; Bachmann, T.; Glaser, E.; Strunk, H.P.; Wesch, W. *Nucl. Instr. And Meth. In Phys. Res. B* 1996, 113, 239-243.

[27] Hallen, A.; Persson, P. O. A.; Kuznetsov, A. Y.; Hultman, L.; Svensson, B. G. *Mater. Sci. Forum* 2000, 338-342, 849.

[28] Héliou, R.; Brebner, J. L.; Roorda, S. *Semic. Sci. Technol.* 2001, 16, 836-843.

[29] Oliviero, E.; Tromas, C.; Pailloux, F.; Declémy, A.; Beaufort, M. F.; Blanchard, C.; Barbot, J. F. *Mat. Sci. And Eng.* B 2003, 102, 289-292.

[30] Nakata, K.; Kasahara, S.; Shimanuki, S.; Katano, Y.; Ohno, H.; Kuniya, J. *J. of Nucl. Mater.* 1991, 179-181, 403-406.

[31] Leclerc, S.; Declémy, A.; Beaufort, M.F.; Tromas, C.; Barbot, J.F. *J. of Appl. Phys.* 2005, 98, 113506-1-113506-6.

[32] Rohde, M. *J. Nucl. Mater.* 1991, 182, 87-92.

[33] Wendler, E.; Heft, A.; Wesch, W.; Peiter, G.; Dunken, H. H. *Nucl. Instr. and Meth. in Phys. Res. B* 1997, 127-128, 341-346.

[34] Pacaud, Y.; Skorupa, W.; Stoemenos, J. *Nucl. Instr. And Meth. In Phys. Res. B* 1996, 120, 181-185.

[35] Zolnai, Z.; Son, N. T.; Hallin, C.; Janzén, E. *J. of Appl. Phys.* 2004, 96, 2406-2408.

[36] Kawasuso, A.; Weidner, M.; Redman, F.; Frank, T.; Sperr, P. ; Krause-Rehberg, R.; Triftshäuser, W.; Pensl, G. *Physica B* 2001, 308-310, 660-663.

[37] Jiang, W.; Thevuthasan, S.; Weber, W. J.; Grötzschel, R. *Nucl. Instr. and Meth. in Phys. Res. B* 2000, 161-163, 501-504.

[38] Rong, Z.; Gao, F.; Weber, W.J.; Hobler, G. *J. of Appl. Phys.* 2007, 102, 103508-1-103508-7.

[39] Lucas, G. ; Pizzagalli, L. *J. Phys. Condens. Matter* 2007, 19, 086208-1-086208-10.

[40] Bockstedte, M.; Mattausch, A.; Pankratov, O. *Phys. Rev. B* 2004, 69, 235202-1-235202-13.

[41] Puff, W.; Mascher, P.; Balogh, A. G.; Baumann, H. *Mater. Sci. Forum* 1997, 258-263, 733-738.

[42] Polity, A.; Huth, S.; Lausmann, M. *Phys. Rev. B* 1999, 59, 10603-10606.

[43] Demenet, J. L.; Hong, M. H.; Pirouz, P. *Scripta Mater.* 2000, 43, 865-870.

[44] Samant, A. V.; Pirouz, P. *Int. J. Refractory Metals Hard Mater.* 1998, 98, 277-289.

[45] Höfgen, A; Heera, V.; Eichhorn, F.; Skorupa, W.; Mö ller, W. *Mater. Sci. And Eng. B* 1999, 61-62, 353-357.
[46] Hojou, K.; Furuno, S. ; Kushita, K. N.; Otsu, H. ; Furuya, Y. ; Izui, K. *Nucl. Instr. and Meth. in Phys. Res. B* 1996, 116, 382-388.
[47] Pacaud, Y.; Stoemenos, J.; Brauer, G.; Yankov, R. A.; Heera, V.; Voelskow, M.; Kögler, R.; Skorupa, W. N. *Instr. And. Meth. in Phys. Res. B* 1996, 120, 177-180.
[48] Yoshii, K.; Suzaki, Y.; Takeuchi, A.; Yasutake, K.; Kawabe, H. *Thin Solid Films* 1991, 199, 85-94.
[49] Wesch, W.; Heft, A.; Heindl, J.; Strunk, H. P.; Bachmann, T.; Glaser, E.; Wendler, E. *Nucl. Instr. and Meth. in Phys. Res. B* 1995, 106, 339-345.
[50] Beaufort, M.F.; Pailloux, F.; Declémy, A.; Barbot, J.F. *J. of Appl. Phys.* 2003, 94, 7116-7120.
[51] Snead, L. L.; Katoh, Y.; Connery, S. *J. of Nucl. Mater.* 2007, 367-370, 677-684.
[52] Snead, L. L.; Nozawa, T.; Katoh, Y.; Byun, T. S.; Kondo, S.; Petti, D.A. *J. of Nucl. Mater.* 2007, 371, 329-377.
[53] Kondo, S.; Katoh, Y.; Snead, L. L. *J. of Nucl. Mater.* 2009, 386-388, 222-226.
[54] Heera, V.; Stoemenos, J.; Kögler, R.; Skorupa, W. *J. of Appl. Phys.* 1995, 77, 2999-3009.
[55] Jiang, W.; Wang, C. M.; Weber, W. J.; Engelhard, M.H.; Saraf, L.V. *J. of Appl. Phys.* 2004, 95, 4687-4690.
[56] McHargue C. J.; Williams, J. M. *Nucl. Inst. and Meth. B* 1993, 80-81, 889-894.
[57] Weber, W. J.; Yu, N.; Wang, L. M.; Hess, N. J. *J. of Nucl. Mater.* 1997, 244, 258-265.
[58] Weber, W. J., Wang, L. M.; Yu, N.; Hess, N. J. *J. of Nucl. Mater.* 1998, 253, 53-59.
[59] Snead, L. L; Nozawa, T.; Katoh, Y.; Byun, T.S.; Kondo, S.; Petit, D. A. *J. of Mat. Sci.* 2007 371, 329-377.
[60] Park, K.H.; Hinoki, T.; Kohyama, A. *J. of Mat. Sci.* 2007, 367-370, 703-707.
[61] Lorenz, D.; Zeckzer, A.; Hilpert, U.; Grau, P.; Johansen, H.; Leipner, H. S. *Phys. Rev. B* 2003, 67, 1721011-1721014.
[62] Tromas, C.; Audurier, V.; Leclerc, S.; Beaufort, M. F.; Declemy, A.; Barbot, J. F. *J. of Nucl. Mat.* 2008, 373, 142-149.
[63] Page, T. F.; Pharr, G. M.; Hay, J. C.; Olivier, W. C.; Lucas, B. N.; Herbert, E.; Riesler, L. *Mat. Sci. Soc. Symp. Proc.* 1998, 522, 53-64.
[64] Menard, M.; LeFlem, M.; Gelebart, L.; Monnet, I.; Basini, V.; Boussuge, M. *Ceramic Eng. and Sci. Proceedings* 2008, 28, 319-326.
[65] Barbot, J. F.; Leclerc, S.; David, M. L.; Oliviero, E.; Montsouka, R.; Pailloux, F.; Eyidi, D.; Denanot, M. F.; Beaufort, M. F.; Declémy, A.; Audurier, V.; Tromas, C. *Phys. Status Solidi a* 2009, 206, 1916-1923.
[66] Gao F.; Weber, W. *J. Phys. Rev B* 2004, 69, 224108-1-224108-10.
[67] Kerbiriou, X. ; Costantini, J. M.; Sauzay, M.; Sorieul, S.; Thomé, L.; Jagielski, J.; Grob, J. J. *J. Appl. Phys*. 2009, 105,. 073513-1-073513-11.
[68] White, C. W.; McHargue, C. J.; Sklad, P. S.; Boatner, L. A.; C. Fralow, C. *Mat. Sci. Reports* 1989, 4, 41-146.
[69] Snead, L.; Hay, J. C. *Nucl. Inst. and Meth. B* 1999, 273, 213-220.
[70] Barbot, J. F.; Beaufort, M. F.; Audurier, V. *Mat. Sci. Forum* 2010, 645-648, 721-724.

In: Silicon Carbide: New Materials, Production … ISBN: 978-1-61122-312-5
Editor: Sofia H. Vanger

Chapter 8

RECENT PROGRESS IN THE PREPARATION TECHNOLOGIES FOR SILICON CARBIDE NANOMATERIALS

S.B. Mishra[1] and A.K. Mishra
Department of Chemical Technology, University of Johannesburg,
Doornfontein 2028, Johannesburg, South Africa

ABSTRACT

Over the years the synthesis of silicon carbide nanomaterials (SiCNM) has been thrust area of research and development for its unique and potential properties. Numerous efforts have been made to generate SiCNM at the large scale and with enhanced purity. SiCNM have been developed in the form of nanosphere, nanorods, nanowires, nanofibers, nanowhiskers and nanotubes using different techniques and methodologies. SiC is particularly attractive due to its excellent mechanical properties, high chemical resistivity, and its potential application as a functional ceramic or a high temperature semiconductor. This chapter deals with preparation technologies of silicon carbide based nanomaterials. An overview of some important applications have also been discussed.

1. INTRODUCTION

Nanosized materials have attracted the scientists across the world due to unique properties and numerous applications. The nanometer is one billionth of a meter and ~ 10,000 times finer than human hair and nanomaterials at this smallest level can takes up shape in the form of nanorods, nanowires, nanoribbons, nanofibers and nanotubes [1-8].

Among the various carbon based nanomaterials, silicon carbide (SiC) nanomaterials are widely investigated due to high strength, good creep, oxidation resistance at elevated temperature, chemical inertness, thermal stability, resistance to corrosion [9-12]. The size, shape and surface composition of these nanostructure materials are the few factors

[1] Corresponding authors: smishra@uj.ac.za.

responsible for these unique and fascinating properties [13-15]. These extraordinary properties allows silicon carbide suitable ceramic for numerous applications that includes semiconducting devices to be used at high temperature and high frequency [14], reinforcement in ceramic composites [15], metal matrix composites [16-18], catalytic support [19-20]. Besides, low thermal expansion coefficient, good thermal shock resistance and porosity of the silicon carbide makes it extremely important ceramic in the field of energy production and environmental protection such as hydrogen permselective membrane support, slurry reuse after polishing and water purification and diesel particulate filters [21-27]. Efforts have been made to develop nanosize hollow spheres of silicon carbide for technical applications in optical, electronic, acoustic and sensing devices ranging from photonic crystals to drug delivery systems [28-30].

2. Properties of Silicon Carbide

Discovered in 1893 by Edword Goodrich Acheson, SiC has unique properties of commercial and industrial importance. SiC in pure form has no colour. However, the industrial products of SiC are found to be black or brown in colour due to iron impurities. The basic structure of SiC is shown in Figure 1a where silicon atom is covalently bonded to carbon atom and Figure 1b show tetrahedral of SiC.

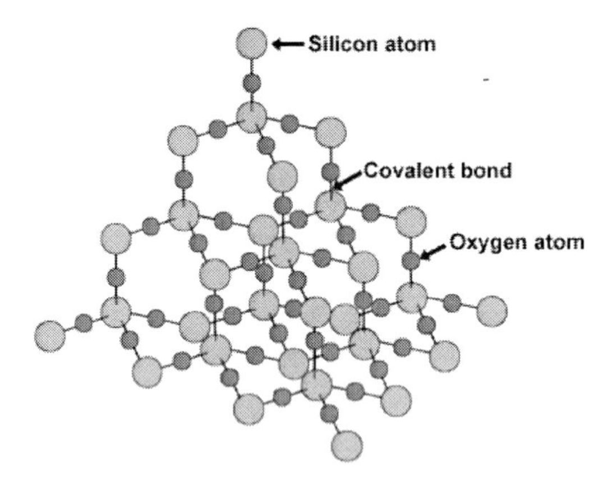

Figure 1a. Structure of silicon carbide.

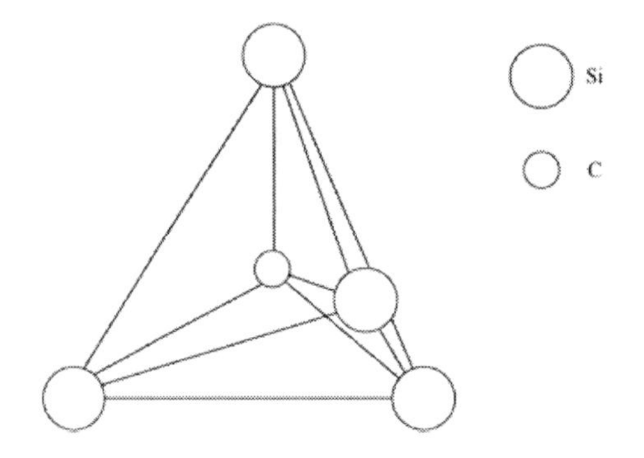

Figure 1b. Tetrahedra showing the formation of SiC.

SiC show one dimensional polymorphism called polytypism and till date, more than 250 polytypes of SiC have been reported. Nearly all the polytypes of SiC have identical planar arrangements of Si and C atoms, with differences in stacking sequence of identical planes and it is this disorderness in the stacking periodicity of similar planes that results in numerous polytypes / crystal structures. Among the variety of polytypes, only three crystalline structures viz. cubic, hexagonal and rhombohedral exist. So, far seventy different crystalline forms of silicon carbide are known [31]. As per conventional nomenclature, SiC polytype is represented by the number of SiC double layers in the unit cell, the appending letter C, H or R for a cubic, hexagonal or rhombohedral symmetry. For example, the 6H hexagonal lattice as shown in Figure 3 has six such layers in the primitive cell with the following succession of the planes. Among these, the most common polymorph of silicon carbide is α-SiC having hexagonal crystal structure is formed at temperature greater than 2000°C. The other polymorph with the growing interest is β-SiC that has face centred cubic crystals and is formed at temperature below than 2000°C. The specific gravity of silicon carbide is 3.2 and has a very high sublimation temperature of approx. 2700°C. SiC has low co-efficient of thermal expansion and is found to be chemically inert in nature.

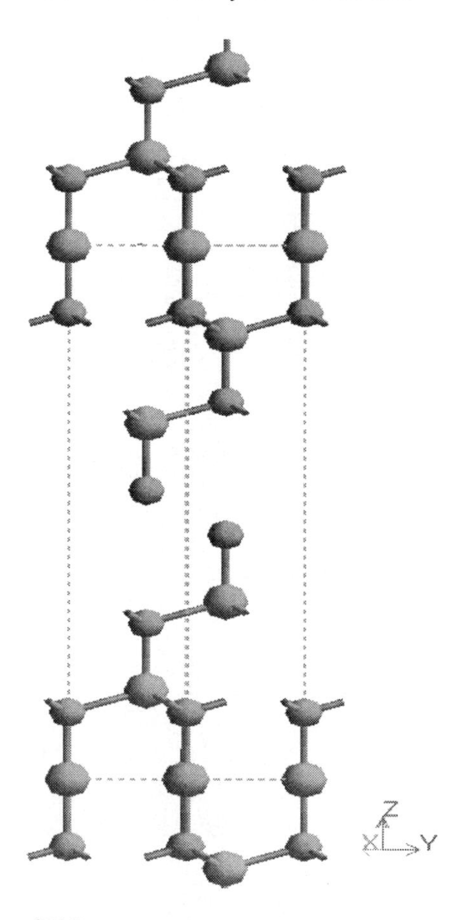

Figure 3. 6H hexagonal lattice of SiC.

The other properties of commercial interest are its high thermal conductivity, high electric field breakdown and high maximum current density. α-SiC has the wide band gap of

3.28 eV (4H) and 3.03 eV (3H) respectively and thus acts as intrinsic semiconductors. Some more details of the various properties of silicon carbide are given in Table 1.

Table 1. Properties of silicon carbide

Property	Value
Mechanical	
Density	3.1 gm/cc
Flexural Strength	550 MPa
Elastic Modulus	410 GPa
Poisson's ratio	0.14
Compressive strength	3900 MPa
Hardness	2800 Kg / mm^2
Mohs Hardness	9.15
Fracture Toughness K_{IC}	4.6 $MPa.m^{1/2}$
Maximum use temperature	1650^{OC}
Thermal	
Thermal conductivity	350-490 $W/m.K^{O}$
Coefficient of thermal expansion	$4 * 10^{6} / {}^{O}C$
Specific Heat	750 $J/Kg.{}^{O}K$
Electrical	
Dielectric Strength	Semiconductor
Volume resistivity	10^{2}-10^{6} ohm.cm
High break down electric field	3-5 MV/cm^{-1}

3. Sources of Carbon and Silica for Producing SICNM

The silicon carbide nanomaterials have been fabricated using various carbon and silica sources by making use of different techniques.

I. Sources of Carbon

Some of the common carbon sources so far used is carbon nanotubes [32-33], pure methane gas [34-35], carbon microfiber [36], pulp mat [37], sucrose [38], phenolic resins [39-41], polyacrylonitrile [42], 1, 2, dimethoxyethane [43], propane [44], carbon nanoparticles [45]. A few other sources of carbon that have been investigated are coal tar pitch [46], polycarbonate [47], polypropylene oxide-polystyrene blend [48], rice husk [49], fuel gas [50], carbon black [51], perhalogenated hydrocarbons [52], wood precursors [53].

II. Sources of Silica

Various types of sources for silica have been investigated for the development of SiCNMs. Among these sources, some of the widely used include silica vapours [32], silica powder [34,51], silicon hydride gas [35], silanes [37, 46, 47-48], silicon chloride [43-44].

4. Preparation Technologies

There are numerous techniques investigated for the synthesis of silicon carbide nanomaterials. The most common used so far includes sol-gel process, carbothermal reduction, autoclave, arc-dischrage, ball milling, chemical vapour deposition, pulsed laser deposition. Some of the other less common techniques radiofrequency magnetic sputtering, pulsed laser deposition, photolithography and chemical vapour infilteration.

I. Sol-Gel Technology

A sol-gel process using metal alkoxides has been widely applied for the synthesis of so-called ideal powders; homogenous, controlled size and high purity. In a typical procedure, the metal alkoxide undergoes hydrolysis and polycondensation process with or without catalyst to produce sol containing the nanoparticles. E.g. to produce silica nanoparticles, the siloxanes or silanes is taken in suitable solvent and hydrolysis is carried out as such or in the presence of acid or base as catalyst. The product so obtained is a sol containing polysilane in nanosize. This can be shown as scheme 1 below.

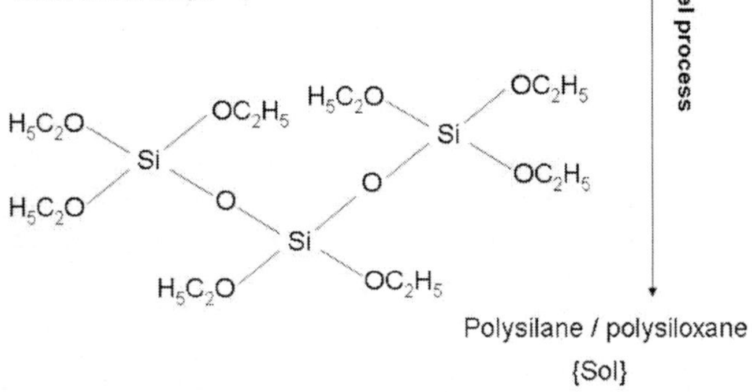

Silane / Siloxane

H_2O
Solvent
Acid or base as catalyst

Hydrolysis
Polycondensation

Sol-gel process

Polysilane / polysiloxane
{Sol}

Scheme 1. Sol containing polysilane.

It is a process that allows the formation nanoparticles produced as sol that is able to produce dense films, dense ceramics, aerogels and fibers. The following scheme 2 shows the various possibilities of sol-gel derived nanoparticles.

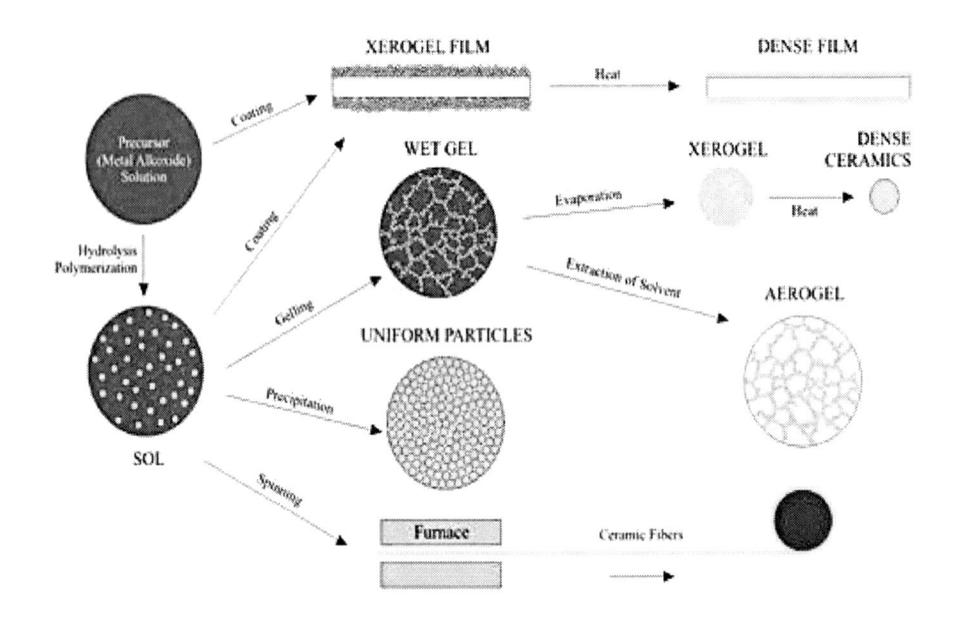

Scheme 2. Various applications of sol-gel process.

Powders synthesized from the precursors have been limited to oxides because of metal-oxygen bonds formed by hydrolysis and condensations. However, many workers have shown that it is possible to synthesise non-oxide powders from organometallics compounds other than metal alkoxides. Sol-gel technology is therefore a platform for developing hybrid systems that can further be processed using to fabricate silicon carbide nanomaterial.

Some of the work where sol-gel technology have been used are discussed here. Monodispersed spherical SiC gel powders were synthesized by hydolysis and condensation of phenyltrimethoxysilane (PTMS) or a mixture of PTMS and tetraethylorthosilicate (TEOS) from a system of silane-H2O-catalyst without use any co –solvent. The monodispersed spherical particle shape was sustained even after 1400 ^{O}C but the initially spherical shape changed to bulk and fiber phase in the PTMS system [76]. In a different study of sol-gel process and polymer blending sol-gel derived silica was blended with coal tar pitch [46], polypropylene-polystyrene blend [48], polycarbonate [47] to yield silicon carbide nanofibers.

Unbleached and bleached soft wood pulps have been used as templates and carbon precursors to produce SiC nanorods. Hydrolyzed tetraethylorthosilicate (TEOS), silicic acid was infiltrarted into the pulps into the pulps followed by a carbothermal reduction to form SiC nanorods [37]. The non-uniform SiC nanorods with thick SiO_2 coating were obtained from unbleached pulp whereas uniform and straight SiC nanorods were formed from bleached pulp. Lignin-silica-titania and lignin-titania hybrid fibers have been prepared by sol-gel processing from lignin, tetraethoxysilane, and titanium tetrakis(2,4-pentenedionate) using a mixture of 2,4-pentanedione and tetrahydrofuran as solvent and sulphuric as catalyst. The hybrid fibers had to be cured to avoid coalescence during heat-treatment. The cures thin fibers could be converted into Si-Ti-C or TiC fibers by firing at 1500 ^{O}C [77].

II. Thermal Synthesis

To develop nanoparticles synthesis with high purity at high process yields and high production rates, thermal routes are widely undertaken [54]. Many researchers have made use of heat, generated by exothermic reactions for the synthesis of nanomaterials [55].This technique is popularly known as combustion synthesis, self propagating high temperature synthesis or thermolysis [56]. Using this technique 1-D nanostructure of cubic phase silicon carbide was efficiently produced from Si containing compounds and halocarbons in calorific bomb [52]. Silicon carbide hollow nanocrystals were fabricated by vapour-solid reactions between carbon nanoparticles and silica vapours in inert atmosphere of argon (Ar) at 1300°C [57]. In an another study gas-solid reactions between solid carbon and SiO vapours at 1200-1400°C yielded SiCNM. Since this method allowed different shapes and sizes, depending on their subsequent uses and having specific surface area up to 150 m^2/g, the process was termed as shape memory synthesis (SMS) [58]. Using this technique (SMS) silicon carbide nanotubes were successfully prepared with medium surface area (30-60 m^2/g) were developed by the reaction between carbon nanotubes and SiO vapours [32]. In an another attempt of SMS , silicon carbide nanotubes were fabricated by the reaction between multi walled carbon nanotubes and SiO vapours at 1200-1250 ^{O}C .This was used as support material for NiS_2-based active phase for the trickle-bed oxidation of H_2S into elemental sulphur [59].

One dimensional silicon carbon nanotubes and nanowires of various shapes and structures were synthesized via the reaction of silicon and multi walled carbon nanotubes at different temperatures. With this methodology a new type of multi walled silicon carbide nanotubes with 3.5-4.5 A^{O} interlayer spacing was identified besides the previously known β-SiC nanowires and the biaxial SiC-SiO_x nanowires [33]. In an another study the pyrolyzed maple and mahogany wood samples as carbon precursors were infiltered with silicon at 1450°C for 30 min in inert atmosphere to get biomorphic silicon carbide [53].Reaction of silicon monoxide with methane at 1300 ^{O}C in an argon atmosphere mixed with hydrogen at a pressure below 10-2 Torr gave rice to first synthesis of 2H-polytype of silicon carbide [34]. Heating at a rate of 15 ^{O}C per minute and holding carbon and silicon powders at a temperature below their eutectic temperature (1404°C) resulted in the exothermic formation of silicon carbide on a sub micrometric scale [60]. Evaporating the solid mixtures of silicon and silicon dioxide in a graphite crucible heated by high-frequency induction system at 1600 ^{O}C resulted in wire like β-SiC nanowhiskers [61].

In a different procedure of vapour-liquid-crystal, mixtures of butane and propane, silicon chloride and hydrogen in the presence of element activator iron were heated at 1250 ^{O}C in an industrial furnace yielding SiC whiskers. The procedure was followed by removal of carbon phases using hydrofluoric acid and annealing to remove unreacted carbon. In a very simple thermal synthesis, scientists used commercially available SiC powder and mixed with aluminium powder (as catalyst). The mixture was heated in tungsten boat at 1700 ^{O}C in a vacuum of 5 * 10^{-2} Torr under inert atmosphere resulting in SiC nanowires [63].

Combustion synthesis was discovered as one of the novel procedures for the production of cubic phase 1-D SiC nanostructures. The mixture of silica containing compounds viz. FeSi, $CaSi_2$ and halocarbons like poly(tetraflouroethylene), syndiotactic polyvinylchloride, hexachloroethane were made into pallets. The combustion process was initiated by an electrically heated reaction promoter attached to the top of the pallet [52].

III. Carbothermal Reduction

Carbothermal reduction is a heat treatment or pyrolysis of carbon and silica at 1400-1500°C in an inert atmosphere of argon gas, to synthesize SiC nanomaterials. At this temperature a number of pyrolytic reactions takes place between carbon and silica as shown below to finally yield the required product.

$$SiO_2(s) + C(s) \longrightarrow SiO(g) + CO(g) \quad (1)$$

$$SiO(g) + 2C(s) \longrightarrow SiC(s) + CO(g) \quad (2)$$

$$SiO(g) + 3CO(g) \longrightarrow SiC(s) + 2CO_2(g) \quad (3)$$

$$3SiO(g) + CO(g) \longrightarrow SiC(s) + 2SiO_2(s) \quad (4)$$

$$CO_2(g) + C(s) \longrightarrow 2CO(g) \quad (5)$$

The overall reaction is

$$SiO_2(s) + 3C(s) \longrightarrow SiC(s) + 2CO(g) \quad (6)$$

As per the above reaction scientist across the world try to develop major yield of SiC nanomaterials in pure form. However, it has still not been reported that carbothermal reduction has been successful in doing so. The real problem that occurs during the process is existence of unburnt carbon / amporphous carbon and unreacted silica along with the SiC. Also, the commonly reported are the stacking faults that generates within the nanostructure.

Carbothermal reduction is widely used to form SiC with desired structures due to broad availability of silica-carbon precursors. SiC nanostructures such as nanofibers, nanoparticles and nanoporous monoliths have been produced by the carbothermal reduction reaction [64]. In an interesting study of carbothermal reduction of carbon-silica nanocomposites, it was observed that the pores present the C-Si nanocomposite, C/SiO_2 ratio and $C-SiO_2$ carbon structure play an important role in the reaction kinetics and SiC nanostructures [65]. It was also observed that varying the reaction conditions (temperature / duration) of carbothermal reduction readily modifies the morphology of SiCNM. Subjecting the mesoporous silica/carbon composite to carbothermal reduction at a temperature of 1250 or 1300 ^{O}C for reduction period of 14h resulted into whiskers and on the other hand when the time period of the reduction was increased to 20h, SiC nanotubes were formed [66]. SiC nanowires were grown at 1400 ^{O}C for 2h in argon flow rate of 2L/min by vapour-liquid-solid process in an absence of metal catalyst from outside. In situ formation of liquid droplet could quickly involve SiO and CO vapours because of its high accommodation co-efficient [67]. Mixed powder of low purity SiO_2 and carbon black were used for carbothermal reduction process to grow SiC nanowires on phenolic resin-coated porous SiC [68].

In the recent attempts the kraft lignin was chosen as carbon source to develop organic-inorganic hybrid using sol-gel technology to fabricate SiC nano materials [69, 70]

IV. Chemical Vapour Deposition

Chemical Vapour deposition is technique where the precursor material is a gas that flows through the reaction chamber containing heated substrate. The gas thus react or decompose upon coming in contact with this heated substrate into nano size solid phase. The following figure 3 a-b show basic set up for the chemical vapour deposition technology and the reaction inside the chamber.

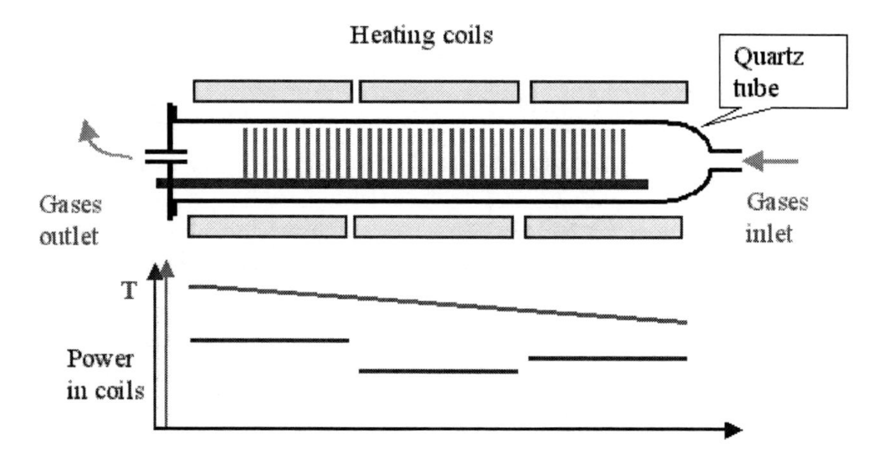

Figure 3a. The basic apparatus for CVD.

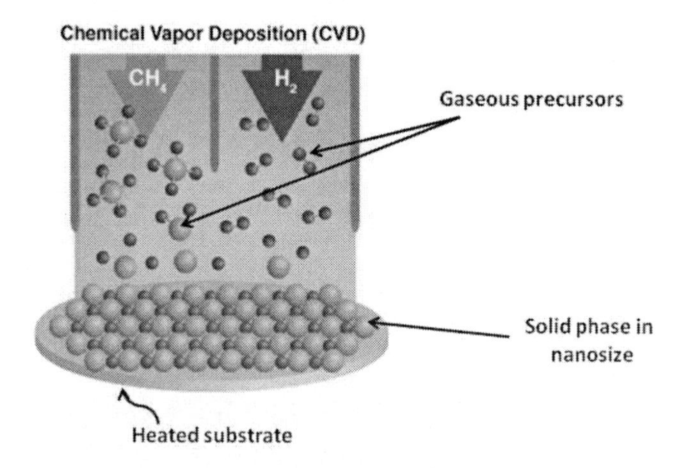

Figure 3b. The process of CVD.

Chemical vapour deposition (CVD) is one of the popular techniques used to produce SiC of various shapes of thin films, powders, whiskers and nanorods using Si-C-H-Cl system [71,72]. SiC nanowires were formed in large quantities with homogenous diameter from methyltrichlorosilane CH_3SiCl_3 (MTS) and hydrogen as precursors without using a metallic catalyst [73]. Figure 1. shows the CVD system for the above stated synthesis of SiC nanowires.

Low pressure chemical vapour deposition (LPCVD) is the process carried out at subatmosphereic pressures. Reduced pressure tends reduces unwanted gas phase reaction and improve film uniform across the wafers [74]. LPCVD improves the rate of production by increasing the mass transfer rate to the susceptor. Multicomponent diffusion and chemical reaction are the crucial factors LPCVD models [75]. Silicon carbide was prepared by LPCVD using hexamethyldisilane as a precursor at temperature between 770-900 ^{O}C, 7sccm flow rate and 0.35 Torr total pressure [74]. The "cold-gas-hot substrate" variant of CVD was chosen to grow SiC in an appratus shown in Fig 2. Disc shaped inductively heated graphite substrates covered with SiC layer were used. Highly purified liquid trichlorosilane ($SiCl_4$) + chloroform (CCl_4) or MTS vaporized in hydrogen acting both as carrier gas and reactant were used as precursors [76].

Plasma-Enhanced Chemical Vapour Deposition (PECVD) process, gases are ionized by an electrical energy source to form plasma. Through the intermediation of highly energized electrons, radical species are formed from the precursor molecule gases. These radicals are chemically very active, and their stability in the gas phase is determined by the rate of recombination and disproportionation reactions [77]. In a study, the technique was used to deposit amorphous SiC films for MEMS reactors, comparing high frequency at 13.56 MHz and low frequency at 380 kHz in order to achieve a low residual stress with a reasonable deposition [35].

V. Miscellaneous

Besides above technologies that have been used to fabricate SiCNMs, some researchers used arc discharge, sintering, ball milling and Yajima process methods to develop these nanomaterials. Researchers used autoclave for reacting 1,2-dimenthoxyethane, silicon chloride and magnesium at 600 °C to develop cubic silicon carbide (3C-SiC) nanoparticles encapsulated in branched wavelike carbon nanotubes. It was concluded that the nanostructure were formed via the sequential deposition growth process (with cores as the templates for the shells).[78] Using ball milling, hollow SiC nanosphere were prepared from carbon sphere and micro grade silicon spheres. The mixture was then heated at variable times at constant temperature to tailor the shell thickness followed by removal of the carbon core by oxidation [79].In a similar study, planetary ball mill was used to produce the equiatomic nanostructured SiC of homogenous distribution of the particles with 0.3μm in size [80].

Some researchers were successful to use direct synthesis approach using silicon monoxide and methane as precursor materials yielding 2H polytype of SiC nanowhiskers with 3C tip. The authors reported the oxide assisted growth process for such a development [81]. Pulsed lased deposition was also explored to develop SiCnms. It was concluded that variation of substrate temperature and substrate material can actually give rise to different morphologies of nanomaterials such as nanoholes, nanosprouts, nanoneedles and nanowires [82]. Using silicon powder and phenolic resin as a starting material, some researchers made use of coat-mix, moulding and high temperature sintering process that also resulted in multiple morphologies of SiC including ordinary SiC, bamboo-like and spindle SiC [83]. Japanese scientist Yajima introduced "Yajima process" in 1975 to fabricate SiC. In this process, sodium is coupled with dimethyl silyl chloride as a monomer in hydrocarbon solvent to produce preceramic polymer that is thermally decomposed to yield SiC nanomaterials. Based on this principle, authors simplified the process, liquid sodium was allowed to react with gaseous phase of silyl methyl chlorides. Two different types of silyl precursors finally resulted into shape variations of the final product [84]. Similarly, some other researchers made use of this technique where they developed polycrystalline SiC at a vapour – solid interface. Here they chose methyl silyl hydrochloride in the vapour phase which reacted with calcium thin film on silica substrate.

Concluding Remarks

Silicon carbide nanomaterials have been widely under investigations using different technologies. There are various parameters that can influence the final product in the nano size. Based on the application of SiC nanomaterials, the researchers can actually fine tune different types of nanosturctures. Among all the technologies, carbothermal reduction becomes an integral part of almost all the preparation methodologies.

References

[1] Tenne, T.; Margulis, L.; Grenet, M. and Hodes, G. *Nature*, 360 (1992) 444.

[2] Ajayn, P.M.; Stephan, O.; Redlich, P.H. and Collix, C. *ibid.* 375 (1995) 564.

[3] Chen, Y.J.; Li, J.B.; Han, Y.S. and Dai, D. J. H. *J. Mater. Sci. Lett.* 21 (2002) 175.

[4] Patel, N.; Kawai, R.; Sugimoto, M. and Oya, A.in Proceedings of Conference ob Carbon, Beijing, Sepetember 15-20, 2002.

[5] Gole, J. in Proceedings of Seminar on Condensed Matter Physics, Connecticut, December 2002.

[6] Pradhan, S.K.; Reucroft, P.J. and Dozier, A. *Carbon 41* (2003) 2873.

[7] Keller, N.; Huu, C.P.; Ehret, G.; Keller, V. and Ledoux, M.J. *ibid.* 41 (2003) 2131.

[8] Shi, W.; Zheng, Y.; Peng, H.; Wang, N.; Lee, C.S.and Lee, S.T. *J. Amer. Ceram. Soc.* 83 (2000) 3228.

[9] Kamakaran, R; Lupo, F.; Grobert, N.; Scheu, T.; Phillipp, N.Y.J and Ruhle, M. *Carbon* 42 (2004) 1.

[10] Pham-Huu, C.; Bouchy, C.; Dintzer, T.; Ehret, G.; Estournes, C.and Ledoux, M. *J. Appl. Catal.*, A 180 (1999) 385.

[11] Okada, K.; Kato, H. and Nakajima, K. *J. Am. Ceram. Soc*. 77 (1994) 1691.

[12] Bao, X.; Nangrejo, M. R.and Edirisinghe, M. J. J. *Mater. Sci.* 35 (2000) 4365.

[13] Eddelstein, A.S. and Cammarata, R.C. Nanomaterials : Synthesis, properties and applications: Institute of Physics: Philadelphia, P.A. 1996.

[14] Thiaville, A.and Miltat J. *Science* 284 (1997) 1939.

[15] Xia, Y.; Yang, P.; Sun, Y.; Wu, Y.; Mayers, B.; Gates, B.; Yin, Y.; Kim, F.and Yan H. *Adv. Mater.* 15 (2003) 353.
[16] Feng, Z.C.; Mascarenhas, A.j.; Choyke, W. J.; Powell, J. A. *J. Appl. Phys.* 64 (1988), 3176.
[17] Becher, P.F.; Hsueh, C.H.; Angelini, P.T.T.N. *J. Am. Ceram. Soc.* 71 (1988) 1050.
[18] Chokshi, A.H. and Porter, J.R. *J. Am. Ceram. Soc.* 68 (1985) C144.
[19] Frevel, L.K.; Saha, C.K. and Petersen, D.R. *J. Mater. Sci.* 30 (1995) 3734.
[20] Pham-Huu, C.; Gallo, P.D.; Peschiera, E. and Ledoux, M. *J. Appl. Catal.*, A 132 (1995) 77.
[21] Boutonet-Kizling, M.; Stenius, P.; Andersson, S.and Frestad, A. *Appl. Catal.*, B 1(1992) 149.
[22] Keller, N.; Pham-Huu C.; Roy, S.; Ledoux, M.J.; Estournes and Guille, J. *J. Mater. Sci.* 34 (1999) 3189.
[23] Fukushima, M. Zhou, Y.; Iwamoto, Y.; Yamazaki, S.; Nagano. T.; Miyazaki, Y.; Yoshizawa, K. and Hirao K. *J. Am Ceram. Soc.* 89 (2006) 1523.
[24] Ledoux, M.J.; Hantzer, S.; Huu, C.P.; Guille, J.and Desaneaux, M.P. *J. Catal.* 114 (1988) 176.
[25] Suwanmethond, V.; Goo, E.P.; Liu, K.T.; Johnston, G.; Sahimi, M. and Tsotsis, T.T. *Ind. Eng. Chem. Res.* 39 (2000) 3264.
[26] Lin, P.-K. andTsai, D.-S. *J. Am. Ceram. Soc.* 80 (1997) 365.
[27] Benaissa, M.; Werckmann, J.; Ehret, G.; Peschiera E.; Guille, J. and Ledoux, M.J. *J. Mater. Sci.* 29 (1994) 4700.
[28] Caruso, F.; Caruso, R.A. and Mohwald, H. *Science*, 282 (1998) 1111.
[29] Sun, Y. and Xia, Y. *Science*, 298 (2002) 2176.
[30] Kidambi, S.; Dai, J.H. and Bruening M.L. *J. Am. Chem. Soc.*, 126 (2004) 2658.
[31] Mirgororodsky, A.P, Sminov, M.B., Abdelmonium, E., Merle, T. and Quintard, P.E. *Phy. Rev.*, B52 (1995) 3993.
[32] Keller, N.; Pham-Huu, C.; Ehret, G.; Keller, V.; Ledoux, M.J. Carbon, 41 (2003) 2131.
[33] Sun, X-H.; Li, C-P.; Wong, W-K.; Wong, N-B.; Lee, C-H.; Lee, C-S.; Teo, B-K. *J. Am. Chem. Soc.* 124 (2002) 14464.
[34] Yao, Y.; Lee, S.T.; Li, F.H. *Chem. Phys. Letts.* 381(2003) 628.
[35] Iliescu, C.; Chen, B.; Wei, J.; Pang, A. *J. Thin Solid Films,* 516 (2008) 5183.
[36] Pham-Huu, C.; Keller, N.; Ehret, G.; Ledoux, M.J. *J. of Catal.* 200 (2001) 400.
[37] Shin, Y.; Wang, C.; Samuels, W.D.; Exarhos, G.J. *Mater. Letts.* 61 (2007) 2814.
[38] Wei, G.C.; Kennedy, C.R.; Harris, L.A. *Bull. Am. Ceram. Soc.* 63 (1984) 1054.
[39] Tanaka, H.; Kurachi, Y. *Ceram.Int.* 14 (1988) 109.
[40] Tanaka H.; Jin, K.Z.;Hirota, K. *Nippon Seramikkusu Kyukai Gakujutsu Ronbunshi* 98(1990) 607.
[41] Ono, K.; Kurachi, Y. *J. Mater. Sci.* 24 (1991) 388.
[42] Sugahara, Y.; Takeda, Y.; Kuroda, K.; Kato, C. *J. Non-Cryst Solids*, 100 (1998) 542.
[43] Xi, G.; Yu, S.; Zhang, R.; Zhang, M.; Ma, D.; Qian, Y. *J. Phys. Chem. B* 109 (2005) 13200.
[44] Li, Z.; Zhang, J.; Meng, A.; Guo, J. *J. Phys. Chem. B* 110 (2006) 22382.
[45] Liu, Z.; Ci, L.; Jin-Phillip, N.Y.; Rühle M. *J. Phys. Chem. C* 111 (2007) 12517.
[46] Raman, V.; Bhatia, G.; Mishra, A.K.; Bhardwaj, S.; Sood, K.N. *Mater. Letts* 60 (2006) 3906.

[47] Raman, V.; Bhatia, G.; Bhardwaj, S.; Srivastava A.K.; Sood, K.N. *J. Mater. Sci. Lett.* 40(2005) 1521.

[48] Raman, V.; Bhatia, G.; Sengupta, P.R.; Srivastava A.K.; Sood, K.N. *J. Mater. Sci.* 42(2007) 5891.

[49] Efremova, S.V.; Korolev, Y.M.; Sukharnikov Y. I. *Doklady Chemistry* 419 (2008) 78.

[50] Vital, A.; Richter, J.; Figi, R.; Nagel, O.; Aneziris C.G.; Bernardi, J.; Graule, T. *Ind. Eng. Chem. Res.* 46 (2007) 4273.

[51] Seo, W-S.; Koumoto, K. *J. Am. Ceram. Soc.* 83 (2000) 2584.

[52] Huczko, A.; Bystrzejewski, M.; Lange, H.; Fabianowska, A.; Cudzilo, S.; Panas, A.; Szala, M. *J. Phys. Chem. B* 109 (2005) 16244.

[53] Singh, M.; Salem, J.A. *J. Eur. Ceram. Soc.* 22 (2002) 2709.

[54] Zhu, W.; Pratsinis, S. E. *AlChE J.* 43 (1997) 2657.

[55] Patil, K.C. Bull. Mater. *Sci.* 16 (1993) 533.

[56] Merzhanov, A.G. in Chemistry of advanced materials; Rao, C.N.R., Ed.; *Blackwell scientific*: Oxford, 1993, p 19.

[57] Liu, Z.; Ci, L.; Jin-Phillip, N.Y.; Rühle. *J. Phys. Chem. C.* 111 (2007) 12517.

[58] Ledoux, M.J.; Guille, J.; Hantzer, S.; Dubots, D. U.S. Patent 4914070, *Pechiney*, 1990.

[59] Keller, N.; Vieira, R.; Nhut, J-M.; Pham-Huu, C.; Ledoux, M.J. *J. Braz. Chem. Soc.* 16 (2005) 514.

[60] Lamerchand, A.; Bonnet, J.P.*J. Phys. Chem.* 111(2007) 10829.

[61] Wu. Y.J.; Wu, J.S.; Qin, W.; Xu, D.; Yang, Z.X.; Zhang, Y.F. *Mater. Lett.* 58 (2004) 2295.

[62] Lutsenko, V.G. *Powder Metal. and Metal Ceram.* 44(2005) 1-4.

[63] Deng, S.Z.; Wu, Z.S.; Zhao, J.; Xu, N. S.; Chen, J.; Chen, J. *Chem. Phys. Lett.* 356 (2002) 511.

[64] Qian, J.M.; Wang, J.P.; Qiao, G.J.; Jin, Z.H.; *J. Eur. Ceram. Soc.* 24 (2004) 3251.

[65] Yao, J.; Wang, H.; Zhang, X.; Zhu, W.; Wei, J.; Cheng, Y.B. *J. Phys. Chem. C* 111 (2007) 636.

[66] Yang, Z.; Xia, Y.; Mokaya, R. *Chem. Mater.* 16 (2004) 3877.

[67] Lee, J-S.; Byeun, Y-K.; Lee, S-H.; Choi, S-C. *J. of Alloys and Compds.* 456 (2008) 257.

[68] Lee, J-S.; Choi, D-M.; Kim, C-B.; Lee, S-H.; Choi, S-C. *J.of Ceram. Process. Res.* 8 (2007) 87.

[69] Mishra, S.B.; Mishra, A.K.; Mamba, B.B.; Krause, R.W. *J. Am. Ceram. Soc.* 92 (2009) 3052.

[70] Mishra, S.B.; Mishra, A.K.; Krause, R.W.;Mamba, B.B. *Mat. Lett.* 63 (2009) 2449.

[71] Zhou, X.T.; Wang N.; Au, F.C.K.; Lai, Peng, H.Y.; Bello, I.; Lee, C.S.; Lee, S.T. *Mater.Sci.Eng.* A 286 (2000) 119.

[72] Huang, H. S.; Liang, B.; Jiang, D.L.; Tan, S.H. *J. Mater. Res.* 31 (1996) 4327.

[73] Fu, Q-G.; Li, H-F.; Shi, X-H.; Li, K-Z.; Wei, J.; Hu, Z-B. *Mater. Chem and Phys.* 100 (2006) 108.

[74] Kleps, I.; Nicolaescu, D.; Stamatin, I.; Correia, A.; Gil, A.; Zlatkin, A. *Appl. Surf. Sci.* 146 (1999) 152.

[75] Roegnick, K.F.; Jenson, K.F. *J. Electrochem. Soc.* 134 (1987) 1777.

[76] Pampuch, R.; Górny, G.; Stobierski. L. *Glass Physics and Chemistry,* 31 (2005) 370.

[77] Hlavacek, V.; Puszynski, J.A. *Ind. Eng. Chem. Res.* 35 (1996) 349.

[78] Xi, G.; Yu,S.; Zhang, R.; Zhang, M.; Ma, D.; Qian, Y.; *J. Phys. Chem. B* 109 (2005) 13200.

[79] 79. Zhang, Y.; Shi, E-W .; Chen, Z-Z.; Li, X-B.; Xiao, B.; *J. Mater. Chem*. 16 (2006) 4141.

[80] 80. Abderrazak, H.; Abdellaoui, M. *Mater. Lett.* 62 (2008) 3839.

[81] 81. Yao, Y.; Lee , S.T.; Li, F.H. *Chem. Phy. Lett*. 381 (2003) 628.

[82] 82. Zhang, H.X.; Feng, P.X.; Makarov, V.; Weiner, B.R.; Morell, G. *Materials Research Bulletin* 44 (2009) 184.

[83] 83. Zhao, H.; Shi, L.; Li, Z.; Tang, C. *Phy. E* 41(2009) 753

[84] 84. Wang, C-H.; Chang, Y-H.; Yen, M.Y.; Peng, C.W.; Lee, C.Y., Chiu, H.T. *Adv. Mater.* 17 (2005) 419.

[85] 85. Wang, C-H.; Lin, H-K .; Ke, T-Y.; Palathinkal, N-H .; Tai, T-J.; Lin, I-N.; Lee, C-Y.; Chiu, H-T. *Chem. Mater*. 19 (2007) 3956.

In: Silicon Carbide: New Materials, Production ... ISBN: 978-1-61122-312-5
Editor: Sofia H. Vanger

Chapter 9

CONVERSION OF SILICON CARBIDE TO CRYSTALLINE FULLERITE

Vadym G. Lutsenko[1]
V. I. Vernadski Institute of General and Inorganic Chemistry, Ukrainian National Academy of Sciences, Kyiv, Ukraine

ABSTRACT

Conversion of carbides to carbon by means of selective removal of metal atoms from carbide lattice allows obtaining porous nanocarbon materials whose phase composition and structure are determined by the properties of initial carbide and conversion conditions. There are no published data on the synthesis of pore-free carbon materials and fullerites by this method. The fundamentally different method for the conversion of SiC to carbon phases by chlorinated derivatives of methane is used. The main point of the method is substitution of carbon atom for silicon atom within the carbide lattice. In the present paper, the unique possibilities of proposed conversion method are demonstrated with silicon carbide (nanowhiskers 3C -SiC, n-type and semi-insulating platelet-shaped single crystals of 6 H -SiC polytype facetted by polar (0001) planes) as an example. By changing the process conditions one can control crystallization of various carbon allotropes which differ in density, and also obtain pore-free transparent mechanically strong coatings exhibiting high adhesion strength, low-porosity materials, and films of fullerite C 42 of different thickness. The above method can be applied for large-scale production of thin pore-free coatings on SiC, for synthesis C 42 -{0001} SiC hetero-structures, and for creation of new strategy in producing fullerites through conversion of carbides with different structure by chlorinated and fluorinated derivatives of methane and other organic compounds.

Keywords: silicon carbide, nanocarbon, fullerite, chlorination

[1] 32/34 Prospect Palladina, 03680 Kyiv, Ukraine, e-mail: vlutsenko33@rambler.ru.

1. Introduction

Conversion of carbides to carbon by means of selective removal of metal atoms from carbide lattice allows obtaining porous nanocarbon materials whose phase composition and structure are determined by the properties of initial carbide and conversion conditions [1-7]. There are no published data on the synthesis of pore-free carbon materials and fullerites by this method [1]. The fundamentally different method for the conversion of SiC to carbon phases by chlorinated derivatives of methane is used [8]. The main point of the method is substitution of carbon atom for silicon atom within the carbide lattice. In the present paper, the unique possibilities of proposed conversion method are demonstrated with silicon carbide (nanowhiskers *3C*-SiC, n-type *6H*-SiC and semi-insulating *4H*-SiC platelet-shaped single crystals facetted by polar (0001) planes) as an example. By changing the process conditions one can control crystallization of various carbon allotropes which differ in density, and also obtain pore-free transparent mechanically strong coatings exhibiting high adhesion strength, low-porosity materials, and films of fullerite C_{42} of different thickness. The above method can be applied for large-scale production of thin pore-free coatings on SiC, for synthesis C_{42}-{0001} SiC hetero-structures, and for creation of new strategy in producing fullerites through conversion of carbides with different structure by chlorinated and fluorinated derivatives of methane and other organic compounds.

2. Experimental

The objects for study were: 1) n-type *6H*-SiC (Russia) and semi-insulating *4H*-SiC (CREE, USA) platelet-shaped single crystals facetted by polar (0001) planes; 2) commercial (Russia) 3C-SiC whiskers [9]; 3) laboratory-made whiskers manufactured by present author by pyrolysis of rice husk [10].

The SiC whiskers are single crystals with growth direction <111> containing parallel coherent micro-twins and stacking faults, in which elements of structure of doped and composite super-lattices are present [9-12]. The content of α-SiC polytypes in commercial SiC whiskers did not exceed 5 wt %, whereas that in laboratory-made whiskers was about 10 wt %. The average whisker diameter was 35 to 600 nm. Just before chlorinating, the surface of whiskers was cleaned from SiO_2 film with HF solution followed by washing in distilled water and drying at 50-60°C [9].

Chlorinating with CCl_4 and $CHCl_3$ was carried out in a quartz tube in the temperature range 600-1050°C for 0.2-5 h. The argon purified of water and oxygen was used as a gas-carrier. Extraction of fullerenes from nanocarbon materials was carried out in a Soxhlets' apparatus with benzene, toluene, and pyridine as solvents.

For study, XRD (DRON-2, Cu K_α), SEM (Superprobe 733, JEOL), TEM (JEM-100CX, JEOL), thermogravimetry (Q-1500D, Hungary), UV-VIS spectroscopy (Specord UV-VIS, Germany), Auger spectroscopy (Jamp-10, JEOL), mass-spectroscopy Autoflex II (Bruker Daltonics Inc., Germany), and chemical analysis have been applied.

3. RESULTS AND DISCUSSION

The conversion process occurs [8] in accordance with the following equations:

$$SiC + CCl_4 = SiCl_4 + C\text{-}C \; (\Delta G^o_{298} = -\,494.9 \text{ kJ mol}^{-1}) \quad (1)$$

$$SiC + CHCl_3 = SiHCl_3 + C\text{-}C \; (\Delta G^o_{298} = -\,298.0 \text{ kJ mol}^{-1}) \quad (2)$$

where ΔG^o_{298} is Gibbs free energy change.

Chlorination of SiC whiskers and 6*H*-SiC single crystals at 680±10°C leads to formation of transparent and mechanically strong nanocarbon coatings on the SiC surface. Also, a thin layer of nanocarbon phase forms that stops the chlorination reaction.

The mass loss of whiskers occurs upon chlorination during the initial stage (10-20 min) of conversion process (Figure 1a). The coating thickness calculated from the mass loss of SiC whiskers is independent of the chlorinating agent (CCl_4, $CHCl_3$) and is about 20 nm. The coatings on the whiskers surface and on planes {0001} 6*H*-SiC are of the same type and exhibit two diffraction maxima at $2\Theta \approx 8.7^o$ and $2\Theta \approx 11.8^o$ (Figure 1b). From the intensity ratio of the above peaks for nanocarbon phases and the line SiC (002)β for planes {0001} 6*H*-SiC it follows that the thickness of coating on the plane terminated by carbon atoms is larger by a factor of 3 than that on (0001) Si plane. The coating on {0001} 6*H*-SiC planes also forms in the initial period of chlorination (10-30 min) (Figure 1c). The nanocarbon phase in the coating is subject to partial dissolution upon extracting for 40 h by boiling pyridine and not by benzene, as indicated by increase in specimen optical transmitting capacity after extracting with pyridine (Figure 1c). Apparently, the coating contains fullerite clusters, which partially dissolve in pyridine. The nanocarbon pore-free and elastic film forms on SiC at 680°C. Because of difference in thermal expansion coefficients of SiC and film, stresses develop at the interface SiC/coating during cooling. As a result, the inter-plane distance (002) SiC (d_{002}) decreases from 0.2527 nm to 0.2510 nm. The removal of coating with a mixture H_2SO_4+HNO_3 or by annealing in air leads to increase in d_{002} up to its initial value. However, optical transmitting capacity of specimens after coating removal does not reach its initial value (Figure 1c) because of micro-roughening of the surface of (0001) planes due to etching of structural defects.

An increase in temperature up to 800±10°C results in virtually linear dependence of carbon layer thickness on chlorination duration. Pores form in carbon coating, through which both chlorinated derivatives of methane and $SiCl_4$ diffuse. Apart from graphite phase with grain size L_a>50 nm (as indicated by broadening of (002) line), the coating contains also a nanocarbon phase or phases with 2Θ = ~19°, ~21°, and ~23.5° (Figure 2a). During conversion, a stress in SiC whiskers arises and causes bending and fracture of some of them, as well as formation of graphite tubes (Figure 2b, c). The coating porosity does not exceed 0.12 cm^3/g (as determined through benzene adsorption). Extracting of carbon materials with benzene and pyridine does not lead to change in solvent color. This allows suggestion that there are no fullerite phases in carbonaceous materials prepared by chlorination of SiC whiskers with chlorinated derivatives of methane at 800±10°C.

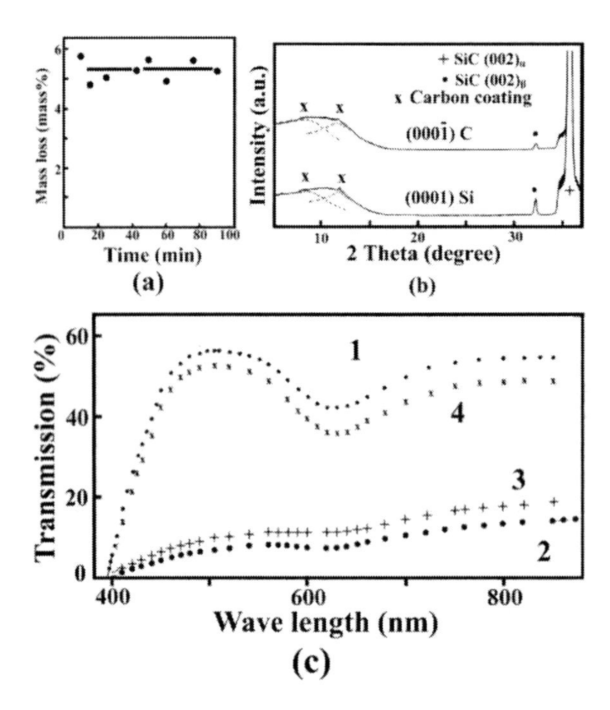

Figure 1. Properties of SiC samples subjected to chlorination by chlorinated derivatives of methane at 680±10oC: (a) dependence of mass loss for SiC whiskers (average diameter 600 nm) on chlorination duration; (b) diffractograms obtained from (0001) planes of 6H-SiC single crystal after chlorination for 60 min; (c) transmission spectrum for 6H-SiC single crystals of n-type: (1) initial sample; (2) after chlorination for 30-100 min; (3) sample 2 after extraction with pyridine for 35 h; (4) sample 3 after removal of carbon coating.

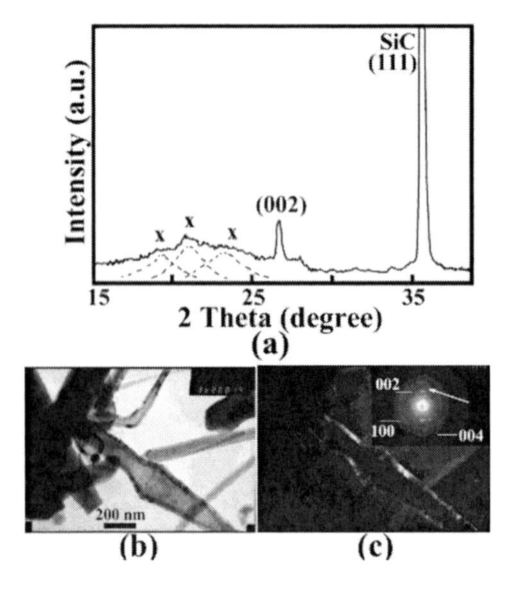

Figure 2. SiC whiskers after chlorination at 800±10oC: (a) diffractogram from whiskers (average diameter 600 nm) after chlorination for 10 min; (b, c) TEM image of carbon tubes produced by total transformation of SiC whiskers to carbon phases (average diameter of initial whiskers 160 nm, chlorination duration 40 min): (b) bright-field image; (c) dark-field image.

Chlorinating at 920±10°C leads to formation of a dense and mechanically strong black coating on the surface of (0001) planes, whose thickness (0.5-30) μm depends on treatment duration. The coating on the plane terminated by Si atoms is thinner by a factor of 3 than that on the (0001) C plane. To this indicate the intensity ratio of lines $2\Theta = 12.3^{\circ}$ and $2\Theta = 24.8^{\circ}$ for coating and the β line of SiC in diffractogram (Figure 3a). The phase composition and size of crystallites in coating are independent of the plane type (Figure 3a). The coating on (0001) Si plane has more perfect microstructure. Boiling of *6H*-SiC single crystals with coating in both benzene and pyridine does not change the solvent color. The morphological features of coating on (0001) C plane are presented in Figure 3b-d.

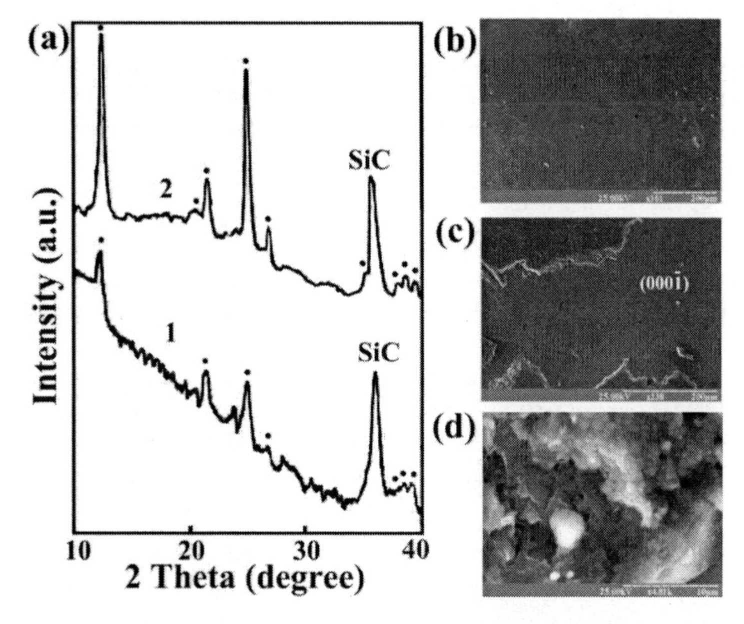

Figure 3. Fullerite coatings on polar planes of *6H*-SiC single crystal of *n* type, obtained by chlorinating with CCl_4 at 920±10°C: (a) diffractograms of (0001) planes (chlorination duration 65 min): (1) (0001) Si plane; (2) (0001) C plane; (b-d) SEM images of coating on (0001) C plane: (b) coating surface; (c) surface of substrate and coating cleavage; (d) enlarged image of the surface of coating cleavage.

The carbon nanofibers prepared by conversion of SiC whiskers at 920±10°C were subjected to boiling in a mixture H_2SO_4+HNO_3 in order to oxidize nanographite phases. Then, the remainder was washed in distilled water, treated with HF solution in order to remove SiO_2 impurity, and again washed in distilled water. The dried remainder contained no less than 98.9-99.4 wt % carbon represented by three phases (Figure 4a), namely: phase with $2\Theta = 10.7^{\circ}$ and half-width of diffraction profile ($\beta_{0.5}$) of about 4°; phase with $2\Theta = {\sim}15^{\circ}$, ~16.5°, and ~22.7° and $\beta_{0.5} \approx 1.4^{\circ}$; phase for which $\beta_{0.5}$ is about 0.3°. Extraction of carbon remainder with benzene does not change the solvent color. Boiling of the remainder in pyridine (which in contrast to benzene dissolves C_{36} [13]) for 60 h leads to change in solvent color first to yellow and then to brown. Evaporation of pyridine results in formation of a solid phase as a mixture of films and whiskers (Figure 4b-e) whose color depends on thickness and varies from brown to black with increase in thickness. To remove pyridine, the crystals and films were dried at 150-180°C. When adding pyridine to dried solid phase, the latter quickly dissolves in pyridine, whereas in toluene and benzene it does not dissolve. Pyridine evaporation causes crystallization of two carbon phases, namely: i) whiskers, and ii) "amorphous" phase with $2\Theta = \sim 15^o$, $\sim 16.5^o$, and $\sim 22.7^o$ and crystallite size of about 3-7 nm (Figure 4a). Comparison of diffractograms for carbon remainder extracted from carbon fibers and for the solid phase that crystallizes upon evaporating pyridine showed that in solid phase there is no phase with $2\Theta = \sim 10.7^o$, which does not dissolve in pyridine. It follows from intensity ratio for the crystalline phase and "amorphous" phase (marked with "X" in Figure 4a) that the latter more readily dissolves in pyridine than does the crystalline phase. So, the relative content of crystalline phase in solid material prepared by evaporation of pyridine is lower than in the initial carbon remainder (Figure 4a). The solid phase prepared from pyridine solution contains a mixture of fullerites C_{84}, C_{70}, C_{60}, C_{50}(?), C_{42}, and C_{30} (Figure 4e). The crystalline phase which forms whiskers (Figure 4a-d) is identical to that of carbon remainder (Figure 4a). Moreover, the carbon films on (0001) 6*H*-SiC planes form a set of diffraction lines, which is identical to that for whiskers (Figs. 3a and 4a). The inter-plane distance d_m for the first low-angle line of whiskers and coatings on 6*H*-SiC planes is 0.71-0.72 nm. This distance is appreciably smaller than that for C_{60} (face centered cubic (fcc), hexagonal closely packed (hcp)) [14, 15] but larger than for C_{36} (hcp) [13]. Therefore, the whiskers and coatings on (0001) 6*H*-SiC planes are phases of fullerite that contain within its molecules less than 60 and more than 36 atoms. Among the lower fullerenes with masses in the range from C_{60} to C_{36}, the C_{42} fullerene has been detected in the mass-spectrum (Figure 4e). An extrapolation of the dependence of d_m for fcc fullerites (C_{84}, C_{70}, and C_{60}) [15-17] on the number of atoms in a molecule to the number of atoms equal to 42 gives a value of inter-plane distance for the (111) line of fcc C_{42} of about 0.7-0.74 nm. The fcc unit cell parameter for C_{42} (whiskers and coatings) calculated after indexing of diffractograms (Figs. 3a and 4a) is 1.24 nm. However, not all of the lines can be indexed via fcc lattice. Probably, the rest of the lines relate to hexagonal and/or rhombohedral polytypes of C_{42}.

Apart from the crystalline C_{42} phase which forms whiskers, the "amorphous" fullerite phase dissolves in pyridine. It is this phase that causes formation of masses corresponding to C_{84}, C_{70}, C_{60}, and C_{30} (Figure 4e). Apparently, that phase contains polymerized fullerite within which there are fullerenes with different number of atoms in a molecule [18]. The phase does not dissolve in benzene and possesses a "memory", i. e. it does not change the initial structure and crystallite size during dissolution in pyridine and subsequent crystallization (Figure 4a). The "amorphous" phase, the whiskers C_{42}, and the phase with $2\Theta = \sim 10.7^o$ (marked with "+" in Figure 4a) exhibit very high chemical stability. In contrast to nanographite, these are not oxidized by hot mixture H_2SO_4+HNO_3. This fact makes it possible to apply the methods of acid treatment for preparing concentrates of fullerite phases and optimizing the process of extraction fullerenes by solvents.

An interaction of chlorinated derivatives of methane with close-packed (0001) planes at 920°C results in formation of single-phase fullerite coatings. Upon conversion of whiskers under the same conditions, the fullerite phase forms also, but its content is less than 10 wt %, i. e. it is represented by small inclusions among other nanocarbon phases. Probably, the C_{42} phase forms upon conversion of close-packed (0001) and (111) planes. The side surface of 3*C*-SiC whiskers, of which sub-microstructure is characteristic with numerous stacking faults, parallel twins [10, 11, 19-22], and elements of structure of doped and composite super-lattices [10-12], is limited by a set of microplanes (100), (110), and (111) [23]. The mosaicity of the side surface of whiskers is the reason for formation of multiphase carbon material with very low C_{42} phase content.

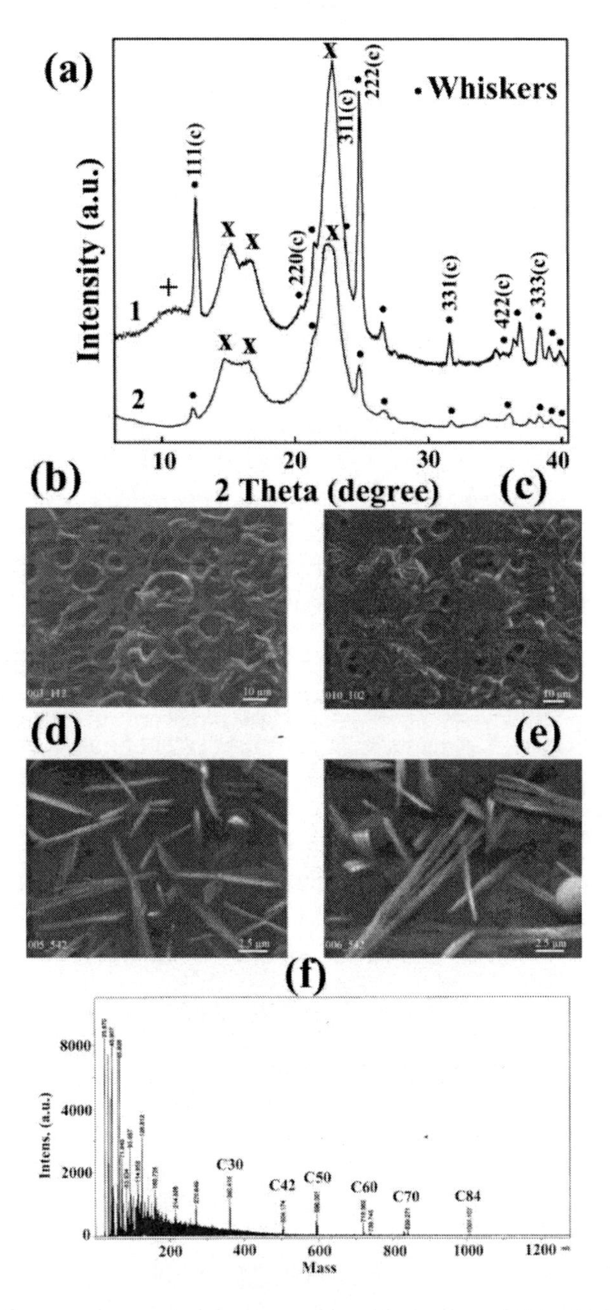

Figure 4. Properties of nanocarbon materials prepared by chlorination of SiC whiskers at 920±10°C: (a) diffractograms of nanocarbon materials extracted from chlorinated SiC whiskers, which were subjected to total conversion: (1) carbon remainder after boiling chlorinated whiskers in a mixture H_2SO_4+HNO_3, treating with HF solution, washing in water, and drying; (2) solid phase crystallized from solution in pyridine prepared by extraction of carbon remainder 1 for 60 h; (b-e) SEM images of solid phase obtained from pyridine extract: (b, c) whiskers and "amorphous" phase; (d, e) C_{42} whiskers; (f) mass-spectra of solid phase obtained from solution in pyridine.

The thin pore-free carbon films are attractive as a barrier layer on the surface of SiC whiskers and nanopowders, which are used for reinforcing [1], for ohmic contacts in structures based on SiC [7, 24, 25], including super-lattices. Finally, the described strategy in conversion of carbides with halogenated derivatives of organic compounds, including cyclic ones, can give a simple method for the production of fullerites and synthesis of nanocarbon structures with unusual properties.

Acknowledgment

I would like to thank Dr. G. S. Oleynik (the Institute for Problems of Materials Science, Ukraine) for the discussion of the results in present paper.

References

[1] Nikitin, A. and Gogotsi, Y. in *Encyclopedia of Nanoscience and Nanotechnology* (ed. Nalwa, H. S.) , V. 7, 553-574 (Amer. Sci. Publ., 2004).

[2] Gogotsi, Y., Welz, S., Ersoy, D. A. and McNallan, M. J. Conversion of silicon carbide to crystalline diamond-structured carbon at ambient pressure. *Nature* 411, 283-287 (2001).

[3] Gogotsi, Yu. G. and Yoshimura, M. Formation of carbon films on carbides under hydrothermal conditions. *Nature* 367, 628-630 (1994).

[4] Kusunoki, M., Suzuki, T., Honjo, C., Hizayama, T. and Shibata, N. Selective synthesis of zigzag-type aligned carbon nanotubes on SiC (000-1) wafers. *Chem. Phys. Lett.* 366, 458-462 (2002).

[5] Maruyama, T. et al. STM and XPS studies of early stages of carbon nanotube growth by surface decomposition of 6H–SiC(000-1) under various oxygen pressures. *Diamond and Related Mater.* 16, 1078-1081 (2007).

[6] Rollings, E. et al. Synthesis and characterization of atomically thin graphite films on a silicon carbide substrate. *J. Phys. Chem. Solids* 67, 2172-2177 (2006).

[7] Zinovev, A.V., Moore, J.F., Hryn, J., Pellin, M.J. Etching of hexagonal SiC surfaces in chlorine-containing gas media at ambient pressure. *Surface Sci.* 600, 2242-2251 (2006).

[8] [8] Lutsenko, V.G. in *Synthesis, Properties and Applications of Ultrananocrystalline Diamond* (D.M.Gruen et al., Eds.), Vol. 192, 289-298 (Springer, The Netherlands, 2005).

[9] Lutsenko V.G. Impurity phases of silicon dioxide in commercial SiC whiskers produced by VLS method. *Mat. Chem. Phys.* , 115, Is. 2-3, 664-669 (2009).

[10] Lutsenko, V.G. Composition and properties of SiC obtained from rice husk. *Superhard mater.* 28, 45-56 (2006).

[11] Lutsenko, V.G. SiC with Quantum Wells and Nanosized Twins, in *Proc. Theodor Grotthuss Electrochem. Conf.*, Vilnius, 5—8 June, 2005 (Juodkazite, J., Selskis, A., eds.) p. 62 (DABA, Vilnius, 2005).

[12] Lutsenko V.G. Silicon carbide whiskers with superlattice structure: A precursor for a new type of nanoreactor. *Acta Materialia,* 56, 2450-2455 (2008).

[13] Piscoti, C., Yarger, J. and Zettl, A. C_{36}: a new carbon solid. *Nature* 393, 771-774 (1998).

[14] Krätschmer, W., Lamb, L. D., Fostiropoulos, K. and Huffman, D. R. Solid C_{60}: a new form of carbon. *Nature* 347, 354-358 (1990).

[15] Jin, Y. et al. Structural and optoelectronic properties of C_{60} rods obtained *via* a rapid synthesis route. *J. Mater. Chem.* 16, 3715-3720 (2006).

[16] M. A. Verheijen, M. A. et. al. The structure of different phases of pure C_{70} crystals. *Chem. Phys.* 166, 287-297 (1992).

[17] Almairac, R., Tranqui, D., Lauriat,J., Lapasset, P.J. and Moret, J. Evidence of phase transitions in C_{84}. *Solid State Commun.* 106, 437-440 (1998).

[18] Zubov, V.I. Interactions between the molecules of different fullerenes. *Fullerenes Nanotubes Carbon Nanostructures* 12, 499-504 (2005).

[19] Nutt, S.R. Defects in Silicon Carbide Whiskers. *J. Am. Ceram. Soc.* 67, 428-431 (1984).

[20] Seo, W.-S., Koumoto, K. and Aria S. Morphology and Stacking Faults of β-Silicon Carbide Whisker Synthesized by Carbothermal Reduction. *J. Am. Ceram. Soc.* 83, 2584-2592 (2000).

[21] Lutsenko, V.G. in *Rock-crushing and Metal-cutting Tools, Technique to their Manufacture and Applications (ed. Novikov, N. V),* Is. 7, 190-196 (Institute Superhard Materials, Kiev, 2004) (in russian).

[22] Gambaz, Z.B, Yushin, G., Gogotsi, Y. and Lutsenko, V.G. Anisotropic Etching of SiC Whiskers. *Nano Lett.* 6, 548-551 (2006).

[23] Geng, L., Zhang, J., Meng,Q.-C. and Yao, C.-K., Side Surface Structure of a Commercial β-Silicon Carbide Whisker, *J. Am. Ceram. Soc.* 85, 2864-2866 (2002).

[24] Weijee, L.U. et al. Ohmic contact behavior of carbon films on SiC. *J. Electrochem. Soc.* 150, G177-G182 (2003).

[25] Zinovev, F.V. et al. Coating of SiC surface by thin carbon films using the carbide-derived carbon process. *Thin Solid Films* 469-470, 135-141 (2004).

INDEX

D

E

F

P

Q

R

S

T

V

U

W

X

Z